MW00814691

# Biological and Medical Physics, Biomedical Engineering

For further volumes:
http://www.springer.com/series/3740

The fields of biological and medical physics and biomedical engineering are broad, multidisciplinary and dynamic. They lie at the crossroads of frontier research in physics, biology, chemistry, and medicine. The Biological and Medical Physics, Biomedical Engineering Series is intended to be comprehensive, covering a broad range of topics important to the study of the physical, chemical and biological sciences. Its goal is to provide scientists and engineers with textbooks, monographs, and reference works to address the growing need for information.

Books in the series emphasize established and emergent areas of science including molecular, membrane, and mathematical biophysics; photosynthetic energy harvesting and conversion; information processing; physical principles of genetics; sensory communications; automata networks, neural networks, and cellular automata. Equally important will be coverage of applied aspects of biological and medical physics and biomedical engineering such as molecular electronic components and devices, biosensors, medicine, imaging, physical principles of renewable energy production, advanced prostheses, and environmental control and engineering.

Volker Schmidt · Maria Regina Belegratis
Editors

# Laser Technology in Biomimetics

Basics and Applications

 Springer

*Editors*
Volker Schmidt
Maria Regina Belegratis
Institute for Surface Technologies
and Photonics
Joanneum Research Forschungsgesellschaft
mbH
Weiz
Austria

ISSN 1618-7210          ISSN 2197-5647 (electronic)
ISBN 978-3-642-41340-7  ISBN 978-3-642-41341-4 (eBook)
DOI 10.1007/978-3-642-41341-4
Springer Heidelberg New York Dordrecht London

Library of Congress Control Number: 2013956322

Printed on acid-free paper

Springer is part of Springer Science+Business Media (www.springer.com)

# Preface

This book deals with the exciting field of biomimetics in combination with laser technology. Biomimetics, the translation from nature-inspired principles to technical applications, is strongly multidisciplinary. Since its invention, the laser has proven many times its versatility. Therefore it is not astonishing that the wide scientific interest entangled with biomimetics has discovered the laser as a fast and reliable processing tool for its purpose. This field offers intrinsically a wide scope of applications for laser-based methods regarding structuring and modification of materials. Plus, the development for novel laser-based processing methods is permanently under development.

This book attempts to give a survey over state-of-the-art laser technology and materials that are used for diverse biomimetic applications. It covers the most important laser lithographic methods and various biomimetics application scenarios ranging from coatings and biotechnology to construction, medical applications and photonics. The term biomimetics is hereby expanded to cover also the field of biotechnology and biomedical applications. Due to the availability of a large spectrum of different laser sources regarding their characteristics such as laser power, wavelength, operation mode, the laser serves as tool and light source for widespread applications.

The book introduces both a laser technology focused approach as well as an application focused approach. It is dedicated to the advanced reader that is already familiar with laser basics and laser technology and to scientists, who may discover a possibility or inspiration to use a laser for their own specific works.

Chapter 1 is dedicated to a short introduction and motivation of the laser application for biomimetics.

Chapter 2 represents a comprehensive review of direct laser writing methods, especially focusing on laser-induced two-photon-based photochemical effects. This method has evolved in recent years as a powerful technology for the realization of micron-, to sub-micrometer resolution structures and gains more and more interest in the field of biomedicine. In this chapter, the basic principles of direct laser writing, a short survey of available techniques, and finally technologies for enhanced performance are described. The review is completed with some examples of direct laser writing in the field of 3D cell culture and tissue engineering.

Chapter 3 deals with direct 3D laser writing of bio-inspired materials related to optical phenomena. This chapter introduces direct laser writing as a flexible technology for the development of 3D microstructures for applications in photonics. The structural designs found in nature often exhibit useful physical phenomena such as photonic bandgaps and circular polarisation stop bands. The investigated biological nanostructures will certainly inspire novel biomimetic materials for photonics applications.

Chapter 4 focuses on selective laser sintering, which is another mature and versatile rapid prototyping method. After the introduction of major rapid prototyping technologies suitable for biomedical applications, a review on laser sintering is made, which includes its working principle and technical benefits as well as materials. A biomimetic application, namely tissue engineering scaffolds and drug or biomolecule delivery vehicles is discussed, showing the great potential also for many other biomimetic and biomedical applications.

Chapter 5 reviews matrix-assisted pulsed laser evaporation for biomimetic applications in drug delivery systems, biosensors and advanced implant coatings. This method emerged more than a decade ago and is dedicated to the transfer of organic materials onto solid substrates, because it represents a minimal harmful approach for transporting and depositing thin films of delicate, heat sensitive molecules, such as organic biomaterials on largely extended active areas. Such films are considered to play an important role in biology, pharmaceutics or sensing applications.

Chapter 6 gives an overview over the process basics, its parameters and the major influences on the quality of complex three-dimensional parts manufactured by Laser additive manufacturing. Its preferred field of application is the one-step manufacturing of complex geometries in low lot sizes, where conventional machining would require a longer overall production time due to a high number of processing steps. In addition, the chapter deals with design guidelines, current applications of laser additive manufacturing and future developments.

Chapter 7 reviews important applications of pulsed laser deposition and recent work in the field of biomimetic coatings. Furthermore, technical limitations and possible solutions are outlined. The general characteristics of pulsed laser deposition relevant to solid-state physics, e.g. the initial ablation processes, plume formation and plume characteristics are discussed as well.

In Chap. 8, laser-assisted bio-printing based on the laser-induced forward transfer is discussed as an emerging and complementary technology in the field of tissue engineering envisaging biomimetics applications. The method allows to print cells and liquid materials with a cell-level resolution, which is comparable to the complex histology of living tissues. Experimental requirements are described and typical multi-component printing, 3D printing approaches and in vivo bio-printing are presented.

Chapter 9 Laser-based biomimetic tissue engineering relies on the controlled and reproducible structuring of biomaterials at micro- and nanoscales by means of laser radiation. Tissue Engineering is defined as the technology aiming to apply the principles of engineering and life sciences towards the development of biological substitutes that restore, maintain or improve tissue function or a whole organ. This chapter reviews current approaches for laser-based fabrication of biomimetic tissue engineering scaffolds. These include laser processing of natural biomaterials synthesized to achieve certain compositions or properties similar to those of the extracellular matrix as well as novel laser fabrication technologies to achieve structural features on artificial materials mimicking the extracellular matrix morphology on various levels. The chapter concludes with the wealth of arising possibilities, demonstrating the excitement and significance of the laser-based biomimetic materials processing for tissue engineering and regeneration.

Chapter 10 reviews material aspects with respect to laser processing of natural materials. The laser is known as a versatile tool and its application is successfully expanded to the processing of natural biomaterials. Various natural biomaterials, such as collagen, agarose, hyaluronic acid and Matrigel, have been treated through laser-induced polymerization, ablation or activation. The presented developments expand and will continue to expand the potential applications of natural biomaterials in biomimetic approaches.

Last but not least, the book ends with Chap. 11 about future perspectives of laser-based research in the wide field of biomimetic applications.

We thank all authors for their valuable contributions and Springer for the pleasant cooperation. Finally, we thank Claus Ascheron from Springer, without whose patience, encouragement and support, this book would not have been possible.

Weiz, Austria                                                                        Maria R. Belegratis
                                                                                          Volker Schmidt

# Contents

# Contributors

**Maria R. Belegratis** Institute for Surface Technologies and Photonics, Joanneum Research Forschungsgesellschaft mbH, Franz-Pichlerstraße 30, 8160 Weiz, Austria, e-mail: maria.belegratis@joanneum.at

**Sylvain Catros** Bioingénierie Tissulaire, INSERM U1026, 146, rue Léo Saignat, Bordeaux 33076, France

**Shaochen Chen** Department of NanoEngineering, University of California, 245B SME Building, MC-0448, San Diego, CA 92093, USA, e-mail: chen168@ucsd.edu

**Peter H. Chung** Department of NanoEngineering, University of California, Atkinson Hall, Room 2314, Gilman Drive #0448, San Diego, CA 92093-9500, USA, e-mail: peterchung@ucsd.edu

**Bin Duan** Faculty of Engineering, Department of Mechanical Engineering, The University of Hong Kong, Pokfulam Road, Pokfulam, Hong Kong

**Claus Emmelmann** iLAS Laser- und Anlagensystemtechnik, Technische Universität Hamburg-Harburg, Denickestr. 17 (L), 21073 Hamburg, Germany

**Sascha Engelhardt** Institute for Laser Technology, RWTH Aachen University, Steinbachstraße 15, 52074 Aachen, Germany, e-mail: sascha.engelhardt@ilt.fraunhofer.de

**Costas Fotakis** Institute of Electronic Structure and Laser, Foundation for Research & Technology-Hellas, 1527, Heraklion 71110, Greece, e-mail: fotakis@iesl.forth.gr

**Min Gu** Faculty of Engineering and Industrial Sciences, Centre for Micro-Photonics and CUDOS, Swinburne University of Technology, Hawthorn, VIC 3122, Australia

**Fabien Guillemot** Bioingénierie Tissulaire, INSERM U1026, 146, rue Léo Saignat, Bordeaux 33076, France

**Bertrand Guillotin** Bioingénierie Tissulaire, INSERM U1026, 146, rue Léo Saignat, Bordeaux 33076, France

**Dirk Herzog** LZN Laser Zentrum Nord GmbH, Am Schleusengraben 14, 21029 Hamburg, Germany, e-mail: dirk.herzog@lzn-hamburg.de

**Jannis Kranz** iLAS Laser- und Anlagensystemtechnik, Technische Universität Hamburg-Harburg, Denickestr. 17 (L), 21073 Hamburg, Germany, e-mail: jannis.kranz@tuhh.de

**Ion N. Mihailescu** Lasers Department, National Institute for Lasers, Plasma and Radiation Physics, 409 Atomistilor Street, 77125 Măgurele, Ilfov, Romania, e-mail: ion.mihailescu@inflpr.ro

**Anthi Ranella** Institute of Electronic Structure and Laser, Foundation for Research & Technology-Hellas, 1527, Heraklion 71110, Greece, e-mail: ranthi@iesl.forth.gr

**Carmen Ristoscu** Lasers Department, National Institute for Lasers, Plasma and Radiations Physics, 409 Atomistilor street, MG-54, 077125 Magurele, Romania, e-mail: carmen.ristoscu@inflpr.ro

**Volker Schmidt** Institute for Surface Technologies and Photonics, Joanneum Research Forschungsgesellschaft mbH, Franz-Pichler straße 30, Weiz 8160, Austria, e-mail: volker.schmidt@joanneum.at

**Gerd E. Schröder-Turk** Theoretische Physik, Friedrich-Alexander Universität Erlangen-Nürnberg, Staudstr. 7B, Erlangen, Germany

**Felix Sima** Lasers Department, National Institute for Lasers, Plasma and Radiation Physics, 409 Atomistilor Street, 77125 Magurele, Ilfov, Romania, e-mail: felix.sima@inflpr.ro

**Emmanuel Stratakis** Institute of Electronic Structure and Laser, Foundation for Research & Technology-Hellas, 1527, Heraklion 71110, Greece, e-mail: stratak@iesl.forth.gr

**Mark D. Turner** Faculty of Engineering and Industrial Sciences, Centre for Micro-Photonics and CUDOS, Swinburne University of Technology, Hawthorn, VIC 3122, Australia

**Min Wang** Faculty of Engineering, Department of Mechanical Engineering, The University of Hong Kong, Pokfulam Road, Pokfulam, Hong Kong, e-mail: memwang@hku.hk

**Eric Wycisk** LZN Laser Zentrum Nord GmbH, Am Schleusengraben 14, 21029 Hamburg, Germany, e-mail: eric.wycisk@lzn-hamburg.de

**Wande Zhang** Department of BioEngineering, University of California, 9500 Gilman Drive, San Diego, CA 92093, USA, e-mail: wande117@gmail.com

**Aping Zhang** Photonics Research Centre, Department of Electrical Engineering, The Hong Kong Polytechnic University, Kowloon, Hong Kong SAR, People's Republic of China, e-mail: aping.zhang@polyu.edu.hk

# Acronyms

| | |
|---|---|
| 2D | Two-dimensional |
| 2PP | Two-photon polymerization |
| 3D | Three-dimensional |
| 3DP | 3D printing |
| ADSC | Adipose-tissue derived stem cells |
| AFM | Atomic force microscopy |
| AFO | Ankle-foot orthose |
| ALP | Alkaline phosphatase |
| AOM | Acousto-optical modulator |
| b-TCP | B-tricalcium phosphate |
| BAEC | Bovine aortic endothelial cell |
| BG | Bioglass |
| BMP | Bone morphogenic protein |
| BSA | Bovine serum albumin |
| CAD | Computer-aided design |
| CAM | Computer-assisted manufacturing |
| Ca-P | Calcium phosphate |
| CCD | Charge-coupled device |
| CGH | Computer generated hologram |
| CHAp | Carbonated hydroxyapatite |
| CICP | Collagen type I production |
| C-PLD | Combinatorial pulsed laser deposition |
| CT | Computer tomography |
| DDD | Drug delivery device |
| DETC | 7-Diethylamino-3-thenoylcoumarin |
| DIC | Differential interference contrast |
| DiLL | Dip-in laser lithography |
| DLP | Digital light processor |
| DLW | Direct laser writing |
| DM | Dichroic mirror |
| DMD | Dynamic micromirror device; Digital micromirror device |
| DMSO | Dimethyl sulfoxide |

| | |
|---|---|
| DNA | Deoxyribonucleic acid |
| DOE | Diffractive optical element |
| DOE | Design of experiment |
| EBL | Electron beam lithography |
| ECFC | Endothelial colony-forming cells |
| ECM | Extra cellular matrix |
| EDS | Electron dispersion spectroscopy; Energy-dispersive X-ray spectroscopy |
| eGFP | Enhanced green fluorescent protein |
| EHS | Engelbreth-Holm-Swarm |
| eV | Electron Volt |
| Excimer | Excited dimer |
| FDA | Food and Drug Administration |
| FDM | Fused deposition modeling |
| FEM | Finite element method |
| FET | Field effect transistor |
| FITC | Fluoresceine isothiocyanate |
| FN | Fibronectin |
| fs | Femtosecond |
| FTIR | Fourier transformed infrared spectrometry |
| GIXRD | Grazing incidence X-ray diffraction |
| GM | Göppert-Mayer |
| GMHA | Glycidyl methacrylate hyaluronic acid |
| GNR | Gold nanorod |
| GRGDS | Glycine-arginine-glycine-aspartic-acid-serine oligopeptides |
| GVD | Group velocity dispersion |
| HA | Hyaluronic acid |
| HA | Hydroxyapatite |
| HDPE | High density polyethylene |
| HMSC | Human mesenchymal stem cell |
| HOP | Human bone-marrow derived osteo-progenitors |
| HUVEC | Human umbilical vein endothelial cells |
| HUVSMC | Human umbilical vein smooth muscle cells |
| IR | Infrared |
| ITX | Isopropylthioxanthone |
| K | Kelvin |
| kHz | Kilohertz |
| LAB | Laser assisted bioprinting |
| LAM | Laser additive manufacturing |
| LASER | Light amplification by stimulated emission of radiation |

| | |
|---|---|
| LAT | Laser ablation texturing |
| LBL | Layer-by-layer |
| LCP | Left circularly polarized |
| LED | Light emitting diode |
| LHD | Left handed |
| LIFT | Laser-induced forward transfer |
| LIL | Laser interference lithography |
| LP | Laser printing |
| LTE | Local thermal equilibrium |
| MAA | Methacrylic acid |
| MALDI | Matrix-assisted laser desorption/ionization |
| MAPLE | Matrix-assisted pulsed laser evaporation |
| MAPTMS | Methacryloxypropyl trimethoxysilane |
| MBE | Molecular beam epitaxy |
| MDM | Multi nozzle deposition manufacturing |
| MHz | Megahertz |
| MIS | Metal-insulator-semiconductor |
| MMP | Matrix metalloproteinase |
| MRI | Magnetic resonance imaging |
| MSC | Mesenchymal stem cell |
| MTS | 3-(4,5-dimethylthiazol-2-yl)-5-(3-carboxymethoxyphenyl)-2-(4-sulfophenyl)-2H-tetrazolium |
| NA | Numerical aperture |
| NIR | Near infrared spectral range |
| Nm | Nanometer |
| OCN | Osteocalcin |
| OCP | Octacalcium phosphate |
| OES | Optical emission spectroscopy |
| OPA | Optical parametric amplifier |
| OPG | Osteoprotegerin |
| PBS | Phosphate buffered saline |
| PBT | Polybutylene-terephtelate |
| PC | Photonic crystal |
| PCL | Poly($\varepsilon$-caprolactone); Polycaprolactone |
| PDLLA | Poly(D, L-lactic acid) |
| PEEK | Polyetheretherketone |
| PEG | Polyethylene glycol |
| PEGDA | Polyethylene glycol diacrylate |
| PEOT | Polyethyleneoxide-terephtalate |
| PG | Progesterone |

| PGA | Poly(glycolide) |
|---|---|
| PHBV | Poly(hydroxybutyrate-co-hydroxyvalerate) |
| PHSRN | Proline-Histidine-Serine-Arginine-Aspargine |
| PI | Photo initiator |
| PLA | Polylactic acid |
| PLCL | Poly(lactide–co-caprolactone) |
| PLD | Pulsed laser deposition |
| PLGA | Poly(D, L-lactide-co-glycolide); Poly(lactic acid-co-glycolic acid) |
| PLLA | Poly(L-lactic acid) |
| PMMA | Poly(methyl methacrylate) |
| PVA | Poly(vinyl alcohol) |
| PVP | Polyvinylpyrrolidone |
| R&D | Research and development |
| RCP | Right circularly polarized |
| RGD | Arginine-glycine-aspartic acid |
| RGDS | Arginine-glycine-aspartic acid-serine |
| rhBMP-2 | Recombinant human bone morphogenetic protein-2 |
| RHD | Right handed |
| RP | Rapid prototyping |
| SAED | Selected area electron diffraction |
| SAM | Self-assembled monolayer |
| SBF | Simulated body fluid |
| SEM | Scanning electron microscope |
| SFF | Solid free form |
| SG | Sol–gel |
| SIMP | Solid-isotropic-material-with-penalization |
| siRNA | Small interfering ribonucleic acid |
| SIS | Small intestinal submucosa |
| SLA | Stereolithography apparatus, stereolithography |
| SLM | Spatial light modulator; Selective laser melting |
| SLS | Selective laser sintering |
| SPA | Single photon absorption |
| SS | Stainless steel |
| SSLS | Surface selective laser sintering |
| STED | Stimulated emission depletion |
| STL | Stereolithography |
| TCP | Tricalcium phosphate |
| TPP | Two photon polymerization |
| TRIS | Tris(hydroxymethyl)aminomethane |
| UHMWPE | Ultrahigh molecular weight polyethylene |

| UV | Ultraviolet |
| VIS | Visible spectral range |
| VN | Vitronectin |
| XRD | X-ray diffraction |
| XTEM | Cross-sectional transmission electron microscopy |
| YAG | Yttrium aluminium garnet |
| $YVO_4$ | Yttrium orthovanadate |

# Chapter 1
# Introduction and Scope of the Book

Volker Schmidt and Maria R. Belegratis

**Abstract** This chapter introduces the scope of the book. It is intended to guide the reader through the book, to find specific information by shortly summarizing information from the following chapters and to build a cross reference to the various applied methods and the corresponding applications. The laser as a powerful light source can be found in nearly any technical application, ranging from consumer electronics (CD, DVD, blu-ray player, scanner), metrology (including environmental monitoring), scientific research (laser development to novel fields in quantum physics, photonics and medicine), arts, industry, information technology to lithography and material processing. It is obvious that the laser meets many requirements from technical challenges inspired by natural evolutionary solutions. Not all of them can be treated in a single book, but a cross section of the powerful combination of both, laser technology and biomimetic thinking, form a powerful approach to novel technical application scenarios as presented in the next chapters, which are considered as guideline and orientation for the reader depending on a laser or application based approach.

## 1.1 Biomimetics as Inspiration for Laser-Based Methods and Applications

The term "biomimetics" commonly defines the understanding of natural structures and functions of biological systems and the corresponding translation of the observed working principles as models for the development of technical systems with enhanced

V. Schmidt · M. R. Belegratis (✉)
Institute for Surface Technologies and Photonics, Joanneum Research Forschungsgesellschaft mbH, Franz-Pichlerstraße 30, 8160 Weiz, Austria
e-mail: maria.belegratis@joanneum.at

V. Schmidt
e-mail: volker.schmidt@joanneum.at

V. Schmidt and M. R. Belegratis (eds.), *Laser Technology in Biomimetics*,
Biological and Medical Physics, Biomedical Engineering,
DOI: 10.1007/978-3-642-41341-4_1, © Springer-Verlag Berlin Heidelberg 2013

properties. On one hand, nature takes advantage of structural features for tailoring and enhancing intrinsic material properties. These structures have been optimized during a long term of evolution. Prominent examples are e.g. structural colors [1], or the wetting behavior of a textured surface [2]. These examples are interesting for many technical applications in design, construction, architecture, robotics, energy management and surface engineering. Here, medical applications are biological lightweight construction and medical implants with biocompatible coatings.

On the other hand, direct processing of natural materials and substituting it by artificial biomaterials that mimic natural tissue are further aspects of learning from nature. By studying mechanisms of cell behavior as well as cell tissue interaction and designing adequate cell environments new aspects are introduced into tissue engineering.

## 1.1.1 Laser Sources

Without the invention of the laser and the unique properties of laser radiation many processes and innovations would not have been possible. Laser technology has experienced a strong development since its first demonstration in the 1960s. Today many different laser types and technical applications for laser radiation are state-of-the-art and new ones are about to be discovered. This short introduction summarizes the main features and concepts of laser processing and surveys the most important application scenarios of the laser as a versatile processing tool, which is presented in the context of biomimetics in the subsequent chapters of this book.

Nature provides many examples how evolution solved environmental issue for the purpose of survival of an organism. Hence, there are many technical problems that can look for inspiration within the framework of biomimetics. The application of laser technology in material science is comparably multifaceted. First of all there are many different laser sources available that are potentially matching a certain application. Secondly, the laser processing is in principle not limited to a certain class of materials. Basically most of known materials show modifications when processed by laser radiation. Consequently applications that are both, inspired by nature and involving laser technology are quite overlapping and a systematic approach to this topic can be quite different.

The selection of an appropriate laser for an envisaged application is mainly determined by the technical specifications of the laser such as wavelength, operational mode, power etc. The wavelength for material processing is demanded by the optical properties of the target material. The wavelength of the laser subsequently determines properties of the optical setup regarding the beam delivery with all included optomechanical components. The operational mode (continuous wave-cw or pulsed) directly influences the interaction regime with the target material and which paths of energy deposition can be triggered by the incident laser power.

Industrial applications are more demanding than scientific applications in terms of acquisition and operational costs, reliability of the laser sources, maintenance requirements, handling, ease of operation, technical complexity, and automation. These factors have great impact on the market value and prize of the final products. Some laser sources have gained specific importance and have made the way from pure research in laboratories to industrial fabrication and applications in material processing (cf. also Table 1.1). The laser types involved in the biomimetic applications and processes described in this book are shortly summarized in the next paragraphs.

### 1.1.1.1 $CO_2$ Lasers

The $CO_2$ laser is a gas laser and is one of the most important lasers in industrial applications regarding high power material processing such as machining, welding, cutting, drilling, and engraving. Although most metals reflect very well its radiation, the processing of metallic workpieces with such lasers is very common. $CO_2$ laser systems are available as pulsed or cw operating systems. Beam power up to 100 kW in cw operation is possible at rather high energy conversion factor (up to 30 %). The $CO_2$ laser represents currently the highest available cw power at reasonable costs and emits radiation at mid-infrared wavelength (10.6 μm). Low power systems are used in research and medicine. Medical applications are motivated by the good absorption of the laser wavelength in water and water containing tissue. The wavelength of the laser requires special optics (ZnSe as lens or window material, Cu as mirror) and cannot be passed through glass fibers, which is sometimes a drawback with respect to beam delivery.

With respect to biomimetics and the presented manufacturing methods, $CO_2$ lasers are used for selective laser sintering (see Chap. 4).

### 1.1.1.2 Excimer Lasers

Laser light sources in the ultraviolet spectral range are either frequency multiplied solid-state lasers or gas lasers. Especially excimer lasers provide pulsed high power ultraviolet to deep ultraviolet emission at a typical repetition rate of a few hundred Hz up to some kHz. The main scientific and industrial applications of excimer lasers are material processing, lithography and medicine, involving methods such as laser ablation, engraving, marking, surface and sub-surface modifications and coatings made via pulsed laser deposition in either projection or direct exposure mode. In projection mode, the laser light is projected via a mask onto a target for UV exposure. In direct exposure mode, the focused laser light directly ablates the material by moving the focused laser across the target. In projection mode the effect of the laser results either in patterned material removal or in exposure of the material without removal, which depends mainly on the laser fluence and applied pulse number.

The active medium of an excimer laser is a gas of electrically excited dimers ("excimer", or more precisely excited complexes), where an excited noble gas atom and a halogen form a noble gas halide, which decays after a short time (typically

**Table 1.1** Important laser types, emission characteristics and field of application

| Lasertype | Gain medium | Emission characteristics (mode of operation, wavelengths, power/pulse energy) | Application |
|---|---|---|---|
| Ar-ion laser | Gas | cw(*) operation $\lambda = 488$ nm, $514.5$ nm some Watt | Spectroscopy, holography, machining |
| He-Ne laser | Gas | cw operation $\lambda = 633$ nm some $0.1$ Watt | Alignment, spectroscopy, holography, Interferometry |
| He-Cd laser | Gas | cw operation $\lambda = 325$ nm some $10$ mW | Lithography interferometry |
| Excimer laser | Gas | pulsed operation (ns) $\lambda = 157$ nm ($F_2$), $193$ nm (ArF), $248$ nm (KrF), $308$ nm (XeCl) some Joule | Lithography, ablation, machining, surgery |
| $CO_2$ laser | Gas | cw, pulsed operation ($\mu$s) $\lambda = 10.6 \mu$m some mW to some kW | Machining, cutting, welding, drilling |
| Nd:YAG | Solid state | cw, pulsed operation (ns, ps) $\lambda = 1064$ nm, $532$ nm (SH**), $355$ nm (TH), $266$ nm (FH) some kW | Material processing, laser pumping, research, surgery |
| Ti:Sapphire | Solid state | pulsed operation (ps, fs) $\lambda = 670 - 1080$ nm some mJ | Spectroscopy, non-linear material processing |
| Nd-Glass | Solid state | pulsed operation (ms, ns, ps) $\lambda = 1062$ nm some $100$ J | High energy multiple beam systems, laser fusion |
| AlGaAs | Semiconductor | cw, pulsed operation ($\mu$s) $\lambda = 780 - 880$ nm some kW (laser diode bars) | Machining, medical, optical discs, laser pumping |
| AlGaInP | Semiconductor | cw operation $\lambda = 630 - 680$ nm | Machining, medical, optical discs, laser pumping |
| InGaAsP | Semiconductor | cw, pulsed operation (ps) $\lambda = 1150 - 1650$ nm | Machining, medical, optical discs, laser pumping |
| GaN | Semiconductor | cw operation $\lambda = 405$ nm | Optical discs, lithography |
| Dye laser | Dye | cw, pulsed operation (ns) $\lambda = 300 - 1200$ nm, depends on used dye | Spectroscopy, research, medical |

(*) cw = continuous wave, (**) SH = second harmonic, TH = third harmonic, FH = fourth harmonic. Adapted from [3]

some ns) into the dissociated state (e.g. Kr*F $\rightarrow$ Kr $+$ F) under emission of UV light. The type of excimer determines the wavelength of the emission. The technically most relevant excimers are ArF, KrF, XeCl, and XeF and there are manifold applications for excimer laser processes such as excimer based optical lithography [4], excimer laser chemical vapour deposition [5], excimer laser micromachining [6] or eye surgery [7]. The high photon energy of the excimer radiation is capable of directly breaking intramolecular bonds of the target material, without only negligible thermal impact on the surrounding material.

### 1.1.1.3 Nd:YAG, Nd:YLF, Nd:Glass Lasers

These solid state lasers provide high power radiation (some kW) with a fundamental wavelength in the near-infrared spectral range (e.g. fundamental wavelength of Nd:YAG laser: 1064 nm). The Nd doped laser crystal is commonly pumped by IR laser diodes (diode pumped solid state laser–DPSSL). Depending on the geometrical shape of the laser crystal, such lasers are often denoted as disc, rod or fiber lasers. Applications for these lasers are machining, medicine or pumping of other lasers such as Ti:Sapphire lasers for the generation of ultrashort pulses. These solid state lasers are use instead of $CO_2$ laser in sintering of metallic and ceramic materials as described in Chap. 4. By using higher harmonics of the emitted radiation, these lasers may substitute excimer lasers in the UV region for laser assisted deposition methods such as matrix assisted pulsed laser evaporation (Chap. 5), pulsed laser deposition (Chap. 7) or laser induced forward transfer (Chap. 8).

### 1.1.1.4 Ti:Sapphire Lasers

The active medium is a titanium doped $Al_2O_3$ crystal (Ti:Sapphire). This crystal emits in the near-infrared spectral region and shows a spectrally broad emission. Because of the uncertainty principle, a short light pulse consists of many spectral components and the active laser must support the stimulated emission for many longitudinal modes of the laser resonator. Hence, Ti:Sapphire laser are very well suited for the generation of short (femtosecond) laser pulses, typically in the wavelength range 750–850 nm, which is achieved by mode-locking of many laser modes with a constant phase relation. This yields a soliton-like propagating pulse within the laser resonator of a very short temporal width. Femtosecond laser pulses facilitate on one hand the examination of very fast transient processes in biology, chemistry and physics and on the other hand proved as a unique tool for lithography and material processing at very high precision. Ultra short laser pulses are mainly demonstrated in Chaps. 2, 3 and 9 with respect to 3D laser lithography and tissure engineering based on non-linear optical processes such as multi-photon absorption.

## 1.1.2 Laser Processing

The use of lasers in a lithographic apparatus relies on the unique properties of laser radiation, such as high monochromaticity, coherence, directed emission of radiation and excellent beam quality along with high focusability. If the laser substitutes the conventional light source and provides high power radiation at a short wavelength with high spectral density, it illuminates in a simple approach similar to conventional lithography a sample through a patterned mask and projection optics. Alternatively, the laser light can be tightly focused and scanned along a defined path, where it modifies the target upon energy transfer from the beam to the material. Additionally, superimposing multiple laser beams yields, owing to its large coherence length and defined polarization, complex intensity patterns at sub-wavelength resolution, which can be transferred into a suitable photosensitive material. Both methods do not require photolithographic masks for patterning a material and are described in Sect. 1.1.3.1.

The concept of laser processing takes advantage of the high definition of the laser radiation with respect to its high intensity, spatial anisotropy (directionality) and spectral properties [8]. The high intensity of a focused laser sources generally alters the target material in a confined region around the incident laser light. Depending on the optical properties of the material, energy is transferred from the laser into a small volume of the material leading to a local increase of the temperature. The evolution of the temperature is governed by the thermal properties of the target material and the amount of deposited energy. Numerous effects can take place in dependence on the deposited net energy (energy deposition in the volume by the laser minus energy diffusion out of the interaction volume) such as phase changing, melting, evaporation or ablation. At low laser intensities, a gentle heating by the laser is generally induced, while at highest laser intensity explosive material ejection and plasma formation is observed. The main difference between high and very high intensity is the duration in which the energy of the laser is deposited in the material. Short (ns) and ultra-short (ps to fs) pulses deposit the energy on a time scale much smaller than diffusion occurs and the heat cannot spread in the material. This leads to a superheated material and eventually to the mentioned effects (Fig. 1.1).

## 1.1.3 Biomimetic Processes Involving Laser Radiation

### 1.1.3.1 Biomimetic Laser-Based Lithography

A straight forward application is to use the laser as *shaping tool* that modifies or creates geometry from or within a target material. The laser acts as a light pencil that draws either in two or three dimensions a structure with a biomimetic shape in a material. The term drawing is quite general and includes processes commonly associated with metals, ceramics or other inorganic materials such as removal (e.g. ablation, etching, milling, drilling, and cutting), joining (welding, sintering), marking or addition (e.g. laser cladding, cusing, laser assisted growth) of material or other

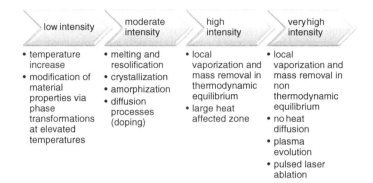

| low intensity | moderate intensity | high intensity | very high intensity |
|---|---|---|---|
| • temperature increase <br> • modification of material properties via phase transformations at elevated temperatures | • melting and resolification <br> • crystallization <br> • amorphization <br> • diffusion processes (doping) | • local vaporization and mass removal in thermodynamic equilibrium <br> • large heat affected zone | • local vaporization and mass removal in non thermodynamic equilibrium <br> • no heat diffusion <br> • plasma evolution <br> • pulsed laser ablation |

**Fig. 1.1** Mechanisms in solid target material at increasing laser intensity

processes like photocrosslinking, photopolymerization, or photoactivation associated with photosensitive organic or biomaterials.

Methods related to laser based shape creation are often termed as rapid prototyping methods and include direct writing methods (direct laser writing–DLW) with a focused laser beam or with multiple interfering laser beams (laser interference lithography–LIL).

Regarding biomimetic applications, the importance of true 3D structuring capabilities of a lithographic method is often emphasized (cf. e.g. Chap. 2). 3D structures can be built by a number of laser based rapid prototyping and fabrication methods (cf. Chaps. 4, 6 and 9) such as (micro-) stereo lithography ($\mu$-SL), selective laser sintering (SLS), or laser additive manufacturing (LAM), which all facilitate a laser-based solid free-form (SFF) fabrication, which are discussed throughout the book chapters. Biomimetic applications of 3D structures are medical implants, such as dental bridges or biologically inspired lightweight structures (cf. Chap. 6).

A common approach for 3D structures is the micro-stereo-lithography ($\mu$-SL) [9], where the shape is built layer-by-layer in a photosensitive material. The pattern in each layer is built either by scanning the focus of a laser or by projecting patterned UV light via masks or a digital mirror device onto the material. The 3D structure is finally built as a layer stack in a repetitive process by vertically translating the material reservoir. This imposes several restrictions on the structure regarding the 3D design (geometrical restrictions due to layer-by-layer processing) and limit the structure resolution. The vertical resolution is limited by the achievable thickness of the individual layers. The lateral resolution is determined from the spot size of the light on the layer. In contrast to the use of conventional UV or VIS lasers in the linear absorption regime of the target material, absorption via inter-band transitions involving more than one photon is a key issue regarding multi-photon based laser lithography, which takes advantage of the strong confinement of the energy transfer of a tightly focused laser beam to a photosensitive material. The energy transfer is responsible for a modification of the material only around the laser focus, which enhances spatial resolution and enables true 3D structuring. The smallest exposed volume element is typically called a voxel and represents the smallest building unit of a 3D structure. μ-SL involving multi-photon based exposure with suitable materials with enhanced spatial resolution was demonstrated for building rather large structures [10].

Multi-photon based laser lithography is a relatively simple method that is inherently capable of 3D structuring, but requires lasers with high peak intensities of the laser pulses. The technological maturity of ultrafast (femtosecond) lasers proved multi-photon-absorption as realistic exposure mechanism with several technical benefits over single-photon exposure. Most important technical benefits of this method are true 3D structuring capabilities and high spatial resolution beyond the diffraction limit (e.g. sub 100 nm structures fabricated with a laser wavelength of 800 nm), which is controlled by the number of applied laser pulses and the laser pulse energy. Finally, this method has found its way to biomimetics as presented in several chapters throughout the book. Recently, Misawa and Juodkazis edited a very comprehensive book about this kind of 3D laser microfabrication [11]. Multi-photon based direct laser writing is a true 3D method that has definitely reached a level of technical perfectness over the last years and may potentially replace other direct writing methods, such as electron-beam lithography in a wide field of applications. It has been applied as a versatile tool in micro- and nano-fabrication anf had been reviewed comprehensively (e.g. [12, 13]) for 3D structures with arbitrary shapes, with freely moving components [14] periodic structures and scaffold structures for photonic and biomedical applications [15, 16], flexible structures for biological cell culture studies [17]. It was also applied to micro replication of biological architectures for cellular scaffolds or custom tissue replacements [18], or in vivo processing of scaffolds with embedded living organisms [19]. Chapter 3 demonstrates biomimetic chiral structures made by 3D laser lithography.

Smooth 2.5D topologies can be achieved with a local variation of the exposure dose with a suitable contrast of the resist. Direct beam writing methods such as direct laser write grey-tone lithography or focused ion beam lithography are capable of fabricating smooth relief structures in photosensitive polymer materials such as commercial positive-tone resists. Due to their inherent capability of varying the exposure dose as a function of the beam position, such methods are often used for the generation of a continuous relief in the target material [20], which could find applications on textured surface with taylored wetting or anti-fouling properties. Often, these structures have a certain periodicity and must cover large surfaces. In such cases, the scanning of a single focused laser beam is a too time consuming process and large areas may not be structured within a reasonable time. For this kind of structures laser interference lithography (LIL) can be used alternatively. LIL is capable of structuring large areas in a single shot exposure (or limited exposure time) without defects and without scanning, but the periodic interference pattern of the laser light limits fabrication to periodic patterns.

The laser provides light of defined wavelength, polarization and coherence, thus enabling coherent superposition of multiple laser beams, whereas the experimental conditions such as laser fluence, film thickness, angle of incidence and polarization of the beams directly correlate to the fabricated patterns.

The beam superposition of multiple laser beams leads to the generation of stable interference patterns, which can be used for patterning films of (usually) positive and negative type photoresist negative type resist, $TiO_2$ gel films [21], hybrid organic-inorganic sol-gel materials [22], biomimetic tissue [23], as well as PEDOT-PSS [24], a conducting polymer, which is important for organic (opto) electronics. The periodicity of the pattern is determined by difference between the wave vectors of the interfering beams and proportional to the wavelength of the laser and the angle between the interfering beams, which are thus crucial and limiting parameters for the achievable spatial resolution. Depending on the number of beams, angle between the beams and polarization, 1D, 2D, and 3D periodic patterns can be fabricated over a large area in a single exposure step. Multiple exposure steps with rotation and translation of a 1D phase mask facilitate complex 3D patterns such as woodpile structures with three beams [25]. The coherence length limits path differences in the optical setup and determines also the maximal area that can be processed in a single exposure step.

Various types of laser sources such as Nd:YAG lasers at 266 nm (fourth harmonic) or 355 nm (third harmonic) [26] are used for interference lithography. Regarding costs, large area LIL with semiconductor lasers seems more attractive. The used AlInGaN laser has a rather low prize and a long coherence length, which is a pre-requisite for processing large sample areas [27]. Recently, LIL was combined with multi-photon polymerization in a four-beam setup for the fabrication of micro lenses. The four beams were generated using a diffractive optical element and a diaphragm to remove undesired laser light from the optical path. The negative type resist was exposed to multiple laser pulses, which facilitates a much faster processing as compared to multi-photon based direct laser writing, which is a sequential voxel-by-voxel build up process. Appropriate hatching or stepping of the exposed area enables the processing over large areas and reduction of the structure degradation at the edge of the exposed are due to the spatial intensity profile of the laser beams [28]. Originally used for regular 2D patterns, interference lithography is increasingly applied to 3D structures.

### 1.1.3.2 Biomimetic Laser-Based Coating and Deposition Methods

The previous section introduced the laser as a direct writing tool for the creation of 2D and 3D biomimetic structures. Mimicking the nature by the generation of bio-compatible coatings and creating environments for living cells is another aspect of biomimetics that can be dealt with laser technology. In this manner, the laser can be used for the deposition of various biomaterials and research in cell biology. Several methods (laser chemical vapour deposition, laser induced transfer methods, pulsed laser deposition etc.) exist, where the laser is used for the patterned deposition of materials on a substrate. Laser chemical vapour deposition (LCVD) was used for the in situ fabrication of micro lenses with precise control of film properties [5]. This process takes place in a reaction chamber, which contains precursor gases and the substrate. The laser is used to locally heat the substrate, which subsequently dissociates the gas precursor and a thin film deposits on the substrate. Using multiple beams or a layered approach to build the structure, it is possible to create also 3D structures. The deposition rate of LCVD depends linearly on the precursor gas pressure and the laser power density and decreases with increasing scanning speed. The deposition rate can be adjusted by these parameters and is much higher than in conventional CVD [29]. Pulsed laser deposition (PLD) methods, ablation mechanisms and applications are discussed in Chap. 7.

Sensitive materials such as biomaterials that are easily destroyed by the laser are often embedded in a matrix material that absorbs the laser energy. This method (matrix assisted pulsed laser deposition–MAPLE) uses a frozen solvent, which is evaporated upon laser irradiation. The material for deposition is evaporated together with the matrix and deposits on a receiving substrate (cf. Chap. 5). Without masks, the material deposition is unpatterned, hence another method, laser induced forward transfer method (LIFT) is often used for laser assisted patterning (cf. Chap. 8). In the LIFT (sometimes also called laser based bio printing–LBP) process, the laser energy is absorbed in a thin film on a transparent substrate, which leads to droplet formation and ejection of the transfer material. Subsequently, the evaporated material precipitates on a second receiving substrate, which is facing the first substrate either in close (micrometer) vicinity. For soft-matter materials such as polymers or biological compound materials, a direct contact between the substrates was found to yield best transfer results regarding resolution and defined edges of transferred pixels [30].

LIFT can be achieved with various types of lasers (UV excimer lasers, Nd:YAG, Ar-ion lasers, fs lasers). The transfer materials are often sensitive to oxygen or humidity, thus requiring a vacuum or inert gas setup. Originally used for the patterned transfer of metal films, it can be applied for a variety of materials including oxides and biomaterials or even more complex multi-layer systems such as a polymer light emitting diode pixel [31] or organic thin film transistors [32]. Such sensitive materials or materials, which are transparent to the incident laser or easily destroyed by the incident laser, can be transferred by using an energy absorbing sacrificial layer (dynamic release layer) between the transfer material and the carrier, which promotes the release of the material. Additionally, the temporal shape of ultrafast laser pulses influences the LIFT process and the achievable resolution on the receiving substrate, which is attributed to fast electron and lattice interactions. It was shown that fs pulses with a short separation (less than 500 fs) show large impact of the deposited pixel size, while the covered area stays constant for longer pulse separations up to 10 ps [33].

Using microsphere arrays as micro lenses, parallel material transfer (parallel LIFT) with an unfocused laser beam can be achieved [34]. Polystyrene beads are on top of a transparent substrate (quartz glass) and focus the incident light onto the single or multi-layered transfer material, which is on the other side of the substrate. Thus, micron to sub-micron holes can be written into the films and corresponding dot patterns on the receiving substrate.

### 1.1.3.3 Biomimetic Laser-Based Biomaterial Processing and Tissue Engineering

From a materials point of view, the laser supports rapid tooling for natural biopolymers (e.g. proteins, polysaccharides) and artificial biomolecules (PLA, PGA, PLGA etc.). Laser radiation offers the possibility to generate 3D biological microstructures (scaffolds) by crosslinking of oxidizable side chains in biomolecules. The purpose is the creation of chemically and physically defined cell environments for applications such as tissue regeneration and gene delivery as described in Chaps. 2, 9, and 10.

In the field of tissue engineering, which is closely related to replace or repair tissue such as bone, cartilage, blood vessels, bladder, skin, muscle etc. the control of cell density and organization is crucial. Laser based method such as laser assisted bioprinting implementing the laser induced forward transfer can deposit patterned films of bioink on substrates. Chapter 8 discusses examples of printed cell types.

## References

1. Steindorfer MA, Schmidt V, Belegratis M, Stadlober B, Krenn JR (2012) Detailed simulation of structural color generation inspired by the morpho butterfly. Opt Express 20:21485–21494
2. Neinhuis C, Barthlott W (1997) Characterization and distribution of water-repellent, self-cleaning plant surfaces. Ann Bot 79:667–677
3. Table 1 adapted from Bäuerle D,(2008) Laser: grundlagen und anwendungen in photonik, technik, medizin und kunst. Wiley-VCH, Verlag GmbH & Co KGaA, Weinheim. ISBN 978-3-527-40803-7
4. Partel S, Zoppel S, Hudek P, Bich A, Vogler U, Hornung M, Voelkel R (2010) Contact and proximity lithography using 193 nm excimer laser in mask aligner. Microelectron Eng 87:936–939
5. Wang Q, Zhang Y, Gao D (1996) Theoretical study on the fabrication of a microlens using the excimer laser chemical vapor deposition technique. Thin Solid Films 287:243–246
6. Chiu C-C, Lee Y-C (2011) Fabricating of aspheric micro-lens array by excimer laser micro-machining. Opt Lasers Eng 49:1232–1237
7. Vossmerbaeumer U (2010) Application principles of excimer lasers in ophthalmology. Med Laser Appl 25:250–257
8. Vainos N (ed) (2012) Laser growth and processing of photonic devices. Woodhead Publishing, ISBN: 978-1845699369
9. Neumeister A, Himmelhuber R, Materlik C, Temme T, Pape F, Gatzen H, Ostendorf A (008) Properties of three-dimensional precision objects fabricated by using laser based micro stereo lithography. JLMN-J Laser Micro/Nanoeng 3(2):67–72
10. Houbertz R, Steenhusen S, Stichel T and Sextl G (2010) Two-photon polymerization of inorganic-organic hybrid polymers as scalable technology using ultra-short laser pulses. In: Duarte (FJ) Coherence and ultrashort pulse laser emission 583. InTech, Rijeka
11. Misawa H and Juodkazis S (eds) (2006) 3D laser microfabrication: principles and applications. Wiley-VCH, Verlag
12. Sun HB, Kawata S (2004) Two-photon photopolymerization and 3D lithographic microfabrication. APS 170:169–273
13. Maruo S, Fourkas JT (2008) Recent progress in multiphoton microfabrication. Laser Photonics Rev 1–2:100–111

14. Sun H-B, Kawakami T, Xu Y, Ye J-Y, Matuso S, Misawa H, Miwa M, Kaneko R (2000) Real three-dimensional microstructures fabricated by photopolymerization of resins through two-photon absorption. Opt Lett 25(15):1110

15. Ovsianikov A, Chichkov B, Mente P, Monteiro-Riviere NA (2007) Two photon polymerization of polymer-ceramic hybrid materials for transdermal drug delivery. Int J Applied Ceram Technol 4(1):22–29

16. Ovsianikov A, Ostendorf A, Chichkov BN (2007) Three-dimensional photofabrication with femtosecond lasers for applications in photonics and biomedicine. Appl Surf Sci 253:6599–6602

17. Klein F, Striebel T, Fischer J, Jiang Z, Franz C, von Freymann G, Wegener M, Bastmeyer M (2010) Elastic fully three-dimensional microstructure scaffolds for cell force measurements. Adv Mater 22:868

18. Nielson R, Kaehr B, Shear JB (2009) Microreplication and design of biological architectures using dynamic-mask multiphoton lithography. Small 5(1):120–125

19. Torgersen J, Baudrimont A, Pucher N, Stadlmann K, Cicha K, Heller C, Liska R and Stampfl J (2010) In vivo writing using two-photon-polymerization. In: Proceedings of LPM2010 - The 11th International Symposium on Laser Precision Microfabrication

20. Fu Y, Ngoi BKA (2001) Investigation of diffractive-refractive microlens array fabricated by focused ion beam technology. Opt Eng 40:511

21. Wang Z, Zhao G, Zhang X, Heguang L, Zhao N (2011) Fabrication of two-dimensional lattices by using photosensitive sol-gel and four-beam laser interference. J Non-Cryst Solids 357:1223–1227

22. Della Giustina G, Zacco G, Zanchetta E, Gugliemi M, Romanato F, Brusatin G (2011) Interferential lithography of bragg gratings on hybrid organic-inorganic sol-gel materials. Microelectron Eng 88:1923–1926

23. Daniel C (2006) Biomimetic structures for mechanical applications by interfering laser beams: more than solely holographic gratings. J Mater Res 21(8):2098

24. Lasagni AF, Hendricks JL, Shaw CM, Yuan D, Martin DC, Das S (2009) Direct laser interference patterning of poly3, 4-ethylene dioxythiophene-polystyrene sulfonate (PEDOT:PSS) thin films. Appl Surf Sci 255:9186–9192

25. Xu D, Chen KP, Ohlinger K, Lin Y (2010) Holographic fabrication of three-dimensional woodpile-type photonic crystal templates using phase mask technique. In: Kim KY (ed) Recent optical and photonic technologies. ISBN 978-953-7619-71-8, p 450

26. Lasagni AF, Roch T, Langheinrich D, Bieda M, Wetzig A (2011) Large area direct fabrication of periodic arrays using interference patterning. Phys Procedia 12:214–220

27. Byun I, Kim J (2010) Cost-effective laser interference lithography using a 405 nm AlInGaN semiconductor laser. J Micromech Microeng 20:55024

28. Stankevicius E, Malinauskas M, Raciukaitis G (2011) Fabrication of scaffolds and microlenses array in a negative photopolymer SZ2080 by multi-photon polymerization and four-femtosecond-beam interference. Phys Procedia 12:82–88

29. Hon KKB, Li L, Hutchings IM (2008) Direct writing technology–advances and developments. CIRP Ann Manuf Technol 57:601–620

30. Palla-Papavlu A, Dinca V, Luculescu C, Shaw-Stewart J, Nagel M, Lippert T, Dinescu M (2010) Laser induced forward transfer of soft materials. J Opt 12:124014

31. Shaw Stewart J, Lippert T, Nagel M, Nüesch F, Wokaun A (2011) Laser-induced forward transfer of polymer light-emitting diode pixels with increased charge injection. ACS Appl Mater Interfaces 3:309–316

32. Rapp L, Nénon S, Alloncle AP, Videlot-Ackermann C, Fages F, Delaporte P (2011) Multilayer laser printing for organic thin film transistors. Appl Surf Sci 257:5152–5155

33. Papadopoulou EL, Axente E, Magoulakis E, Fotakis C, Loukakos PA (2010) Laser induced forward transfer of metal oxides using femtosecond double pulses. Appl Surf Sci 257:508–511

34. Othon CM, Laracuente A, Ladouceur HD, Ringeisen BR (2008) Sub-micron parallel laser direct-write. Appl Surf Sci 255:3407–3413

# Chapter 2
# Direct Laser Writing

Sascha Engelhardt

**Abstract** Direct laser writing has emerged in recent years as a powerful technology for the realization of micron to sub-micrometer resolution structures in the field of biomedicine. The technology is based on the nonlinear optical effect of two-, or multi-photon absorption, inducing photochemical effects in a defined volume. These photochemical effects can be utilized for the fabrication of microstructures, as well as for a defined 3D chemical surrounding. In this contribution, the basic principles of direct laser writing are described, followed by an explanation of process relevant aspects and a short survey of available techniques and technologies for enhanced performance. In the last part of this chapter, some examples of direct laser writing in the field of 3D cell culture and tissue engineering are given.

## 2.1 Introduction

In his satirical novella "Flatland: A Romance of Many Dimensions", published 1884, Edwin A. Abbott describes a two-dimensional world, Flatland, inhabited by simple geometrical figures. When visited by a sphere of the 3D spaceland, the protagonist has difficulties to grasp the existence of such a world. In his world, only a two-dimensional representation of the sphere, a circle, can be perceived. In the end the protagonist accepts the existence of 3D-beings and hypothesis about even higher dimensional worlds [1].

In analogy, much research concerning cell biology has been restricted to a two-dimensional world over the last years. However, tissue is a 3D object, and cells as part of the tissue are 3D by nature. Studying cells in a petri dish is like studying the sphere, from the point of view of a flatlander. Aspects of their nature will be hard

S. Engelhardt (✉)
RWTH Aachen, Institute for Laser Technology, Steinbachstraße 15, 52074 Aachen, Germany
e-mail: sascha.engelhardt@ilt.fraunhofer.de

S. Engelhardt
Fraunhofer Institut for Laser Technology, Steinbachstraße 15, 52075 Aachen, Germany

V. Schmidt and M. R. Belegratis (eds.), *Laser Technology in Biomimetics*,
Biological and Medical Physics, Biomedical Engineering,
DOI: 10.1007/978-3-642-41341-4_2, © Springer-Verlag Berlin Heidelberg 2013

to understand, or even impossible to analyze if the third dimension is not accounted for. In recent years, 3D cell culture systems address this issue and have revealed remarkable difference in cell behavior when compared to standard flat cell culture. One possibility to generate arbitrary 3D biomimetic structures is direct laser writing. Here, a focused laser beam is translated through a material bath. In a confined volume in the focal region, a photochemical reaction occurs which is used for the fabrication of the 3D structures. These structures may be physical structures as a result of a curing process, or chemical structures due to a photochemical activation process. The 3D capability of direct laser writing has its origin in a nonlinear behavior. In most cases direct laser writing is associated with a two- or multi-photon process [2]. In this case, nonlinear absorption processes occur when the photon density is high. In direct laser writing these high photon densities are only present in the vicinity of a tightly focused laser beam with short pulse duration. Thus, the reactive volume is restricted to just a few femtoliters and high resolution 3D microstructures can be generated.

In this book chapter, the basic principles of direct laser writing will be described. A typical setup for direct laser writing is introduced in Sect. 2.2 before the process of two-photon absorption is described in more detail in Sect. 2.3. The photochemical processes, which are used in direct laser writing, are subject of Sect. 2.4. Section 2.5 will focus on specific aspects of direct laser writing, which have to be considered while applying the technology. Section 2.6 deals with some technological advances, which address the main drawbacks of direct laser writing, before an overview of biomimetic applications of the method is given in Sect. 2.7.

## 2.2 Experimental Setup

Depending on the application, several possibilities exist for direct laser writing setups. However, the required basic components are quite similar (several reviews exist, e.g. [3–9]). In order to initiate nonlinear optical effects, which are responsible for 3D direct laser writing, high photon densities are necessary. These photon densities are normally realized by tightly focusing a short- or ultrashort (fs–femtosecond) pulsed laser beam with a high NA objective (NA–numerical aperture, 0.5 and above). The laser wavelength is in the visible (VIS) or near infrared range (NIR) range. A common laser source is a mode-locked Ti:Sapphire ultrafast oscillator, which provides laser pulses with typical pulse durations in the range of tens to hundreds of femtoseconds, at a center wavelength of approximately 780 nm. These laser systems may either work as an amplified Ti:Sapphire systems usually operating in the kHz range or long cavity oscillators working at low MHz frequencies. The intensity of the laser beam has to be tuned finely, since the direct laser writing process sensitively depends on it, especially when the process is aimed at a sub-diffraction limited resolution. Although neutral density filters can be used to tune the laser intensity, they do not have the ability for such fine tuning. Therefore other approaches, such as a combination of a $\lambda/2$-waveplate and a polarizing beam splitter, acousto-optical modulators (AOM) or Pockels cells are often used.

Since a mode-locked laser source provides a continuous pulse-train and hence cannot be triggered, the on- and off-state of the laser beam is either controlled by a mechanical shutter or an optical device, such as an AOM. The advantage of the AOM is its fast response time (ns range), which is several magnitudes smaller compared to mechanical shutters (ms range). However, mechanical shutters are more easily implemented and are less susceptible to imprecise alignment. Additionally, AOMs are optimized for a certain wavelength and therefore not suited if a second laser source with a different wavelength is needed in the same setup. In general, AOMs should be utilized if the fast response times are mandatory for the application, and mechanical beam shutters if ease of use and versatility of the setup are of interest.

Beam expansion and collimation, realized with a telescope, is necessary to slightly overfill the back aperture of the focusing objective, so that the entire NA of the objective can be used and thus the resulting focal point is diffraction limited.

Besides these essential components, an optical isolator, for example a Faraday rotator is often implemented to avoid interference from reflected light, which can cause problems in the direct laser writing process. A cost efficient, but less effective method to achieve this goal is the insertion of a slightly misaligned neutral density filter. When femtosecond laser pulses pass a dispersive medium such as glass, the pulse duration broadens due to group velocity dispersion (GVD). The pulse broadening leads to a reduction in laser intensity, which in consequence reduces the two photon absorption efficiency. The optical component which causes most GVD in a standard direct laser writing setup is the objective, which consists of a complex lens system, often with a total thickness of several centimeters. This effect can be counteracted by introducing inverse GVD, which can be accomplished by a pair of prisms, or by diffractive mirrors.

3D-laser writing is accomplished by a computer controlled translational stage, which is able to translate 3D computer aided designed (CAD) data into axis movement. Piezo driven stages are often used because of their high accuracy. However, for piezo stages, total translation lengths are limited to a few hundred micrometers. If larger structures are desired, linear stages with less accuracy or stitching becomes necessary.

In order to identify the right focal position, most setups include an online monitoring system (Fig. 2.1).

Beside this basic setup, several other setups have been proposed which mostly have the aim of decreasing process time, or expand the possible structure sizes. An overview of these approaches is given in Sect. 2.6 of this chapter.

## 2.3 Two Photon Absorption

A molecule can be excited from its ground state $E_g$ to an energetically higher state $E_e$ if the molecule absorbs a photon with energy equal or larger than the energetic gap between the molecular states $\Delta E = E_e - E_g = \hbar\omega_a$ (Fig. 2.2). Since the transition of the electronic states occurs much faster than the correlated change in nuclear motion

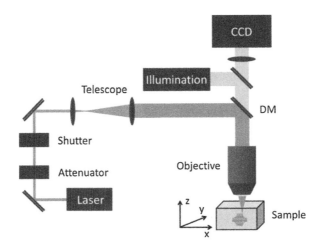

**Fig. 2.1** Optical pathway of a basic setup for direct laser writing. The laser beam of a suitable laser system, delivering ultrashort laser pulses can be attenuated by neutral density filters, a combination of a λ/2-waveplate and a polarizing beam splitter, an AOM or a Pockels cell. The beam status is controlled by an optical or mechanical shutter. A telescope expands the laser beam to a diameter where it slightly overfills the back aperture of a focusing objective. A dichroic mirror (DM) separates the incoming laser beam from an imaging system, consisting of an illumination source and a tube lens, imaging the focal region on a camera, such as a CCD chip for online monitoring. A translational stage moves the photosensitive material in all three spatial directions relative to the focal position

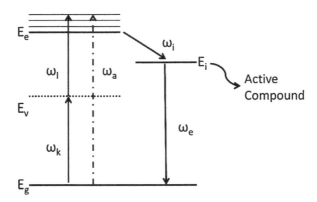

**Fig. 2.2** Energy level diagram for the single and two photon absorption process. Both processes can lead to the generation of an active compound for photochemical processes

(Franck-Condon principle), the molecule undergoes internal conversion until reaching an intermediate state $E_i$, from which the molecule can relax to its ground state by the emission of a photon with lower energy $\hbar\omega_e$, which has a longer wavelength (red shifted) compared to the absorbed photon. Two photon absorption describes a similar process, where $\Delta E$ is realized by the almost coincidental absorption of two photons with $\Delta E = \hbar\omega_k + \hbar\omega_l$. The physical derivation of this process was already described 1931 by Maria Göppert-Mayer through second order perturbation theory [10]. In this chapter, a more qualitative explanation will be given. If a photon of energy $\hbar\omega_k < \Delta E$ interacts with the molecule, no real transition from $E_g$ to $E_e$ is allowed. However, for a short period of time $\Delta t \leq \frac{\hbar}{2\,\hbar\omega_k}$, given by the energy time uncertainty principle, the molecule can be elevated to a virtual state $E_v$. During this time, a second photon has only to overcome the energy gap $\hbar\omega_l = \Delta E - \hbar\omega_k$ and a real transition can occur. This transition is only possible, if the combined energy of the two incident photons is sufficient to bridge the gap between the ground and the excited state of the molecule:

$$\hbar\omega_k + \hbar\omega_l - \hbar\omega_i - \hbar\omega_e = 0$$
$$\hbar\omega_k + \hbar\omega_l = \hbar\omega_a \qquad (2.1)$$

Of practical importance is the degenerate case $\hbar\omega_k = \hbar\omega_l = \frac{1}{2}\hbar\omega_a$, where two photons of the same energy are absorbed. In terms of technical implementation this means that only a single laser source is required to initiate two photon absorption processes. From this simple picture the activation rate can be approximated.

The number of photons interacting with the molecule per time is given by $\sigma N$, where $\sigma$ is the absorption cross section of the target molecule, and N denotes the number of photons per area and time. The number of photons initiating the two photon absorption process in the time frame $\Delta t$ is $\sigma \Delta t N^2$. Thus the two photon transition rate $\frac{\partial n}{\partial t}$ per area can be written as:

$$\frac{\partial n}{\partial t} = \frac{\sigma^2 \Delta t}{2} M\,N^2$$
$$= \frac{\sigma_2}{2} M\,N^2 \qquad (2.2)$$

with $M$ as the density of absorbing molecules. $\sigma_2$ represents the two photon absorption cross section. The two photon initiation rate is directly proportional to the two photon absorption cross section, the density of absorbing and the squared photon flux (cf. 2.2).

In this simple picture of two photon absorption, the two photon absorption cross section follows the single photon absorption cross section at half the wavelength $\sigma_2(\lambda) = \sigma^2(\lambda/2)\Delta t(\lambda)$. Indeed, this trend can be observed for many absorbing molecules [11–13] and can serve as a basic thumb rule for choosing appropriate absorbing molecules.

However, for quantitative predictions concerning $\sigma_2$ this qualitative model is too simple. The enhanced green fluorescent protein (eGFP), which is a commonly used chromophore in cell biology, has a maximum absorption cross section at 489 nm of approximately $2 \times 10^{-16}\,cm^2$. The corresponding maximum two photon absorption cross is situated at 920 nm and is approx. 40 GM (1 GM $= 1 \times 10^{-50}\,cm^4$ s/photon) [11]. Based on the simple model $\sigma^2\Delta t$, results in a value for $\sigma_2$ that is two orders of magnitude larger than the value actually measured.

Additionally a significant blue shift of 29 nm between the one and two photon case can be observed. The first deviation has its origin in the far more complex nature of two photon absorption. Actually for molecules with an inversion center, or other symmetric molecules, different transition rules apply for one- and two photon induced excitation. For centrosymmetric molecules, a one photon allowed state $E_e$ cannot be reached directly by a two photon process. In this case only transitions between states with different parity can be reached by one photon absorption and with equal parity by two photon absorption [14].

The blue shift was theoretically explained for the green fluorescent protein [12]. A number of Frank–Condon active vibronic modes lead to a stretching, or bending around a central atom. These vibrations lead to a significant non-Condon influence on the excitation process, which in consequence lead to the observed blue shift.

In summary, the described very simple model can give a good qualitative explanation of the two photon excitation process, with regard to basic behavior of absorbing molecules. A more precise model of the excitation process has to include the actual structure of the molecule and its symmetries, as well as considering vibrational cross talks that influence absorption behavior.

This aim can be achieved by looking at the interaction of photons with the molecule that can be described by the polarization $\overrightarrow{P}$, represented by a power series of the electrical field $\overrightarrow{E}$:

$$\overrightarrow{P} = \varepsilon_0 \left[ \chi^{(1)} \overrightarrow{E} + \chi^{(2)} \overrightarrow{E}^2 + \chi^{(3)} \overrightarrow{E}^3 + \cdots \right] \tag{2.3}$$

$\varepsilon_0$ is the electric constant and $\chi^{(i)}$ denotes the ith order susceptibility. Two photon absorption belongs to the third order nonlinear polarization, since three electrical fields (described previously $\omega_k$, $\omega_l$ and $\omega_i$) interact to give a fourth electrical field, the nonlinear polarization given by $\omega_e$. Thus, two photon absorption is governed by the imaginary part of the third order susceptibility $\chi_{im}^{(3)}$, whereas the real $\chi_{real}^{(3)}$ part is associated with nonlinear refraction. The third order susceptibility is a representation of the more complex description of the absorbing molecule. The relationship between $\sigma_2$ and $\chi_{im}^{(3)}$ is given by [4]:

$$\sigma_2 = \frac{8\pi^2 \hbar \omega_k^2}{n^2 c^2} \chi_{im}^{(3)} \tag{2.4}$$

$n$ is the first order refractive index of the medium and $c$ is the vacuum speed of light. For direct laser writing, the two photon absorption cross section is one possibility to enhance the process efficiency, since it directly influences the absorption rate. By manipulating $\chi_{im}^{(3)}$ many substances with huge two photon absorption cross sections have been synthesized in the last years, resulting in two photon cross sections in the range of 102 GM to 103 GM (e.g. [14–16]).

## 2.4 Chemical Processes for Direct Laser Writing

### 2.4.1 Photopolymerization

Polymerization describes a chain reaction in which monomers are translated into a polymeric network (Fig. 2.3). Polymerization can be initiated by several trigger mechanisms like heat or radiation [17, 18]. In the case of photopolymerization, electromagnetic radiation is used, mostly in the form of UV-photons, to deliver the necessary energy for the activation of a photosensitive additive called photoinitiator (PI) [18]. There exist two distinct photopolymerization mechanisms which are employed for direct laser writing, radical photopolymerization and cationic photopolymerization [3]. Free radical photopolymerization uses radicals, which are generated after the excitation of the photoinitiator to start a polymerization chain reaction. These radicals are either a direct result of homolytical cleavage of the photoinitiator (type I photoinitiator), or an energy transfer to a coinitiator molecule (type II photoinitiator). Radicals $R^\circ$ react with monomers $M$, for example by addition to double bonds of acrylates,

$$R^\circ + M \rightarrow RM^\circ,$$

which then results in a chain propagation in the form of

$$RM^\circ + M \rightarrow RM_2^\circ \rightarrow \cdots \rightarrow RM_n^\circ.$$

Therefore a single radical can be responsible for a crosslinking chain reaction and the quantum yield for photopolymerization can easily reach values of 102–103 [19, 20]. The chain propagation continues, until two radicals interact to form a stable species, either by a second radical recombining with the growing polymer chain or a second growing polymer chain via electron transfer. Additional parasitic effects, such as radical quenching by singlet oxygen, limit the polymer propagation [19]. For a good material system, the polymer network forms readily after laser irradiation and stops when the light source is turned off. This behavior implies good process control and low dark polymerization and can only be achieved if activation-, propagation- and termination rates are in balance. The sudden change in material density due to polymerization leads to a change in the refractive index of the material. On the one hand, this behavior allows in-process observation of the polymerized structures, but on the other hand, the refractive index change can negatively influence the laser propagation through the forming structure. As a result many fabrication schemes utilize a layer-by-layer approach to minimize beam distortion, although one of the advantages of direct laser writing is its capability for arbitrary 3D structuring [5, 9, 21]. In general the fabrication path and the position of the focusing objective with respect to the forming structure have to be chosen carefully in order to minimize this effect.

Cationic photopolymerization can circumvent this problem. Here, the photoinitiator forms a strong Brønsted acid [19], which is capable of polymerizing epoxides or vinyl esters by ring-opening reactions. The following polymerization mechanism resembles that of radical polymerization; however, cationic polymerization possesses in general lower curing speeds as well as an increased dark polymerization. The catalytic nature of the reaction can also lead to quantum yields larger than one. For many substances, such as the commonly used SU-8 [22–26], the photo irradiation only activates the material and the polymerization is carried out subsequently, for example by thermal treatment of the material. Thus, there is no refractive index change during irradiation, which facilitates the realisation of 3D structures.

However, specimen induced spherical aberration caused by the refractive index mismatch of specimen and the medium between specimen and exit pupil of the focusing objective leads to focus distortion and a loss of focal intensity. This effect increases with increasing penetration depth of the focus into the specimen. In Sect. 2.5.3 the effect of spherical aberration on direct laser writing is discussed in more detail.

## 2.4.2 Photocrosslinking

Photocrosslinking is closely related to photopolymerization, but the involved materials do not undergo chain propagation (Fig. 2.3). It describes the photoinduced bond formation between two molecules or macromolecules. The photoinduced crosslinking of biomolecules, such as proteins, is described by photocrosslinking. Here, the photon energy is absorbed by a photosensitizer, which is first excited to a singlet state, followed by intersystem crossing to a triplet state [27, 28]. This excited photosensitizer can then directly interact with the biomolecule via hydrogen abstraction (type I reaction), or with ground state oxygen to form singlet oxygen, a highly reactive oxygen species (type II reaction). Both mechanisms are known to induce protein unfolding, fragmentation, aggregation and many other disruptive processes [29]. However, for direct laser writing the aggregation of proteins offers the possibility to generate 3D protein microstructure. This aggregation is driven by crosslinking of oxydizable side chains in proteins, such as tryptophan, tyrosin, histidine or cysteine [30–34]. As in the case of radical polymerization, crosslinking results in a refractive index change. In general, the quantum yield is smaller than one, because of the lack of chain propagation. Another aspect is the lifetime of the photosensitizer. While most photoinitiators decompose after irradiation to form free radicals, photosensitizers of type II can excite 103–105 singlet oxygen molecules before photobleaching occurs [27]. Additionally, proteins contain inherent photosensitizers in the form of aromatic amino acids, which show a strong linear absorption at wavelengths of 260–280 nm and are therefore susceptible for two photon absorption with visible light [35–37]. Each of these aromatic amino acids, such as phenylalanine, tyrosine or tryptophan exhibit only weak two photon absorption capabilities, but each protein contains several of them. Hence, it is possible to perform two-photon initiated crosslinking of bovine serum albumin (BSA) without an additional photosensitizer by using frequency doubled Nd:YAG laser sources emitting at 532 nm [38, 39].

The precise photochemical reaction pathways for protein crosslinking are not yet fully understood, because of the sheer protein size and the resulting multitude of possible binding mechanisms [29]. Several studies were performed to understand the underlying mechanisms for protein aggregate formation based on DNA binding, or synthetic polymers functionalized with different oxidizable groups (e.g. [30–33, 40–42]). These studies were able to identify several possible reaction mechanisms. For example it has been shown, that aggregates may form through radical-radical termination reactions such as the formation of di-tyrosine of two tyrosine derived phenoxyl radicals [32, 43], or through addition reactions of carbonyl function containing oxidized histidine with nucleophiles, such as lysine, arginine and cysteine side chains [44, 45].

## 2.4.3 Photoactivation

The two chemical processes described above, primarily lead to physical changes in the pristine material. The formation of covalent bonds between monomers or macromolecules changes solubility of exposed regions, which results in stable 3D-structures. Additionally, the density of crosslinking has a direct influence on mechanical characteristics, such as elasticity or hardness [19]. Besides physical changes, chemical modification through laser irradiation is also a possibility that has been pursued extensively in recent years (see for example the reviews [46, 47]). There exist several methods to achieve this goal; however, all these techniques are based on the photoactivation of a chemical moiety on an existing surface or polymer network (Fig. 2.3). In most cases hydrogels are used as a 3D material backbone in which the photoactivation leads to a 3D chemical microenvironment. This chemical functionalization can be primarily induced using one of two pathways [46, 47]. In the first case, the hydrogel which contains susceptible side groups is incubated with photosensitive molecules. Upon irradiation these photosensitive molecules are activated so they can form a covalent bond with side groups of the hydrogel [48–50]. In the following sections, this process is denoted photofunctionalization. The desired chemical functionality is in this case defined by the photosensitive molecule. In the second case, the hydrogel contains reactive side groups, which are protected by a photolabile group, acting as a cage for the actual functional group. Upon exposure to light these protective groups are cleared off, and are able to leave the reactive side group of the hydrogel [51, 52] which delivers the desired chemical functionality. In both cases additional subsequent wet chemical steps can be performed, in order to further bind functional moieties, such as proteins or adhesive peptides covalently to the generated chemical active sides [53] (Fig. 2.3).

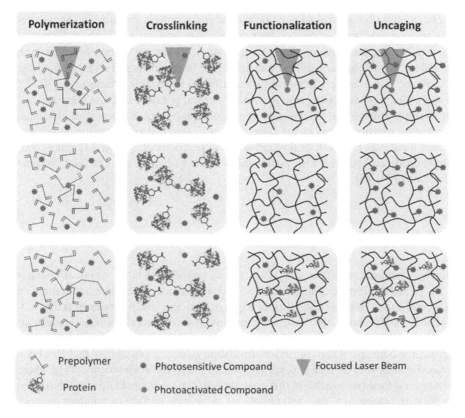

**Fig. 2.3** The four main photochemical reactions used for direct laser writing: photopolymerization, photocrosslinking, photofunctionalization and photouncaging (inspired by [54]). All four reactions begin with the activation of a photosensitive compound. In the case of photopolymerization, this active compound leads to a chain reaction, which crosslinks prepolymers. Crosslinking resembles polymerization, but without the chain reaction. Functionalization and uncaging describe chemical processes, which result in a functional moiety embedded in an existing network by addition of a functional group, or removing of a protective group. Details see text

## 2.5 Principles of Direct Laser Writing

Direct laser writing utilizes a focused laser beam to initiate a photochemical reaction as described in the previous section. By scanning the focus through all three spatial directions, arbitrary structures can be directly written inside the sample.

In order to manipulate a substance in three dimensions via laser irradiation, the photo induced reaction has to be localized to a finite 3D volume. One possibility to achieve such a volume is the quadratic two photon absorption dependency of the photon flux (2.2). Two main prerequisites for direct laser writing have to be met: First, since the probability of linear absorption is far greater than of nonlinear absorption, it is mandatory that the substance used for direct laser writing has a small linear

absorption cross section for the laser wavelength. If the material is not transparent
to the laser wavelength the formation of cavities and other unwanted effects will
dominate the process. Additionally, according to the Lambert-Beer-law, the photon
flux will exponentially decrease with the penetration depth, making 3D manipulation
nearly impossible. Second, in order to initiate two photon absorption, a localized high
photon density is necessary. This feature is normally achieved by using ultrashort
pulsed laser sources in combination with a tightly focusing microscope objective. In
the vast majority, Ti:Sapphire oscillators, emitting near-IR laser pulses with a pulse
width of approximately 100 fs are used for direct laser writing [3, 5, 7, 9]. Because
of their small pulse width these lasers are able to deliver massive photon fluxes, while
obtaining a minimum of thermal heating. Additionally, many UV absorbers used for
photopolymerization are designed to work for a wavelength of 300–400 nm, which
is suited for NIR irradiation at 800 nm. However, laser sources with much longer
pulse widths or even cw-laser sources have been applied for direct laser writing
[38, 39, 55–58]. For an improved understanding of the laser source prerequisites, a
closer look on the two photon initiated transition rates is necessary. The transition rate
$\partial n/\partial t$ depends linearly on the two photon absorption cross section $\sigma_2$, the number
density of absorbing molecules $M$ and quadratically on the photon flux $N$. The
two photon absorption cross section $\sigma_2(\lambda)$ strongly depends on the employed laser
wavelength $\lambda$ and is generally higher for smaller wavelengths, since many of the
employed photosensitive compounds exhibit their maximum two photon absorption
cross section in the visible range [13]. For the comparison of laser sources, a constant
$\sigma_2$ and $M$ is assumed. Additionally a quantum yield $\phi$ has to be considered, since
not every transition leads to a molecule being able to initiate a chemical reaction.
When diffusion effects are neglected, the total photon flux in the case of pulsed laser
irradiation can be approximated by integration over a single pulse and multiplication
with the laser repetition rate $v$:

$$\frac{\partial R_i^{pulsed}}{\partial t} = \frac{\sigma_2 \Phi M}{2} \int\limits_{pulse} N(t)^2 dt \upsilon$$

$$\frac{\partial R_i^{cw}}{\partial t} = \frac{\sigma_2 \Phi M}{2} \cdot N(t)^2 \qquad (2.5)$$

$\partial R_i^{pulsed}/\partial t$ denotes the initiation rate for pulsed laser irradiation and $\partial R_i^{cw}/\partial t$ the
corresponding initiation rate for the continuous wave case. If the pulse duration $\tau_p$
is small compared to the time between two pulses $\upsilon^{-1}$, the photon flux $N(t)$ can be
written as $N_0$ with the average laser power $P$ and the beam radius $r$:

$$N_0 = \frac{P}{\upsilon \pi r^2 \tau_p \hbar \omega} \qquad (2.6)$$

The beam radius can be approximated by the resolution limit of the focusing objective
to $r \approx 0.61\lambda\ NA^{-1}$, with the numerical aperture $NA$. Thus the initiation rate for
pulsed and continuous irradiation is given by:

$$\frac{\partial R_i^{\text{pulsed}}}{\partial t} = \frac{\sigma_2 \Phi M}{2} \cdot \frac{P^2 NA^4}{\upsilon \tau_p \pi^4 0.55 \lambda^2 \hbar^2 c^2} = K_M \cdot K_L^{\text{pulsed}}$$

$$\frac{\partial R_i^{\text{cw}}}{\partial t} = \frac{\sigma_2 \Phi M}{2} \cdot \frac{P^2 NA^4}{\pi^4 0.55 \lambda^2 \hbar^2 c^2} = K_M \cdot K_L^{\text{cw}} \qquad (2.7)$$

With the material parameter $K_m$ and the laser parameters $K_L^{\text{pulsed}}$ and $K_L^{\text{cw}}$. The advantage of pulsed laser sources can be seen directly by comparing these two equations. If the laser sources emit at the same wavelength, the difference between pulsed and continuous irradiation can be summarized to $K_L^{\text{pulsed}} = (\upsilon\tau_p)^{-1} K_L^{\text{cw}}$. For a typical experimental setup using a fs-laser source, with $\tau_p = 100$ fs and $\upsilon = 80$ MHz this results in a value of $12.5 \times 10^5$, so that $\frac{\partial R_i^{\text{pulsed}}}{\partial t} \gg \frac{\partial R_i^{\text{cw}}}{\partial t}$, which explains the wide use of fs-laser sources for direct laser writing. Since these laser sources are normally expensive, researchers were looking for alternative laser sources. Passively Q-switched frequency doubled Nd:YAG laser have a pulse duration of typically 600 ps and a repetition rate of 10 kHz. Thus $(\upsilon\tau_p)^{-1}$ results in $1.6 \times 10^5$, which is only one order of magnitude smaller than a normal fs-laser source. Because the wavelength is smaller, this difference in magnitude can be easily compensated, or even overcompensated by the wavelength dependency of $\sigma_2$ and the reduced focus width depending linearly on the applied wavelength. This effect manifests itself through lower necessary average laser power, while maintaining the total irradiation time [55]. A direct comparison of continuous irradiation to the two above mentioned pulsed laser solutions for a wavelength independent $\sigma_2$ shows the close proximity of the initiation rate for different average laser powers of the fs- and ps-laser sources, whereas continuous wave excitation has an approximately 6 orders of magnitude smaller rate of initiation (Fig. 2.4). This huge difference can be compensated by a higher average laser power and through the material parameter $K_M$ [56].

Direct laser writing is influenced by a multitude of different mechanisms, of which the rate of initiation is but one. For example, it takes approximately 1–100 ms dwell time before the two photon absorption process manifests itself through a chemical reaction such as local hardening [59, 60]. This translates into $10^4$–$10^6$ pulses for the fs-system and into 10–1000 pulses for the ps-system and represents the fact that in order for the chemical reaction to occur, a certain threshold value of the number of transformed molecules has to be present in the focal volume.

The previously described mathematical model for the initiation rate neglects any diffusive processes. In this model, the number of initiated molecules growing during the pulse width of the applied laser beam is constant during the time between pulses given by the repetition rate. However, this model is only valid if the diffusion length $\langle r_{Dif} \rangle$ or the diffusion time $\tau_{Dif} = \upsilon^{-1}$ is small compared to the illuminated focal region. The diffusion length $\langle r_{Dif} \rangle$ of a single molecule can be calculated by the Einstein–Smoluchowski relationship [61, 62]:

$$\langle r_{Dif} \rangle = \sqrt{2 \cdot d \cdot D(T) \cdot \tau_{Dif}} \qquad (2.8)$$

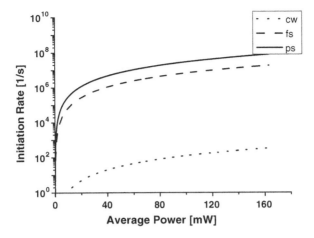

**Fig. 2.4** Comparison of the initiation rate for fs-, ps-pulsed and cw laser sources. Although having much longer pulse durations, $\mu$-chip ps-lasersystems result in a nearly identical rate of initiation compared to fs-sources. The reason lies in the reduced repetition rate, and thereby higher pulse energy and in the smaller wavelength compared to standard fs-sources. The smaller wavelength additionally reduces the irradiated volume, which is why the initiation rate in this graph is even higher for ps-irradiation compared to fs-irradiation. Continuous irradiation however, results in a 6 order of magnitude smaller rate of initiation, when diffusion is neglected and two-photon absorption is considered

where $d$ is the number of spatial directions and $D(T)$ is the temperature dependent diffusion constant

$$D(T) = D_0 \cdot e^{-\frac{E_A}{RT}} \tag{2.9}$$

with $D_0$ as the maximum diffusion constant, $E_A$ as the activation energy for diffusion, $R$ the gas constant and $T$ as the temperature. Typical values are $D_0 = 10^{-7} \text{ m}^2 \text{ s}^{-1}$ and $E_A = 3 \times 10^3 \text{ J mol}^{-1}$ [63], thus at $T = 300$ K, $\langle r_{Dif} \rangle$, results in a value smaller than 1 $\mu$m for approximately $\upsilon_c = 2 \times 10^5$ Hz. Assuming an approximate focus diameter of 1 $\mu$m, the above described initiation model can be well applied for repetition rates exceeding $\upsilon_c$. However, the actual kinetics is far more complex. For a Ti:Sapphire oscillator, which emits 100 fs laser pulses at a repetition rate of 80 MHz the diffusion time $\tau_{Dif}$ is 12.5 ns. The material is a monomer resin, containing a few weight percent photoinitiator, with an absorption spectrum suitable for two photon absorption. The first pulse interacting with the monomer resin will elevate a certain amount of photoinitiator molecules by two photon absorption from a singlet ground state into an excited singlet state. This transition occurs in a timeframe of $10^{-15}$ s and is almost instantaneous. Following the Frank–Condon principle, the excited photoinitiator undergoes internal conversion by vibrational relaxation on a timescale of $10^{-14}$–$10^{-11}$ s, followed by intersystem crossing ($10^{-11}$–$10^{-8}$ s) to a triplet state [64]. In most cases, such as photopolymerization and photocrosslinking, this triplet state is the basis for the subsequent chemical process. In the case of type I radical photopolymerization, the excited photoinitiator forms radicals $\alpha$-cleavage, which then starts the polymerization chain reaction. Besides the propagation and termination rate, diffusive processes, heat accumulation and concentration gradients influence the kinetics of the polymerization. Most of the processes can be modeled by a series of partial equations:

$$\frac{\partial \vec{Q_i}}{\partial t} = K \cdot \Delta \vec{Q_i} + \sum_i (\vec{S_i} - \vec{D_i}) \qquad (2.10)$$

$\vec{Q_i}$ denotes the parameter of interest, which can be the monomer-, radical-, photoinitiator-, or inhibitor concentration, or the accumulated heat, $K$ is the corresponding diffusion constant $\vec{S_i}$ and $\vec{D_i}$ and describe relevant sources and drains to be considered. In the case of radical concentration $\vec{R}$, the equation results in [65]:

$$\frac{\partial \vec{R}}{\partial t} = D \cdot \Delta \vec{R} + \Phi \sigma_2 \vec{P} N^2 - \left[ 2k_t \vec{R}^2 + (1-f) k_p \vec{M} \cdot \vec{R} + k_z \vec{Z} \vec{R} \right] \vec{e} \qquad (2.11)$$

with the molecular diffusion constant $D$, the termination rate for radical combination $k_t$, the radical trapping by monomers described by $(1-f) k_p$, where f is a number between 0 and 1 describing the amount of radicals which initiate polymerization, with the propagation rate constant $k_p$ and $k_z$ the inhibition rate constant for the inhibition with a radical catcher $\vec{Z}$, such as solved oxygen. This rate equation describes the kinetic effects that are governed by diffusion, the two-photon induced generation of radicals and three main sources for radical termination, radical combination, radical trapping and radical quenching. Although the different rate equations can be easily stated, the actual solution for the polymerization kinetics is rather cumbersome, because of the temporal and spatial dependencies of the involved molecular species, which are also connected by the temperature dependent diffusion. However, a numerical approach was pursued using standard material parameters and two different laser sources [65, 66]. Both laser sources had pulse durations of 150 fs, emitted at a wavelength of 800 nm and an intensity of $5 \times 10^{11}$ W/cm$^2$, but one laser source had a repetition rate of 80 MHz and the other of 1 kHz. Comparing the results obtained by the numerical calculation with the above described model for two photon absorption the numerical solution for the 1 kHz laser system yielded a radical density in the order of magnitude of $10^{-7}$–$10^{-6}$ mol/cm$^3$ after a single pulse. Taking into account the intensity and photoinitiator density are the same; the simplified model yields an identical radical density. Diffusive processes in the numerical model are negligible for the 80 MHz laser system, which is in good agreement with the random walk model described earlier, where the average distance of a single radical molecule is below the typical focus size for repetition rates exceeding $10^5$–$10^6$ Hz. For the 1 kHz system the numerical calculations result in a loss of radicals in the center of the focus to approximately 27 % in the laser-off period. When using the random walk model for a 1 kHz system, the loss in radical density can be estimated by the average distance $\langle r_{Diff} \rangle \approx 10$ μm, given for the diffusion coefficient $D(T)$ and the diffusion time $\tau_{Diff}$, which results in a loss of 25–12 %, depending on whether a 2- or 3D system is considered. However, with prolonged process time, the kinetic effects become more dominant. The diffusion constant, as well as the rate constants are temperature dependent and grow with increasing temperature. During the polymerization process, the temperature rises due to vibrational relaxation of photoinitiator molecules, as well as due to the exothermic nature of polymerization. It can easily reach several hundred degrees Celsius in the processing time. However, the simple model which neglects temperature dependent diffusion is the basis for several voxel growth models that can be found in literature and experimental data can be nicely explained by these models [21, 38, 59]. All of these models postulate a critical photon flux, or radical density necessary for polymerization to occur. The kinetic effects, present during polymerization mainly reduce the efficiency of conversion of produced radicals to a polymer network. This reduced efficiency is approximated by the models that imply higher necessary threshold values. Thus, the general form of the equation is still valid, while parasitic effects are incorporated through the experimental threshold value.

### 2.5.1 Voxel Formation

The above described initiation model can be used to calculate the size of the influenced volume. This volume is normally called volume pixel or voxel and can be applied to all the described direct laser writing methods. In the case of photopolymerization and photocrosslinking, voxels denote the insoluble volume that is generated by irradiation at a single spot for a certain amount of irradiation time, whereas for photofunctionalization it describes the volume where the photochemical effect is located. Most research on voxel size and shape in dependence on the process parameters has been performed for photopolymerization and crosslinking, since the voxel can be accurately measured by scanning electron microscopy [3, 5, 9]. However, to a great extent, the knowledge gained for this type of voxels can be transferred to photofunctionalization, because many of the physical characteristics correspond to all three methods.

The basis for voxel models is the assumption that the desired photochemical reaction, i.e. photopolymerization, photocrosslinking, or photofunctionalization, occurs above a certain threshold value. This threshold value can be incorporated into the photon flux, intensity, energy or radical density. This nonlinear material behavior has its origin in the chemical kinetics of the system. Parasitic effects, like quenching of radical or active groups have to be overcompensated in order for the desired chemical reaction to occur (Fig. 2.5). Experimentally this effect manifests itself through a threshold value. In fact, this nonlinear material behavior is observed for two photon– as well as for one photon based laser processes and can even be applied to 3D writing using UV laser sources [6, 56, 67].

The size of the voxel is derived by including the depletion of photoactive molecules into (2.2) [21]:

$$\frac{\partial R(t)}{\partial t} = (P_0 - R(t))\, \sigma_2 \chi N^2 \tag{2.12}$$

where $R$ is the radical density at the time $t$ and $P_0$ is the initial concentration of initiating molecules. For a time independent photon flux $N(r, z, const.)$ the spatial photon flux $N(r, z)$ can be written for a Gauss shaped laser beam at $r = 0$ and $z = 0$ as the center of the focus, as:

$$N(r, z) = N_0 \left( \frac{\omega_0}{\omega(z)} \right)^2 e^{-2r^2/\omega(z)^2} \tag{2.13}$$

where $N_0 = N(0, 0)$, $r$, is the radial distance from the center axis of the beam $\omega_0$, is the beam waist and the beam width at position z, $\omega(z) = \omega_0 \sqrt{\left(1 + \left(\frac{z}{z_R}\right)^2\right)}$, with the Rayleigh length $Z_R$. The solution for the voxel diameter $d(N_0, t)$ and length $l(N_0, t)$ can now be calculated by integration of the rate equation over the irradiation time t and the condition $R(r, z) \geq R_{Th}$ and results into:

$$d(N_0, t) = \omega_0 \sqrt{\ln \left( \frac{\sigma_2 \chi t N_0^2}{\ln\left(P_0/(P_0 - R_{Th})\right)} \right)} \tag{2.14}$$

$$l(N_0, t) = Z_R \sqrt{\sqrt{\frac{\sigma_2 \chi t N_0^2}{\ln(P_0/(P_0 - R_{Th}))}} - 1} \tag{2.15}$$

There are several parameters that can be used to lower the voxel sizes. The quadratic dependency of the photon flux, or intensity, has a direct influence on the voxel size, compared to the single photon case. Additionally, the threshold value $R_{Th}$ allows only photochemical reactions to occur if certain intensity is reached. The combination of quadratic dependency and threshold behavior allows voxel sizes prohibited by the Abbe-criterion for the employed focusing objective. For small enough irradiation times, the accurate choice of these parameters allow voxel diameters in the range of 80–100 nm, which are often found in literature [24, 60, 68]. Smaller voxel sizes are more difficult to achieve, because if the threshold value closes in on the maximum intensity, the process is becoming more susceptible to inadequate laser parameters like pointing stability or inhomogeneity of the material [69].

The two most frequently used parameters to influence the voxel size are the average laser power, i.e. the photon flux (P-scheme), and the irradiation time (T-scheme). The P-scheme and the T-scheme result into different growth behaviors [59]. In general, the growth characteristic in the P-scheme is steeper than in the T-scheme. Additionally, it was found experimentally with a fs-laser source emitting at 780 nm and a focusing objective of NA 1.4 for urethane acrylates, that the aspect ratio of the voxels in the

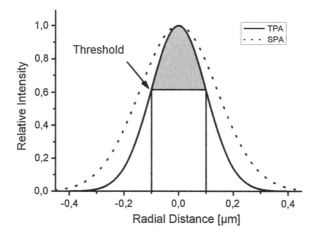

**Fig. 2.5** The intensity distribution for the radial distance to the focus center for the one- and two photon absorption case. Considering the same wavelength, the quadratic intensity leads to reduction of the effective focus size. The non-linear material behavior introduces a threshold intensity, which has to be reached for the photochemical process to be initiated, resulting in a sub-diffraction limited resolution

P-scheme growths more rapidly than in the T-scheme [59]. Even more, the aspect ratio in the T-scheme is almost constant and even slightly declining for long exposure times of 500 ms. This behavior was explained by introducing two separate voxel formation models. In the P-scheme, diffusion is neglected and the formed voxel resembles isolines of photon fluxes exceeding the threshold flux, denoted focal spot duplication. In the T-scheme, voxel formation starts from an isoline, such as in the P-scheme, but its growth is mainly governed by diffusion. This model is denoted voxel growth. While in focal spot duplication the aspect ratio is expected to grow because of the spatial nonuniformity of the point spread function, the aspect ratio in the voxel growth model will be almost constant, since diffusion effects are uniform in all spatial directions [59].

A crucial parameter for the voxel geometry is the NA of the focusing objective. For most applications, an almost spherical voxel shape is desired, to generate high resolution structures in all three spatial directions. However, the aspect ratio increases according to $1/NA$, thus the aspect ratio of small NA objectives is significantly larger.

Additionally the lateral resolution is proportional to $1/NA$, making high NA immersion objectives the first choice when high resolutions are desired. However, using smaller NA objectives can have its merits depending on the application. From a practical point of view, no immersion medium is necessary for objectives with a NA smaller than the refractive index of air, which makes it easier to exchange samples, or fabricate larger structures. Additionally working distances of several millimeters are common for low-NA objectives, whereas high performance immersion objectives often have only a working distance of several hundred micrometers. The combination of a large working distance and easy handling leads to a reduced affinity of errors due to improper use of the optics.

## 2.5.2 Determining the Size of Voxels

The influence of the different parameters on the size of the voxel can be readily measured, since the photosensitive material acts as a 3D film, which captures the point spread function. For this reason, voxels have to be fixed to a surface, mostly a glass surface, to prohibit the accidental removal during post-processing steps. This necessity leads to truncation effects, which have to be considered [70]. Two methods are widely used to eradicate this truncation effects, the ascending Voxel method [70] and the suspended bridge method [71]. For the ascending voxel method, an array of voxels is generated, where each voxel has an increased distance to the surface, starting inside the glass and ascending to a position where the voxels are no longer fixed to the surface and flips over unveiling its length. These arrays can now be measured by scanning electron microscopy, where the lateral and vertical voxel size of the last adhered voxel is determined [70]. Another method utilizes light microscopy to determine the number of voxels generated with the ascending voxel method that are fixed to the glass surface, thus determining the voxel length [38]. This method can be applied if post-process shrinkage of the voxels is an issue, such as in the case of protein crosslinking. The suspended bridge method relies on the fabrication of support structures, between which lines are being fabricated. The vertical and lateral size of the lines can then be measured by scanning electron microscopy. The connection of the two methods is the dwelling time of the initiating laser beam, which can be directly controlled using the ascending voxel method, and is related to the writing speed for the suspended bridge method.

## 2.5.3 Spherical Aberration

One of the biggest advantages of low NA objectives is the reduced spherical aberration that has its origin in refractive index mismatch [72]. Considering a typical experimental setup for direct laser writing, the sample material with a refractive index of $n_1$ is situated between two glass slides of refractive index $n_2$, separated by spacers with a thickness $d$. At the interface of glass and sample the refractive index mismatch leads to reflection and refraction of the focused laser beam. Reflection occurs to a small percentage for all angles; however, large angle illumination leads to total internal reflection for high NA objectives, which in consequence leads to a reduction of the effective numerical aperture. Thus the NA is restricted by the smallest refractive index, which is for most immersion objectives the refractive index of the sample material. In the case of protein crosslinking the refractive index is close to the refractive index of water and the effective NA of the objective is limited to 1.33. Light that passes the refractive index boundary will be refracted according to Snell's law. The different path length of the paraxial part and the boundary part of a focused laser beam will lead to two main effects. First, the actual focal position will be different compared to the nominal focus position for $n_1 = n_2$ [73, 74]. The resulting focal shift will be towards the glass slide facing the objective if $n_2 > n_1$. The second effect is a decrease in intensity given by a broadening of the focal spot. The extent of such a spherical aberration on the shape of the point spread function has been calculated previously for confocal and two-photon fluorescence microscopy [72, 75–77], which faces identical problems, when imaging deep into a specimen. When exiting the aperture of the objective, the wavefront has a converging spherical shape and according to the Huygens–Fresnel Principle, each point is the start of secondary spherical wavelets [72, 77]. The application of Fermat's principle from each of the secondary wavelets to a point in the focal region delivers the resulting point spread function (for more details: Hell et al. [72]). The numerical solutions demonstrate the strong influence of refractive index mismatch $n_1 - n_2$, the NA and the nominal focusing depth on the extent of spherical aberration. The resulting focus shift depends almost linear on the nominal focal position, which can be corrected by appropriate scaling of the 3D structure. However, one important issue is the loss of the axial symmetry with respect to the main maximum of the focus. In confocal microscopy this leads to a difference in the resulting image when focusing above or below a fluorescent specimen. The loss of axial and vertical resolution and more importantly the drastic reduction of laser intensity have a huge impact on direct laser writing. The lateral and axial full width at half maximum of the point spread function increases by a factor of 1.25 and 1.5 respectively for a NA of 1.33 and an oil ($n_2 = 1.518$) immersion objective, focusing in glycerol ($n_1 = 1.47$) at a nominal focus position of 50 μm [77]. However, the intensity of a fluorescence signal, which indicates the absorption rate of the photoactive dye, dropped to 30 % of the original value. Compared to confocal microscopy this effect is less drastic in direct laser writing, since the point spread function of the observation system is negligible. But even a smaller drop in intensity cannot be easily compensated in direct laser writing, since the resolution depends sensitively on the applied laser intensity.

In order to reduce the amount of spherical aberration, a proper selection of the optical components is necessary. One of the main goals is to reduce the difference of refractive indices. Most photoactive resins have a refractive index of 1.4 to 1.5, but in the case of protein crosslinking a refractive index close to the refractive index of water is common. In most applications, the photoactive material is placed between glass slides of a certain thickness. Thus, in the best case an immersion objective with an immersion medium having a refractive index close or equal to that of the material which is corrected for the applied glass thickness should yield the best results. Specialized objectives exist, which are designed for water, glycerin or oil immersion. Thus, depending on the material the best matching objective should be chosen. If this possibility however, does not exist, a reduction of the NA can lead to a better resolution and lower intensity variations compared to high NA objectives, depending on the nominal focal position (Fig. 2.6). Another possibility is the reduction of the nominal focal position. This technique is wide spread among laboratories performing direct laser writing. For high resolution structures, the structure is built from the bottom to the top, with the laser beam entering the material from the bottom. By this means, the realization of high structures poses a problem, since spherical aberration increases with increasing structure height. Additionally, refractive index

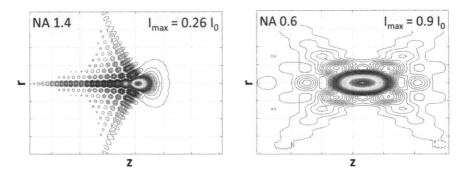

**Fig. 2.6** Calculated intensity distribution of the focal region for objectives with numerical aperture of 1.4 and 0.6 at a nominal focus position of 100 μm. The focus of the objective with NA 0.6 shows relatively low distortion and the maximum intensity is still 90% of the original intensity. The objective with NA 1.4 shows pronounced effects of spherical aberration with the characteristic intensity tails. Since much of the available intensity is situated in secondary maxima the maximum intensity has dropped to 26% of the original intensity

changes in the material after laser irradiation, which are common for radical polymerization or protein crosslinking, lead to light scattering which has an additional negative effect on resolution and on the intensity distribution. Thus, cationic polymerization is widely used for the realization of high resolution structures, because crosslinking and the subsequent change in density and refractive index occur after the laser writing process in a post exposure baking step.

## 2.5.4 Viscosity

For the generation of complex 3D structures, it is imperative that during processing, the position of the fabricated structure exactly correlates with the position given by the 3D model of the structure. If a structure is generated inside a material with low viscosity, loose ends will float away and deformation occurs. Depending on the viscosity, the geometry and mechanical properties of the crosslinked structure these deformations can occur almost instantaneously. High viscosity materials or even solid materials reduce this floating effect and allow larger and more complex structures. However, most materials used for photopolymerization and photocrosslinking are in a liquid state and drifting effects have to be considered. One possibility to solve this issue is a smart design and fabrication strategy of the structure, for example by implementing support structures [38] (Fig. 2.7). Many cationic based photopolymerization materials are in a solid state during irradiation. The widely used Epon SU-8 is a good example of such an epoxy based material, which is used for many challenging applications such as generating photonic crystal [78]. The solid and exposed epoxy resin has to undergo a post exposure bake step, during which thermal energy is used to form the polymerized structures. Due to the accumulated thermal energy

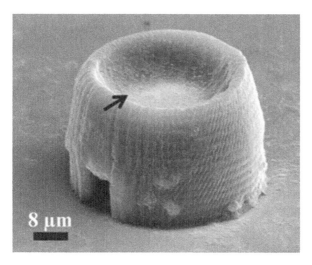

**Fig. 2.7** Free hanging protein membrane (indicated by *arrow*), that could be fabricated due to the barrel shaped polymeric support structure [55]

during light exposure, it is however, possible to circumvent postbaking, resulting in a resolution that is twice as good as the resolution achieved after postbaking, most probably due to reduced dark reactions [78]. Another effective method to increase the viscosity is by generating oligomers by a pre-illumination of the resin with UV light. By this means, a free hanging gear-wheel was realized [79].

## 2.5.5 Shrinkage

For high resolution applications, shrinkage of the fabricated microstructures poses a huge challenge. During crosslinking, or polymerization, the density of the material is increased by the transition from the liquid to the solid state, whereby the generated structures tend to shrinkage [80]. Additional shrinkage occurs after the removal of residual pre-polymer and subsequent drying [81]. This additional shrinkage has its origin in the crosslinking density. If the material is not fully cured, i.e. has a small crosslinking density, the material network resembles a molecular sponge-like structure. After removing non-crosslinked material, small cavities form, which can collapse upon drying [81]. Additional deformation can occur, when capillary forces that are present during drying exceed the mechanical strength of the fabricated structures [82, 83]. The fabricated structures are normally fixed to a solid surface, either by covalent bonds or other adhesive processes. With increasing distance to the solid surface, shrinkage leads to an increasing volume loss, resulting in a characteristic Eiffel-tower like deformation (Fig. 2.8). Most drastically this effect can be observed for crosslinked proteins, which form a hydrogel, incorporating considerable amounts

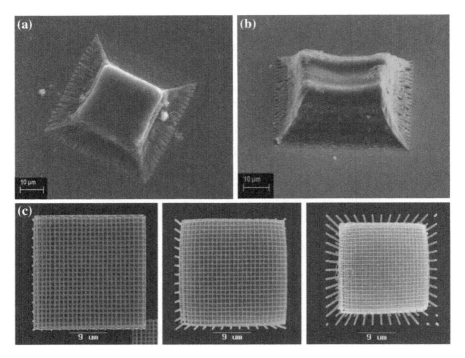

**Fig. 2.8 a, b** Protein-cubes, made of bovine serum albumin, crosslinked by fs-laser irradiation show drastic shrinkage after drying. Due to the fixation to the glass surface, a characteristic pyramid-like shape of the shrunken cubes arises. **c** Top view of woodpile structures fabricated with laser power from *left* to *right* of 8, 5.5 and 4.5 mW. Shrinkage due to post-processing leads to a size reduction and therefore an increasing resolution for decreasing laser powers. The reason is incomplete polymerization processes, which leave not crosslinked material in the structure [84].

of water. If these structures are dried and the water evaporates, protein structures tend to drastic deformation.

This shrinkage behavior depends strongly on the crosslinking density and the irradiation parameters. In general shrinkage is more severe for values that are close to the initiation threshold value, since here the smallest crosslinking density is present [81, 84]. Besides the obvious drawbacks for 3D structuring, shrinkage has to be considered, when the effect of laser parameters on voxel-formation is studied. For example, a direct size measurement with a scanning electron microscope will result in too small values for voxel size [81]. Here, a voxel counting method, described in Sect. 2.5.2 will yield more reliable results.

To minimize the effects of shrinkage, many commercial materials, such as Epon SU-8 are optimized for low shrinkage.

Another strategy to compensate shrinkage, is by adapting the 3D design of the structure appropriately. It could be shown, that the shrinkage of woodpile like structures could be almost eliminated, by including the inverse shrinkage volume [80].

On the other hand, the increased shrinkage can be used to enhance the possible resolution, if the design is chosen appropriately [84].

## 2.5.6 Damaging Effects

The process window is defined as the difference of the laser intensity threshold for the onset of the photochemical reaction and the intensity above which damaging effects, like cavity and bubble formation occur. The damage threshold for most materials is reached by intensities which are only a few times larger than the minimum necessary intensity [9, 55]. This small process windows and the associated cavity formation due to small inhomogeneities in the material can render a stable process almost impossible, because a single cavity usually destroys the whole generated structure.

Basically, cavities are forming when the energy that is present in the focal region surpasses the energy necessary for evaporation of the material. Depending on the material system and the laser parameters there exist several mechanisms which result in cavity formation [85]. First, the shear accumulation of thermal energy during processing can be a reason. In practice, this effect can be observed while working below the intensity damage threshold, but pausing at a fixed position for a considerable time, for example by reducing the writing speed. One possible origin for this thermal energy is the linear photo absorption, which is normally negligible for single pulses, but may have a significant impact after pulses, characteristic for MHz based laser systems. However, it has been calculated for water that even several seconds of irradiation with a fs-laser system, with an average power of 100 mW and a repetition rate of 80 MHz, focused through an objective with a NA 1.2, the temperate increases approximately by 3 K [86]. Thus, if the linear absorption cross section of the material resembles water, linear absorption can be ruled out for fs-based systems. For longer pulse durations this effect becomes more dominant. Additionally, many photochemical reactions are exothermic in nature, such as radical based photopolymerization, and can add to the thermal built-up [87].

However, the far more dominant effect leading to bubble formation is optical breakdown. Optical breakdown describes a process similar to the well-known electrical breakdown, where a non-conducting medium such as air can become highly conducting, if the electrical field strength exceeds the dielectric strength of the medium. For tightly focused high intensity laser irradiation, optical breakdown leads to a drastic increase of the absorption coefficient and to the formation of a plasma, which couples more energy into the material and leads to bubble formation [85, 87–92].

The optical breakdown process starts with the presence of an ionized quasi-free particle, such as an electron. For the theoretic description of an optical breakdown, the material is treated as an amorphous semiconductor, since the bound electronic states of the material resemble an insulator, while the quasi free electrons represent a conductive state [92, 93]. This quasi-free electron interacts with the potential of a molecule or ion and can absorb more photons by inverse bremsstrahlung [85].

This absorption leads to an increase of the kinetic energy of the electron, until the energy is sufficiently high to generate additional quasi-free electrons via impact ionization, which leads to a cascade effect and the formation of high quasi-free electron densities (Fig. 2.9). If the quasi free electron density exceeds a certain threshold, it leads to damaging effects, such as cavitation bubble formation, thermomechanical stress or shockwaves [94].

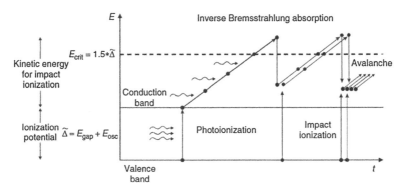

**Fig. 2.9** Schematic overview of the optical breakdown process [85]. This effect is the main reason for cavitation bubble formation during direct laser writing

The starting point for the optical breakdown is the presence of a quasi-free electron which acts as a seed for the following cascade ionization. This quasi-free electron can be generated if the laser intensity exceeds the ionization potential, which can be accomplished mainly by multi-photon ionization or quantum tunneling, impact or cascade ionization and thermionic emission. The latter is in general not important for laser parameters normally used for direct laser writing [85]. Depending on the laser parameters, either multi-photon ionization or impact ionization is the dominant effect. The kinetics of the optical breakdown can be described by a rate equation of the quasi free electron density $\rho$ [85]:

$$\frac{\partial \rho}{\partial t} = \eta_{mp} + \eta_{casc}\rho - g\rho - \eta_{rec}\rho^2 \qquad (2.16)$$

Sources for quasi free electrons are primarily multi-photon ionization $\eta_{mp}$ and cascade ionization $\eta_{casc}\rho$, whereas losses are described by quasi free radical diffusion $g\rho$, with the diffusion constant $g$ and electron-ion recombination $\eta_{rec}\rho^2$ with the recombination rate $\eta_{rec}$. During a single pulse, multi-photon ionization is dominant in the sub-picosecond to picosecond range, since the multi-photon ionization rate is proportional to $I^k$, where denotes the intensity and k the number of photons that are involved in the ionizing process. With increasing pulse durations however, cascade ionization becomes more dominant, because it is proportional to $I$. Here, the initial quasi-free electron is associated with material impurities, linear absorption or even two photon based activation.

In general, the presence of a photo absorbing substance increases the probability for the existence of a quasi-free seed electron via two-photon absorption or thermionic emission, which can lower significantly the minimum intensity necessary for optical breakdown [95]. For direct laser writing however, these molecules are fundamentally important to start the photochemical reaction. This is one of the reasons, why the damage threshold for direct laser writing is often very close to the initiation threshold for the photochemical reaction.

## 2.5.7 Mechanical Properties

The mechanical properties of the construct play an important role for many bio-mimetic applications. The elasticity and the hardness of a 3D biomimetic tissue graft should mimic its biological model. Some examples of the mechanical variety are brain tissue, which has a typical Young's modulus of approximately 1 kPa, muscle tissue of 10 kPa and collagenous bone of 100 kPa. It could be shown that mes-enchymal stem cells that possess the ability to differentiate in a variety of cell types react on their mechanical environment. A range of hydrogel substrates with Young's moduli of 0.1–1, 10 and 34 kPa, induce the production of neurogenic-, myogenic- and osteogenic-markers respectively [96]. Thus, stem cells recognize their mechan-ical environment and react accordingly. This prominent example demonstrates the importance of mechanical properties while studying cell behavior.

Direct laser writing offers several strategies to tackle this issue, because of its material diversity and possible control over the photochemical processes. First, the material that is processed can be adapted. By mixing two photosensitive resins, each having a different Young's modulus, the resulting elasticity of the copolymer can be easily tuned. An example are the two resins Sartomer 499 and Sartomer 368, which have a Young's modulus of 0.1 and 1.2 GPa respectively, mixing them leads to an almost linear transition from one to the other Young's modulus [97]. Besides copolymerization, direct adaptation of the resin formulation can be pursued. The elasticity depends directly on the formed polymer network, namely the crosslinking density. According to the theory of rubber elasticity, the Young's modulus $E$ depends linearly on both, the crosslinking density $D$ and the temperature $T$, if $T$ is well above the glass transition temperature [98]:

$$E = 3RTD \qquad (2.17)$$

$R$ denotes the gas constant. For this reason, different strategies to influence crosslink-ing densities have been pursued and a broad range of Young's modulus is possible. The formulation of a prepolymer can include different substances, such as crosslink-ing agents, filler materials, chain transfer agents, or reactive diluents [99–102]. The number of these active groups dictates the crosslinking density of the resulting poly-mer network, which in consequence changes the elasticity. Additionally, the chain length of the molecules influences the crosslinking density. For biological appli-cation polyethylenglycol is a common filler molecule, since it demonstrates good biocompatibility and many chain lengths are commercial available [103, 104]. By this material adaptations, the Young's modulus can be adapted from the GPa to the kPa regime [105–108].

Another possibility to influence the crosslinking density is the irradiation itself. A higher irradiation dosage, leads to a higher crosslinking density until saturation is reached. This strategy is not applicable for synthetic polymers used for biologi-cal applications, since the unreacted active compounds of the photosensitive resins are mostly toxic [109]. The situation is different when using proteins to form the crosslinked structure. In order to obtain a "close-to-nature" material system, no further changes to the protein are normally desired. Additionally, there are no problems with non-crosslinked residues, since the crosslinking responsible oxidizable side groups occur naturally. During two-photon induced protein crosslinking, the voxel growth can be described by a simple rate equation (2.12). This model describes the perimeter of the voxels by defining a critical radical density. With increasing photon flux or irradiation time, the perimeters and therefore the voxels are growing accordingly. The crosslinking density; however, is the quantity which is changed inside the boundary of the voxels when the critical radical density is reached. Thus, the same physical models can be applied to the crosslinking density as for the voxel growth.

The T-scheme describes the uniform voxel growth by an accumulation of radicals and oligomers with ongoing irradiation time. Following this scheme, the crosslinking density and therefore the Young's modulus increases with irradiation time. This behavior can be observed for two photon induced protein microstructures. Albumin, Lysozyme as well as Avidin have been crosslinked using laser dwell times of approx. 7–35 ms/$\mu$m$^2$ and the Young's modulus increased from approximately 0.01–10 MPa [110]. Remarkably, the three different proteins show an almost identical behavior upon irradiation. The Young's modulus, as well as its increase, is almost identical for each of the proteins, which imply the importance of the photochemical crosslinking on the crosslinking density. Beside crosslinking density, protein conformation is essential for the actual Young's modulus. Deviations from the isoelectric point, the pH value of the medium where the protein carries no net electrical charge, can cause the protein to destabilize and finally unfold because of excess charge. Indeed, the highest Young's modulus could be measured for protein structures incubated at the isoelectric point [110]. The more flexible the protein network, the more water can be incorporated into this hydrogel. Thus, it is not surprising that the same mechanism can be also observed for the swelling behavior of crosslinked protein structures. Swelling can be exactly controlled up to the point that micro actuation devices made of proteins can be realized [111]. An additional mechanism which leads to similar results is a temperature increase. With increasing temperature proteins start to destabilize, unfold and finally denaturate. Each step leads the protein further away from its compact structure, and thereby increases the protein's flexibility. This temperature driven effect can be directly observed through a decrease of the Young's modulus with each destabilizing step [110].

## 2.6 Advanced Setups for Direct Laser Writing

### 2.6.1 Process Speed

3D direct laser writing as described in Sect. 2.2 is based on generating a highly localized focused laser beam, which can be translated in three dimensions through a material. Depending on the technical implementation of the direct laser writing, this method may encounter several inherent limitations regarding crucial parameters such as process speed, structure size and resolution, which can be addressed by implementing more complex setups as discussed in the following.

When considering direct laser patterning for application fields such as the generation of scaffolds for tissue engineering, the main drawback of the method is its low writing speed. In general writing speeds in the order of magnitude of tenth of micrometer to millimeter per second can be achieved, when using translational stages and commercial available photosensitive materials. The fabrication time $t_{Fab}$ of a 3D structure is proportional to:

$$t_{Fab} \propto \frac{V_{St} \cdot F}{d_V \cdot h_V \cdot v} \tag{2.18}$$

where $V_{st}$ is the volume of the final structure, $F$ is a filling factor, giving the percentage of crosslinked versus non-crosslinked parts of the 3D-structure, $d_V$ is the voxel diameter, $h_V$ is the voxel height and $v$ is the writing speed. In order to minimize fabrication time, small filling factors are advantageous. The smallest filling factor can be achieved if only the outer boundary of a closed volume is being crosslinked, while the inner volume is crosslinked subsequently by UV-irradiation [21, 112]. However, this method can only be applied for inelastic materials that can withstand the mechanical stress of rinsing. Scaffolds for tissue engineering mostly do not have a large percentage of closed volumes. Their structure is sponge-like, or woodpile-like, allowing cell ingrowth while providing a mechanical backbone. In the case of bone tissue engineering, typical porosities are between 0.2 and 0.9 [113] and therefore their respective filling factors are between 0.1 and 0.8. As can be seen from these values, a reduction of the filling factor is not a feasible route for scaffold production.

Besides the filling factor the resolution, or the voxel diameter and height, is an important factor for the fabrication time and macroscopic scaffolds generated by direct laser writing are often accompanied with a sacrifice of resolution [114]. One possibility to address this issue is to move the focus faster through the sample, i.e. to increase the writing speed. The writing speed is influenced by a rate equation (cf. (2.11)) where laser parameters such as intensity and repetition rate play a crucial role, besides material parameters. In order to preserve the desired high resolution of the arbitrary 3D structures, acceleration phases, oscillations and beam control are of great importance. One elegant possibility is to use galvanometric scanners where the sample and most of the mass can reside at a fixed position, while the focus is being moved by two galvanometric driven mirrors (Fig. 2.10b). Because almost no mass has to be accelerated highly precise beam deflection, at high writing speeds can be accomplished. Although writing speeds of several meters per second are technically feasible, this method imposes high demands on the material. If a focus of 1 μm diameter is moved with 100 μm/s, the dwell time for each focus area is approximately 10 ms. The dwell time reduces to 100 μs for a writing speed of 1 cm/s and for a writing speed of 1 m/s to 1 μs. This reduction in dwell time $t$ has to be compensated by increased pulse energy $E_P$, so that $E_P^2 \cdot t$ is constant. Thus, compared to a writing speed of 100 μm/s, the pulse energy has to be increased

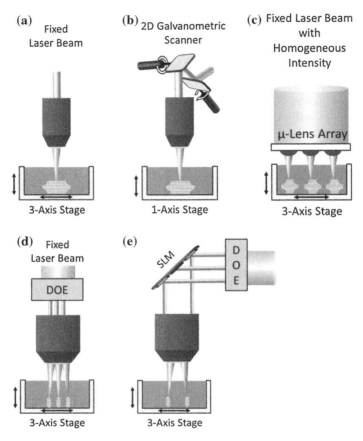

**Fig. 2.10** Principle drawings of examples of direct laser writing process schemes, depicting the main differences. **a** The most common setup utilises a fixed laser beam, focused by a microscope objective. Focus translational is achieved by a 3 axis stage. **b** If a galvanometric scanner is used, the horizontal movement of the focus is caused by the rotational movement of scanner mirrors. **c** A $\mu$-lens array can be used to generate an array of foci with the aim to fabricate many structures in parallel. **d** Apart from $\mu$-lens arrays diffractive optical beam splitters (DOE) can be used for the generation of such multifocus arrays. **e** Single foci of such multifocus applications can be controlled by spatial light modulators (SLM), such as digital micromirror devices

by a factor of 10 for 1 cm/s and by a factor of 100 for 1 m/s. Since the formation of cavitation bubbles scales with the pulse energy and the damage intensity threshold for common materials is only several times the crosslinking threshold intensity, a considerable increase in process speed using fast beam deflection is necessarily linked to the development of highly efficient photosensitive material systems. Photoinitiators having large two photon absorption cross sections, such as bis dialkylamino- or diarylamino-substituted diphenylpolyenes and bis(styryl)benzenes [22, 115], or having strong photochemical reactivity, such as triple-bond-containing 1,5- bis(4-(N,

N-dimethylamino)phenyl)penta-1,4-diyn-3-one [16] have been developed in order to solve this problem (for further information on the topic see [14]).

One way to increase processing time without the need of faster beam deflection is parallelization. The main idea is to have not one, but many foci inside the sample, which leads to a reduced process time, which is proportional to the number of foci.

Microlens arrays are capable of generating such multifocal arrays and can be easily integrated in a direct laser writing setup (Fig. 2.10c). The number of structures that can be built simultaneously can be as high as several hundred and is mainly limited by laser beam properties [116, 117]. Each structure should be equal, which necessitates a highly uniform laser intensity distribution over the entire microlens array. This uniformity can be achieved by using top-hat beams, or largely overfilling the array with a Gaussian beam, which yields good uniformity [116, 117]. Another possibility for the generation of multifocus arrays is the use of diffractive beam splitters (Fig. 2.10d). In this case, a periodic grating generates a set of separated beams, which depend on the period of the grating and the wavelength of the incident laser beam. These separated beams exhibit the same temporal-spatial characteristics and can be subsequently focused by an objective to form a set of well-defined focal spots suited for direct laser writing. The number of foci accessible with this method is generally considerably smaller compared to microlens arrays [118].

Both, microlens arrays and diffractive beam splitters, are well suited for fabricating repetitive units of a microstructure at a given separation distance defined by the optical properties of the setup. In order to use the speed advantage of multifocus applications for the generation of a single arbitrary 3D structure, each of the foci has to be controlled individually. This control is possible by imaging the multifocus array on a controllable spatial light modulator (SLM), such as a digital light processor (DLP) [119] or a dynamic micromirror device (DMD) [120] (Fig. 2.10e). By this means, each focal spot can be individually turned on and off. This controllable array of focal spots can be focused in the sample by a high NA objective. As before, the separation distance of each focus is defined by the optical setup and motion of a translational stage may be necessary for good fabrication results.

An elegant method which is free of the limitations of a set array of focal spots is based on holographic interference patterns. These patterns can be realized by a reflective SLM displaying a computer generated hologram (CGH) that acts as a phase hologram [121]. The main advantages of this approach are the arbitrary positioning of the focal spots and the ability to adjust the intensity of each focus individually.

## 2.6.2 Enhanced Resolution

Another effort in direct laser writing research is the increase in spatial resolution well beyond the diffraction limit. State of the art direct laser writing setups can reach lateral feature sizes of approximately 100 nm driven by the interplay of two photon induced formation of reactive species such as radicals and scavenging processes such as quenching effects due to oxygen or chain termination.

Theoretically, by finely tuning the excitation intensity, even smaller structures may be generated. However, at reduced laser intensity fluctuations in laser power show an increasingly pronounced influence on the generated structures. As a result of the reduced laser intensity, the crosslinking density is reduced as well, leading to mechanical weaker structures. Although crosslinking may not be sufficient for the formation of a microstructure, oligomers as well as reactive species will form in the irradiated areas of the prepolymeric bath. These substances are changing the kinetics of the crosslinking process, since less effort is necessary to finalize the crosslinking process, leaving already irradiated areas in the vicinity of the actual crosslinking volume more susceptible to irradiation. This memory effect of the material hinders the fabrication of closely spaced high resolution structures in the fabrication process. Typical minimum distances of crosslinked polymer are approximately 450–600 nm [56, 122, 123], depending on the employed direct laser writing method. Although lines of approximately 80 nm have been reported, this limitation to the spacing restricts the actual resolution capabilities significantly.

A solution to these issues is to stop or to deactivate reactive species from forming outside their intended boundaries. A possible technology is the so called stimulated emission depletion (STED), already known from and widely used in fluorescence microscopy [124, 125]. This method uses two distinct laser beams, one for excitation and one for stimulated emission. The excitation beam is a focused laser beam standardly used in confocal microscopy and in direct laser writing. The second beam however, passes through a phase mask to form a doughnut-like beam shape to encompass the focus of the excitation beam. The light absorbing molecule is excited by the excitation beam from a ground state S0 to an excited state S1, from which it undergoes intersystem crossing to a triplet state T1, which results in the generation of reactive species. The depletion laser beam disrupts this process by transferring S1 back to S0 through stimulated emission before the molecule can undergo intersystem crossing. As a direct result, the achievable resolution is no longer restricted by the Abbe limit and much higher resolutions are feasible [126–128]. This STED approach poses high demands on the materials used in direct laser writing. Normally applied photoinitiators are designed for high quantum yields of the intersystem crossing from S1 to T1 in order to increase the effectiveness for the formation of active species, i.e. high intersystem crossing rates and low S1-S0 transition rates. However, the S1-S0 transition mediated emission is important for an efficient STED process and high transmission rate are desirable. Thus, suited materials should show sufficiently large oscillator strength of the S1-S0 transition and large excitation lifetime, to allow stimulated emission to occur, while maintaining sufficient generation of reactive species for the photochemical activity. Given a two photon excitation at 810 nm and a depletion wavelength of 532 nm, isopropylthioxanthone (ITX) and 7-diethylamino-3-thenoylcoumarin (DETC) have been reported to fulfil these requirement [129].

Using DETC as an initiator, crosslinked lines with a separation distance of 175 nm, which is slightly below the Abbe limit of the used optical setup, were realized [130].

Beside a separation of excitation and stimulated emission through different wavelengths, it is also possible to separate activation and deactivation through intensity. Certain photoinitiators, such as malachite green HCL [131, 132] are able to undergo excitation upon high intensity near-infrared irradiation, while emission is stimulated at the same wavelength but at drastically reduced intensities, such as laser light emitted by a continuous wave laser system. The basis for this effect is the band structure of the photoinitiator. Most probably, the excited initiator experiences an intersystem crossing until reaching a relatively stable and long lived intermediate state, from which radical formation can occur. This state can undergo single photon induced stimulated emission with the same wavelength as the two photon absorption. Thus, a balance between deactivation and activation occurs, which can be shifted by low intensity irradiation towards deactivation. An effective way of achieving high deactivation rates while maintaining low activation rates is the use of long pulsed or even cw-laser sources. This beam sources can deliver high energy but low intensity laser beams, thus combining high deactivation rate with no, or minimal two-photon induced crosslinking. By applying malachite green carbinol base, structure sizes as small as 40 nm [131, 132] were generated, which is a drastic enhancement of typical achievable structure size and resolution compared to standard direct laser writing.

The balance of activation and deactivation depends on the intrinsic properties of the photoinitiator, on the kinetic transition rates and the laser intensity. A photoinitiator that shows stimulated emission at the same wavelength as two photon activation will reach equilibrium between activation and deactivation without additional deactivation beam. If the irradiation increases, some photoinitiators shift the balance from activation towards deactivation. This behavior results in a reverse growth characteristic compared to standard initiation molecules. Normally, the size of the crosslinked structures increases with decreasing writing speed, since active species have more time to accumulate. However, for reverse growth initiators, such as malachite green HCL, increased writing speed means less deactivation and hence larger structure sizes. By combining photoinitiators with normal and reverse growth characteristics it is possible to implement a material system that is independent of the writing speed, effectively counteracting writing speed inhomogeneities due to acceleration and deceleration of the translational system [133].

## 2.6.3 Large Structures

Although direct laser writing offers the possibility of generating 3D structures with sub-micrometer resolution, the technology has significant shortcomings in producing large structures, such as structures on the centimeter scale. The reason for this limitation is threefold: (1) The limited writing speed, which is normally below 1 mm/s makes the generation of large scale structures, even if possible, time consuming to a point where it is no longer feasible. (2) High numerical aperture objectives, which are necessary to achieve high resolutions, have a limited working distance, giving a simple physical boundary to generate large scale vertical structures. (3) Spherical aberration leads to a strong intensity dependency of the penetration depth, therefore adding to point 2.

For increased process speed several approaches can be pursued as described in Sect. 2.6.1. Using a standard direct laser writing setup, high writing speeds of several mm/s are feasible when employing materials with a sufficient dynamic range. However, the size of the crosslinked features is more difficult to control since small deviations in laser intensity or the material lead to a significant effect. Thus, structures produced at high writing speed are often produced with lower resolution where deviations are less pronounced [114, 134].

Working distance can be maximized up to several millimeters and spherical aberration can be reduced significantly by using low numerical aperture objectives, at the cost of resolution, and an aspect ratio far greater than one [134, 135]. However, the achieved resolution is comparable to the resolution that is possible using UV-based μ-stereolithography [136, 137] and the overall benefit of direct laser writing has to be evaluated case by case due to the higher costs of a direct laser writing machine based on multi-photon absorption.

One possibility to address this issue, while maintaining reasonable resolution, is to decouple vertical translation and focusing depth. Decoupling can be achieved by adding a translational stage that moves the object and not the focal position vertically. By this means 3D structures can be realized with a fixed focal position leading to a setup that is no longer restricted by working distance and spherical aberration (Fig. 2.11a). Such a setup was used to generate structures of several millimeters height [138] and allows for more degrees of freedom compared to a setup where only the focus is moved inside a material bath. Another possibility to overcome the restrictions imposed by spherical aberration is to utilise refractive index matched photosensitive materials (Fig. 2.11b). The material then acts as the immersion medium, eradicating specimen induced spherical aberration and allowing structures higher than permitted by the working distance of the objective. This method is called dip-in-lithography (DiLL) and was developed by Nanoscribe GmbH (Nanoscribe GmbH, Eggenstein-Leopoldshafen, Germany). The main drawback of this method is its dependency on the refractive index of the photosensitive material and the limitation this restriction imposes on material choice (Figs. 2.12, 2.13).

Since the fabrication occurs in a material bath, structures that consist of more than one material are only feasible using subsequent process steps. This is a common drawback of laser-based additive manufacturing techniques and indeed is thought of one of the huge challenges in stereolithography [137]. One class of additive manufacturing techniques that does not have the same restriction is 3D printing, such as 3D plotting, or 3D inkjet printing [139]. Here, the structural features are not realized by selective irradiation, but by a selective positioning of material, which is subsequently fixed by a suitable method such as photopolymerization, heat, or chemical crosslinking. Although these systems work fine for rapid prototyping purposes, the achievable resolution is restricted by the minimum droplet size of the material positioning process. Inkjet printing utilizes the contraction of piezoelectric elements to deposit material through an array of small nozzles. Since the nozzle diameter and the pressure build-up can be well regulated and designed to be small, inkjet based printers can achieve a resolution in the order of magnitude of tenth of micrometers, compared to hundreds of micrometers for plotting based systems. However, this

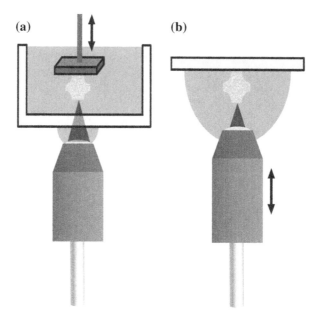

**Fig. 2.11** Two methods to circumvent the limited fabrication height imposed on the process by the limited working distance of the objective and specimen induced spherical aberration. **a** The distance between objective and fabrication plane is fixed, while the sample is pulled out of the material bath by a translational stage. **b** The refractive index of the photosensitive material matches the refractive index of the immersion medium (e.g. is close to glass). Both methods eliminate intensity variations of the laser beam in different structuring heights, which results in a homogenous fabrication quality over larger heights

resolution is still two to three orders of magnitude lower than that is achievable by direct laser writing. By combining inkjet printing and direct laser writing, two main advances are feasible [140, 141]. For macroscopic hierarchical structures, such as a blood vessel system, the overall processing time can be reduced significantly. In the case of a blood vessel system, the diameter of necessary vessels ranges from several millimeters, for dispersing blood throughout the system evenly, to several micrometers, where the exchange of nutrients occur through capillaries. This approach can reduce the volume that has to be generated by direct laser writing to the absolute necessary. Thus, the superior process speed of inkjet printing governs, depending on the structure, is used to reduce the necessary production time.

Besides an increase in production time by omitting direct laser writing where it is not necessary, the material positioning technique of inkjet printing can be used for the fabrication of structures consisting of more than one material.

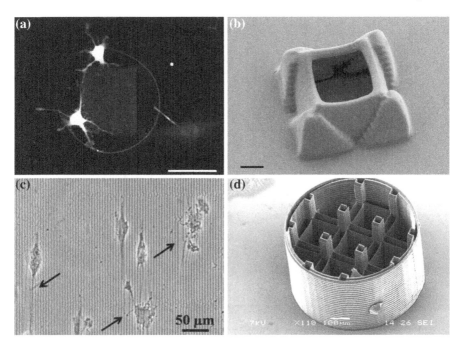

Fig. 2.12  a Neuronal cells encapsulated by a 1 μm thick BSA/fibrinogen cylinder migrate towards an adhesive fibrinogen scaffold situated in the center [170] (scale bar 50 μm). b Micromaze fabricated by direct laser writing of BSA (scale bar 10 μm) [120]. c Chondrocytes cultivated on modified gelatin fibres with a diameter of approx. 1 μm and a distance of 4 μm show strong alignment [55]. The *arrows* mark spots, where the cell induced force leads to a delamination of the gelatin lines from the glass substrate. d 3D structure of modified gelatin with a low concentration of 0.005 % photoinitiator fabricated with direct laser writing [171]

## 2.7  Direct Laser Writing in 3D Cell Culture and Tissue Engineering

The study of cells in a 2D environment, namely a petri dish, has revealed significant scientific insights, it has become more clear over the last couple of years, that the third dimension plays a major role for cell adhesion, proliferation and signaling pathways [142]. A cell cultivated in a petri dish, can detect its mechanical surrounding, which is an inelastic surface, surrounded with a liquid. Additionally the cell detects chemical cues on the surface, mostly in the form of a protein coating and other cells in proximity residing also in this 2D world. However, in a living organism the circumstances are dramatically different. Cells interact in all three spatial directions with their neighboring cells and the constituents of the extracellular matrix. For example the formation of focal adhesions, large dynamic protein complexes connecting the cytoskeleton to the extracellular matrix is triggered along the whole cell body and is significantly different when studied in 2D and 3D [143]. Additionally,

**Fig. 2.13** **a** Human sarcoma cells cultivated on a woodpile structure fabricated with direct laser writing of acrylates and their motility and **b** migration speed analyzed by fluorescent live cell imaging [97]. **c** Fibroblasts adhere selectively to polymer cubes that promote cell attachment but not to the PEG-DA structural backbone [152]

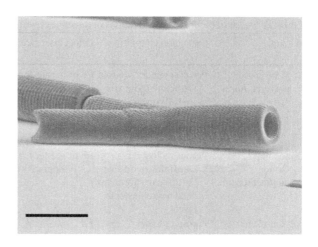

focal adhesions do not only attach the cell to a surface, they act as communication channel, conveying mechanical and chemical information, which implicates the significance of a 3D cell culture. Two examples that demonstrate the importance of the 3D environment are, human breast endothelial cells that have been cultured in 2D behave like tumor cells, but revert to normal growth characteristics when cultured in a 3D environment [144], and the different structure, localization and function of fibroblastic cell-matrix adhesions in in vivo 3D cell cultures compared to several 2D counterparts, which leads to an enhanced biological activity [143].

Direct laser writing offers the possibility to create chemically and physically defined cell environments, which can be designed to fit the needs of the experiment or application. Let us consider for example focal adhesions, protein clusters that have a major impact on the interaction of cells to their environment. A large quantity of these focal adhesions is only 0.2 $\mu m^2$ in size [145]. Direct laser writing is able to produce structure sizes that are in the same order of magnitude. Combined with its inherent 3D capability and material diversity, direct laser writing offers the possibility to realize 3D cell culture systems that could be used to study this aspect in a reproducible way. Although, the application of direct laser writing in this field of research is not widely spread yet, some application examples are presented here to demonstrate its potential for 3D cell culture and tissue engineering (A summary of published studies can be found in table 2.1).

Basically, two distinctive routes can be identified for the generation of such 3D cell cultures [47]. The first route uses photopolymerization of synthetic polymers, or–crosslinking of proteins to generate a physical object. These structures have defined mechanical and structural properties. Chemical properties are defined through the surface chemistry of the applied material, or by subsequent coating procedures. The second route employs photoactivation inside a hydrogel, which results in a defined 3D chemical structure.

**Table 2.1** A list of biomimetic applications of direct laser writing

| Material | Structure/cell type | Direct laser writing method | Notes | Reference |
|---|---|---|---|---|
| Acrylate based polymer, fibrin | Porous cylinder scaffold structure (pore size ~100 μm)/human pulmonary microvascular endothelial cells | PP of master for Soft Lithography of biomolecules | Good cell viability, adhesion and alignment in pores | [146] |
| Ormocomp®, Ormocer®, polylactid acid | Porous cylinder scaffolds (pore size ~100 μm)/fibroblast and neuroblastoma cells | PP of master for Soft Lithography | Cells coat entire scaffold | [147] |
| Poly(ethylene glycol) diacrylate | Porous cylinder scaffolds (pore size ~25 μm)/fibroblasts | PP | Washing protocols reduce cytotoxicity from photoinitiators | [104] |
| Methacrylamide modified gelatin | 3D mesh scaffold (mesh size ~250 μm)/Adipose-derived stem cells | PP | Thin gelatin mesh due to dark reactions inside pores improves cell viability. Differentiation into adipogenic lineage after 7 days | [114] |
| BSA, biotinylated calf intestinal alkaline phosphatase | Protein fibres microparticles and scaffolds | PC | Chemical gradients within cell culture, enzymatic nanoreactors | [148] |
| BSA | Protein microchambers/motile *E. coli* | PC | Pressure measured that *E. coli* apply to a protein membrane if confined | [110] |
| BSA | Lines as confinement/neuroblastoma cells | PC | Outgrowth of neuritis can be spatially manipulated by BSA lines | [149] |
| Collagen I, II and IV | Lines/fibroblasts | PC | Cells orientate parallel to lines | [150] |

(continued)

**Table 2.1**  (continued)

| Material | Structure/cell type | Direct laser writing method | Notes | Reference |
|---|---|---|---|---|
| Photopolymerizable triacrylates | Woodpile like scaffold with pore sizes from 12 to 110 $\mu$m/fibrosarcoma cells | PP | Observation of cell migration | [97] |
| Ormocomp® | Pillars connected by beams/cardiomyocytes | PP | Measurement of cell forces in 3D | [151] |
| Ormocomp®, polyethylene glycol diacrylate (PEG-DA) | Protein repellent pillars connected with beams, Ormocomp® cubes on repellent beams/fibroblasts | PP | Cells adhere selectively to Ormocomp® cubes | [152] |
| Ormocer® | Polymer support pillars, self-assembled peptide fibrills | PP and photobiotin/avidin mediated self-assembled fibril growth | Peptide fibrils with diameter $\sim$<1 $\mu$m and length $\sim$5 $\mu$m | [153] |
| Ormocer® | Mesh scaffold | PP, binding of photobiotin/avidin via UV | Scaffold with defined biochemical surface | [154] |
| Triblock copolymer | Mesh scaffolds, circular scaffolds and line arrays/fibroblasts | PP | Material cytocompatible and biodegradable | [155] |
| Photocurable polymer (Accura SI10) | Cubic scaffold, mesh size 200 $\mu$m/hepatocytes | PP | Hepatocytes cultured on scaffold showed higher values for liver specific functions compared to flat culture | [135] |
| Acrylate base resin, methacrylamide modified gelatin (Gel-Mo), BSA, fibronectin | Polymer-protein hybrid structures, lines/chondrocytes | PP, PC | Free hanging protein membranes on polymer support, strong adhesion of cells to Gel-Mo | [55] |

(continued)

**Table 2.1** (continued)

| Material | Structure/cell type | Direct laser writing method | Notes | Reference |
|---|---|---|---|---|
| Polytetrahydrofura-nether-diacrylate resin | Branched vascular scaffold/endothelial cells | PP | Material with tunable elasticity, inner layer of endothelial cells | [99] |
| Ormocer® | Fibers scaffolds/fibroblasts | PP Single shot fibers (possible aspect ratios of 180:1) through self-focusing and seeded growth | Long, high aspect ratio fibers. Measurement of contractile forces of cell by bending | [156, 157] |
| BSA, hyaluronic acid | Different protein microstructures, such as lines and spirals/neuronal cell and glia cells | PC BSA was crosslinked inside hydrogel and coated with laminin via biotin/avidin reaction | Combination of topographical and chemical cues inside hydrogel. Guided cell growth achieved | [53] |
| Polyethylene glycol diacrylate (PEGDA) | 3D cell guiding patterns inside hydrogel/Fibroblasts | PA Functionalization of biodegradable PEGDA with peptide containing diacrylates | Encapsulation of live cells possible. Guided cell growth achieved | [158] |
| Polyethylene glycol diacrylate (PEGDA) | Biomimetic replica of vascular systems/Endothelial cells, mesenchymal progenitor cells | PA Functional peptide patterns (RGD, IKVAV) | Cells resemble structure of natural capillary system | [159] |
| PEG-precursor including vinyl groups and nitrobenzyl ether moieties | Functional lines, functional degraded tubes/Fibroblasts | PA Photofunctional-ization process and photocleavage separated by wavelength | Combination of controlled chemical cues and physical stimuli through degradation | [50] |
| Polyethylene glycol diacrylate (PEGDA) | Functional triangles, lines and meshes | PA Subsequent fabrication of chemical active sides consisting of up to three materials | Combination of growth factors to study synergetic effects on cells | [160] |

*PP* photopolymerization, *PC* photocrosslinking, *PA* photoactivation

## 2.7.1 Photocrosslinking of Proteins for Biomimetic 3D Structures

With the discovery that proteins can be crosslinked by direct laser writing, it was proposed that these crosslinked protein matrices may serve as scaffolds for regenerative medicine, or advanced 3D cell culture systems [161].

These structures combine the chemical environment of natural tissue, with the ability to govern their structure with a sub-$\mu$m resolution. However, during crosslinking the proteins are subjected to highly reactive molecular species (e.g. singlet oxygen) and high photon dosages. Additionally their intrinsic structure is being changed by the crosslinking process itself, for example by bond breakage of sulfide bridges as well as thermal effects [29, 150, 162, 163]. Taking into account all these effects that may negatively influence protein structure and protein functionality, the question arises if it is indeed possible to maintain protein functionality during direct laser writing. Since a direct comparison of the protein structure before and after crosslinking is not easily achieved, this issue was engaged by studying the conservation of certain protein functions after crosslinking. One of the first experiments in this direction was carried out using alkaline phosphatase and a bovine serum albumin solution [163]. After crosslinking a Michaelis-Menten kinetic analysis of the enzymatic activity of the alkaline phosphatase was performed, with the result that the activity reached normal values. Thus, the laser irradiation had no visible effect on the enzymatic activity of the enzyme itself. Additionally it was shown that crosslinked proteins were susceptible to subsequent enzymatic activity [150], indicating that the proteins are not changed in a way that makes them no longer recognized by enzymes.

Fibronectin and fibrinogen matrices were produced and stained with protein specific dyes [162]. In this case diffusion parameters were measured. It could be shown, that the fluorescent signal increased with increasing photon dosage and the diffusion rates decreased. Thus, the targeted functional parts of the proteins are still active after crosslinking and the increase in fluorescent signal is associated with an increase in crosslinking density. If the laser irradiation would have a negative effect on protein functionality, the fluorescent signal should decline with increasing photon density and not increase. These results are strengthened by the fact that crosslinked avidin demonstrates the same tendency using biotin as a marker [39, 57] and even retains its sensitivity to hydrogen ion activity [39].

Although the underlying mechanism of photoinduced protein crosslinking is not yet fully understood, the results so far demonstrate a remarkable conservation of protein functionality during direct laser writing.

Protein microstructures produced by direct laser writing can be used to study the interaction of single cells with their extracellular matrix in a controlled, yet biomimetic way (Fig. 2.12). For this purpose, it is important to differentiate influences that arise due to topographical features and those that are chemical in nature. The sheer presence of 3D microstructures can alter cell behavior significantly, for example in the form of contact guidance [164, 165] or even structure induced cell behavior (e.g. [166, 167]). Additionally to these physical influences, structured chemical cues that are purely 2D can lead to the same effect. These patterns are often generated using patterning techniques, such as micro contact printing [168], or direct laser writing [169].

Direct laser writing of proteins necessarily combines both influences in a single structure. This entanglement of influences has to be considered when interpreting cell experiments. One possible route to distinguish the underlying causes is to crosslink different types of material. Crosslinked bovine serum albumin (BSA), a protein that normally does not promote cell adhesion, is an excellent cell guiding material. Cells cultivated on BSA line structures change their shape according to the underlying structure [172].

The biocompatible nature of BSA even allows for in-situ cell guiding of cells. Here, cells are cultured in a petri-dish or other suitable vessel and the cell medium is exchanged with a BSA solution. Since BSA can be crosslinked without an additional, mostly toxic photosensitizer BSA microstructures can be generated in the vicinity of a living cell, without damaging effects. These BSA structures then serve as an in-situ guiding tool. The technique was demonstrated using nerve cells and a μ-chip laser source emitting at 532 nm [39].

Chemical guidance by crosslinked proteins that promote cell adhesion was observed for proteins such as fibronectin, fibrinogen or collagen. In contrast to BSA microstructures, cells demonstrated a strong chemical connection to the protein structure by the presence of focal adhesions [55, 150, 172].

One example where the strength of direct laser writing of proteins is utilized for a specific biological application is the generation of a basement membrane model to study cell migration and adhesion of ovarian cancer cells [173]. The basement membrane, a dense mesh of crosslinked proteins, plays a major role during metastasis of many types of cancer [174]. Of course, the complexity of natural basement membrane is not fully captured by the biomimetic model. In fact, the basement membrane model generated by direct laser writing consisted of a single type of protein, laminin, which is just one of the main components of the ovarian basement membrane [175, 176]. Single laminin fibers with a diameter of 600 nm, a height of 2 μm and a separation distance of 10 μm are written on a glass surface on a homogenous BSA passivation layer. Ovarian cancer cell lines plated on such a scaffold demonstrate distinctive contact guidance parallel to the laminin fibers, as well as an increase in migration [173]. Although these first results are promising, the main advantage of in vitro models generated by direct laser writing of proteins over other methods, such as collagen gels [142], is their easy structural adaption, the amount of materials that are processable and the ability to generate in vitro assays which are excellently suited for the analysis of the cell behavior.

Although the idea of forming extracellular matrix proteins to a structure that possesses feature sizes comparable to those found in nature is appealing, there is a major drawback of the method. Complex 3D structures need a certain amount of mechanical stability. Most crosslinked proteins however, are quite mechanically weak. Stability can arise from the material itself, for example of a protein mixture of fibronectin and BSA, where BSA increases the mechanical stability [170], or by increasing the crosslinking density by the introduction of artificial side groups, such as gelatin with methacrylamides [55, 114, 171, 177]. Using the latter, 3D structures can be generated that support proliferation and migration of adipose derived stem cells, and even differentiation into the adipogenic lineage can be observed after seven days in culture [114].

Another possibility to overcome stability issues is a combination of soft protein structures and more rigid polymer support structures [38, 153]. Here the polymer support structure provides the necessary mechanical stability, so that the functional protein structures can be generated with processing parameters that give the best biomimetic results in terms of functionality and elasticity. This technique was used to generate nanometer-sized peptide fibril between polymer supports [153], and free hanging protein membrane assays [38, 55].

## 2.7.2 Photopolymerisation of Synthetic Polymers for Biomimetic 3D Structures

Although direct laser protein writing is intrinsically biomimetic, synthetic polymers have been used to a large extend. Their main advantage over proteins is their chemically well-defined nature and their susceptibility to adaptations. In their simplest form, polymers serve as a physical backbone, which carries topographical and mechanical information (Fig. 2.13). Chemicals cues are subsequently added by adsorption of proteins or peptides. Besides, or better because its simplicity, this method is widely spread for the fabrication of scaffolds for tissue engineering [178] and defined biomimetic cell environments [156, 179]. The fabrication of small scale 3D biomimetic grafts such as small vascular systems is possible. By using biocompatible photosensitive materials, with mechanical characteristics close to nature, small biomimetic vessels with an inner diameter of 20 μm and a wall thickness of several micrometers can be realized [99]. Woodpile-like structures, with different mesh sizes ranging from 12 to 100 μm were used to measure cell motility in a 3D environment [97]. Human fibrosarcoma cells were tracked with a fluorescent live cell imaging process and 3D migration patterns of single cells recorded. The mean migration speed of these cells in the scaffolds was higher than the speed observed in a two-dimensional cell culture, while the probability of cell movement to occur was lower.

Hepatocytes cultured on cubic polymer structures demonstrated higher liver specific activity compared to a flat cell culture [135]. The difference between two-dimensional and 3D cell culture arises due to structural properties and is not linked to a chemical interaction with the scaffolds. The higher cell motility is most probably due to a higher degree of freedom for cell movement [97]. The higher liver specific functionality of hepatocytes cultured on cubic scaffolds arises due to a locally increased cell concentration, which in consequence leads to a higher concentration of soluble factors maintaining hepatocyte phenotype [135].

The mechanical forces that heart muscle cells apply during their beating cycle can be measured by the fabrication of single cell in vitro assays [151]. In this case, a

well-defined scaffold of Ormocomp® pillars was generated and connected by small
diameter beams in a spider web-like fashion. The myocytes adhered to the beams,
consequently applying a force during contraction. Due to the well-known and defined
geometry of the connecting beams, measuring the deformation leads to a mean cell
force involved of approximately 48 nN [151]. These single cell assays can be further
specialized by combining materials with different chemical properties. By combining
a protein repellent polymer for the pillars and beams with Ormocomp® cubes situated
on the beams, cell adhesion can be specifically limited to the Ormocomp® cubes
[152].

Another possibility to combine structural definition and chemical functionalization is to use the biotin-avidin binding mechanism. Cubic scaffolds made of Ormo-
cer® were treated with photobiotin, which was bound to the polymer surface by
UV-irradiation of an excimer laser source [153, 154]. By this means, avidin labeled
substances can be linked to the biotin binding sites, which allows for a wide variety
of biological active substances to be used.

## 2.7.3 Photoactivation of Hydrogels

All these processes rely on photocrosslinking or photopolymerization to generate
a physical 3D structure with significant mechanical and topographical properties.
Another possible rout is to create a 3D chemical environment inside a cell compatible surrounding. In general such a surrounding is realized by hydrogels, which
resemble closely the natural environment of cells. Of course the transition between
photocrosslinking and photofunctionalization is gradual.

Hyaluronic acid is a widely used hydrogel for 3D cell culture and tissue engineering applications (cf. e.g. [180–184]). Such hyaluronic acid hydrogels can be soaked
with a water based photocrosslinkable substance, such as biotinylated bovine serum
albumin [53]. By this means, 3D chemical active sides can be fabricated by direct laser
writing inside the hydrogel. Avidin labeled substances can be linked subsequently to
the biotin sides of the crosslinked albumin structures. Such an approach combines
physical and chemical stimuli for cells in a controlled manner and it was shown that
dorsal route ganglion cells and hippocampal neural progenitor cells are able to recognize these structures and demonstrate guided outgrowth [53]. A comparable method
employs polyethylene glycol diacrylates (PEGDA) [158]. Hydrogels formed from
the biocompatible PEGDA can be made biodegradable by the incorporation of a
collagenase sensitive peptide sequence. Such hydrogels can be cured by the addition
of a photoinitiator upon UV irradiation, even with the presence of live cells clustered
in fibrin gel. Subsequently, the peptide Arg-Gly-Asp-Ser-Lys (RGDSK) containing
PEGDA were added and crosslinked by direct laser writing to remaining free diacrylate groups [158]. Cultured fibroblasts showed a strong tendency to migrate along the
fabricated chemical cues [158]. Based on confocal images of microvascular structures, biomimetic replicas could be formed consisting of an RGDS patterned PEGDA
hydrogel, cultured with human umbilical vein endothelial cells and mesemchymal

progenitor 10T1/2 cells. These cells, organize into capillary-like tubule structures, when cultured in the same, yet unpatterned material [185], indicating the biomimetic strength of the approach.

This fabrication strategy is not restricted to a single functional group. By thoroughly removing any unbound residues of the active agent, and repeating functionalization steps, different functional groups can be formed to patterns [159, 160].

Additional, by varying laser intensity, or repeating irradiation, the concentration of such active compounds can be finely tuned [160, 186]. Such a spatial concentration gradient was observed to promote ingrowth of neuronal progenitor cells [186].

Hydrogel based direct laser writing can also be used to combine physical and chemical functionalities. For this means, hydrogels based on a PEG backbone containing vinyl groups and nitrobenzyl ether moieties were developed [48, 50]. Depending on the applied wavelength different reactions can be triggered. Single photon irradiation with visible light, or the respective two photon wavelength absorbed by the photoinitiator eosin Y triggers a thiol-containing biomolecule to covalently bind. This reaction was applied to bind the cell adhesive active peptide groups RGD and PHSRN. The nitrobenzyl ether moieties, incorporated in the PEG-backbone undergo photocleavage by UV irradiation, or the appropriate two photon wavelength, resulting in degeneration of the hydrogel network. Due to the difference in wavelength, both mechanisms can be used in the same sample. Fibroblasts migrated in channels, formed by the cleavage process, only if functional RGD peptides were present [50].

## 2.8 Conclusion

In just a decade, direct laser writing has matured in a way that many laboratories apply this technology for diverse applications. The availability of turn-key femtosecond laser oscillators, and even commercially available complete direct laser writing setups, has facilitated its use in biology and medicine. It can be expected that the first main impact of direct laser writing in a biological context, will be the realization of controlled 3D cell microenvironments as in-vitro test systems, due to the unique process properties:

- A large diversity of applicable materials, ranging from inelastic to elastic synthetic polymers, hydrogels, to biological components, such as extracellular matrix proteins.
- Controllable physical properties, such as elasticity.
- Highly defined structural features and inherent 3D capability.
- Geometrical freedom, since direct laser writing is not a layer-by-layer fabrication technology.
- Controllable chemical characteristics.

These properties allow the realization of highly controlled test systems, where physical, chemical and structural influences on cell behavior can be analyzed separately, or in combination. These in-vitro test systems may help to answer biological questions, where three-dimensionality is a key issue, such as cancer cell migration, interaction of cells with their extracellular matrix environment or angiogenesis.

However, a deeper understanding of the crosslinking mechanisms and the relationship between process parameters and the resulting structural properties is still necessary. This need becomes most obvious in the case of protein crosslinking. Although basic conservation of protein properties has been demonstrated, the detailed mechanisms leading to the cross-linked construct are yet to be revealed. Yet, such an understanding is crucial for the realization of a highly controlled in-vitro test system.

A major issue which limits the widespread use of the technology today is its limited speed. A typical writing speed in the order of magnitude of mm/s is not sufficient to realize macroscopic scaffolds for tissue engineering, or the number of in-vitro test necessary for biological analysis, on a commercially feasible timescale. Here, a combination of process development and adapted high performance materials may yield the solution.

# References

1. Abbott EA (2010) Flatland: a romance of many dimensions. Merchant Books, New York
2. Deubel M, von Freymann G, Wegener M, Pereira S, Busch K, Soukoulis CM (2004) Direct laser writing of three-dimensional photonic-crystal templates for telecommunications. Nat Mater 3:444–447
3. LaFratta CN, Fourkas JT, Baldacchini T, Farrer RA (2007) Multi-photon fabrication. Angewandte Chemie-International Edition 46:6238–6258
4. Lee KS, Kim RH, Yang DY, Park SH (2008) Advances in 3D nano/microfabrication using two-photon initiated polymerization. Prog Polym Sci 33:631–681
5. Li LJ, Fourkas JT (2007) Multi-photon polymerization. Mater Today 10:30–37
6. Maruo S, Ikuta K (2000) Three-dimensional microfabrication by use of single-photon-absorbed polymerization. Appl Phys Lett 76:2656–2658
7. Maruo S, Fourkas JT (2008) Recent progress in multi-photon microfabrication. Laser Photonics Rev 2:100–111
8. Ovsianikov A, Mironov V, Stampfl J, Liska R (2012) Engineering 3D cell-culture matrices: multi-photon processing technologies for biological and tissue engineering applications. Expert Rev Med Devices 9:613–633
9. Sun HB, Kawata S (2004) Two-photon photopolymerization and 3D lithographic microfabrication. nmr-3D Analysis. Photopolymerization 170:169–273
10. Göppert-Mayer M (1931) Über Elementarakte mit zwei Quantensprüngen. Annalen der Physik 401:273–294
11. Blab GA, Lommerse PHM, Cognet L, Harms GS, Schmidt T (2001) Two-photon excitation action cross-sections of the autofluorescent proteins. Chem Phys Lett 350:71–77
12. Kamarchik E, Krylov AI (2011) Non-Condon effects in the one- and two-photon absorption spectra of the green fluorescent protein. J Phys Chem Lett 2:488–492
13. Schafer KJ, Hales JM, Balu M, Belfield KD, Van Stryland EW, Hagan DJ (2004) Two-photon absorption cross-sections of common photoinitiators. J Photochem Photobiol A Chem 162:497–502
14. Rumi M, Barlow S, Wang J, Perry JW, Marder SR (2008) Two-photon absorbing materials and two-photon-induced chemistry. Photoresponsive Polym I 213:1–95
15. Moon JH, Yang S (2010) Chemical aspects of three-dimensional photonic crystals. Chem Rev 110:547–574

16. Pucher N, Rosspeintner A, Satzinger V, Schmidt V, Gescheidt G, Stampfl J, Liska R (2009) Structure-activity relationship in D-pi-A-pi-D-based photoinitiators for the two-photon-induced photopolymerization process. Macromolecules 42:6519–6528
17. Fouassier PF, Rabek JF (eds) (1993) Fundamentals and methods. Elsevier Applied Science, London
18. Fouassier PF, Rabek JF (eds) (1993) Polymerisation mechanisms. Elsevier Applied Science, London
19. Cowie JMG, Arrighi V (2007) Polymers: chemistry and physics of modern materials. CRC Press, New York
20. Neumann MG, Schmitt CC, Ferreira GC, Correa IC (2006) The initiating radical yields and the efficiency of polymerization for various dental photoinitiators excited by different light curing units. Dental Mater 22:576–584
21. Serbin J, Egbert A, Ostendorf A, Chichkov BN, Houbertz R, Domann G, Schulz J, Cronauer C, Frohlich L, Popall M (2003) Femtosecond laser-induced two-photon polymerization of inorganic-organic hybrid materials for applications in photonics. Opt Lett 28:301–303
22. Witzgall G, Vrijen R, Yablonovitch E, Doan V, Schwartz BJ (1998) Single-shot two-photon exposure of commercial photoresist for the production of three-dimensional structures. Opt Lett 23:1745–1747
23. Teh WH, Durig U, Drechsler U, Smith CG, Guntherodt HJ (2005) Effect of low numerical-aperture femtosecond two-photon absorption on (SU-8) resist for ultrahigh-aspect-ratio microstereolithography. J Appl Phys 97:054907
24. Juodkazis S, Mizeikis V, Seet KK, Miwa M, Misawa H (2005) Two-photon lithography of nanorods in SU-8 photoresist. Nanotechnology 16:846–849
25. Yin XB, Fang N, Zhang X, Martini IB, Schwartz BJ (2002) Near-field two-photon nanolithography using an apertureless optical probe. Appl Phys Lett 81:3663–3665
26. Liu YH, Nolte DD, Pyrak-Nolte LJ (2010) Large-format fabrication by two-photon polymerization in SU-8. Appl Phys A Mater Sci Proces 100:181–191
27. Derosa MC, Crutchley RJ (2002) Photosensitized singlet oxygen and its applications. Coord Chem Rev 233:351–371
28. Ochsner M (1997) Photophysical and photobiological processes in the photodynamic therapy of tumours. J Photochem Photobiol B Biol 39:1–18
29. Pattison DI, Rahmanto AS, Davies MJ (2012) Photo-oxidation of proteins. Photochem Photobiol Sci 11:38–53
30. Spikes JD, Shen HR, Kopeckova P, Kopecek J (1999) Photodynamic crosslinking of proteins. III. Kinetics of the FMN- and rose bengal-sensitized photooxidation and intermolecular crosslinking of model tyrosine-containing N-(2-Hydroxypropyl)methacrylamide copolymers. Photochem Photobiol 70:130–137
31. Spikes JD, Shen HR, Kopecek J (1999) Effects of pH on the kinetics of the FMN-and rose bengal (RB)-sensitized photooxidation of tyrosine and of tyrosine with the amino and/or carboxyl groups blocked. Photochem Photobiol 69:84S
32. Shen HR, Spikes JD, Smith CJ, Kopecek J (2000) Photodynamic cross-linking of proteins. V. Nature of the tyrosine-tyrosine bonds formed in the FMN-sensitized intermolecular cross-linking of N-acetyl-L-tyrosine. J Photochem Photobiol A Chem 133:115–122
33. Shen HR, Spikes JD, Smith CJ, Kopecek J (2000) Photodynamic cross-linking of proteins. IV. Nature of the His-His bond(s) formed in the rose bengal-photosensitized cross-linking of N-benzoyl-L-histidine. J Photochem Photobiol A Chem 130:1–6
34. Wang W, Nema S, Teagarden D (2010) Protein aggregation-pathways and influencing factors. Int J Pharm 390:89–99
35. Rehms AA, Callis PR (1993) 2-Photon fluorescence excitation-spectra of aromatic-amino-acids. Chem Phys Lett 208:276–282
36. Guzow K, Szabelski M, Rzeska A, Karolczak J, Sulowska H, Wiczk W (2002) Photophysical properties of tyrosine at low pH range. Chem Phys Lett 362:519–526
37. Birch DJS (2001) Multi-photon excited fluorescence spectroscopy of biomolecular systems. Spectrochim Acta Part A Mol Biomol Spectrosc 57:2313–2336

38. Engelhardt S, Hu YL, Seiler N, Riester D, Meyer W, Kruger H, Wehner M, Bremus-Kobberling E, Gillner A (2011) 3D-microfabrication of polymer-protein hybrid structures with a Q-switched microlaser. J Laser Micro Nanoeng 6:54–58

39. Kaehr B, Ertas N, Nielson R, Allen R, Hill RT, Plenert M, Shear JB (2006) Direct-write fabrication of functional protein matrixes using a low-cost Q-switched laser. Anal Chem 78:3198–3202

40. Verweij H, Vansteveninck J (1982) Model studies on photodynamic cross-linking. Photochem Photobiol 35:265–267

41. Dubbelman TMAR, Vansteveninck AL, Vansteveninck J (1982) Hematoporphyrin-induced photooxidation and photodynamic cross-linking of nucleic-acids and their constituents. Biochimica et Biophysica Acta 719:47–52

42. Balasubramanian D, Du X, Zigler JS (1990) The reaction of singlet oxygen with proteins, with special reference to crystallins. Photochem Photobiol 52:761–768

43. Lam MA, Pattison DI, Bottle SE, Keddie DJ, Davies MJ (2008) Nitric oxide and nitroxides can act as efficient scavengers of protein-derived free radicals. Chem Res Toxicol 21:2111–2119

44. Agon VV, Bubb WA, Wright A, Hawkins CL, Davies MJ (2006) Sensitizer-mediated photooxidation of histidine residues: evidence for the formation of reactive side-chain peroxides. Free Radical Biol Med 40:698–710

45. Tomita M, Irie M, Ukita T (1969) Sensitized photooxidation of histidine and its derivatives: products and mechanism of reaction. Biochemistry 8:5149–5160

46. DeForest CA, Anseth KS (2012) Advances in bioactive hydrogels to probe and direct cell fate. Ann Rev Chem Biomol Eng 3(3):421–444

47. Kasko AM, Wong DY (2010) Two-photon lithography in the future of cell-based therapeutics and regenerative medicine: a review of techniques for hydrogel patterning and controlled release. Future Med Chem 2:1669–1680

48. DeForest CA, Polizzotti BD, Anseth KS (2009) Sequential click reactions for synthesizing and patterning three-dimensional cell microenvironments. Nat Mater 8:659–664

49. Baskin JM, Prescher JA, Laughlin ST, Agard NJ, Chang PV, Miller IA, Lo A, Codelli JA, Bertozzi CR (2007) Copper-free click chemistry for dynamic in vivo imaging. Proc Natl Acad Sci USA 104:16793–16797

50. DeForest CA, Anseth KS (2011) Cytocompatible click-based hydrogels with dynamically tunable properties through orthogonal photoconjugation and photocleavage reactions. Nat Chem 3:925–931

51. Yu HT, Li JB, Wu DD, Qiu ZJ, Zhang Y (2010) Chemistry and biological applications of photo-labile organic molecules. Chem Soc Rev 39:464–473

52. Aujard I, Benbrahim C, Gouget M, Ruel O, Baudin JB, Neveu P, Jullien L (2006) o-Nitrobenzyl photolabile protecting groups with red-shifted absorption: Syntheses and uncaging cross-sections for one- and two-photon excitation. Chem A Eur J 12:6865–6879

53. Seidlits SK, Schmidt CE, Shear JB (2009) High-resolution patterning of hydrogels in three dimensions using direct-write photofabrication for cell guidance. Adv Funct Mater 19:3543–3551

54. Kasko AM, Wong DY (2010) Two-photon lithography in the future of cell-based therapeutics and regenerative medicine: a review of techniques for hydrogel patterning and controlled release. Future Med Chem 2:1669–1680

55. Engelhardt S, Hoch E, Borchers K, Meyer W, Kruger H, Tovar GEM, Gillner A (2011) Fabrication of 2D protein microstructures and 3D polymer-protein hybrid microstructures by two-photon polymerization. Biofabrication 3:025003

56. Thiel M, Fischer J, von Freymann G, Wegener M (2010) Direct laser writing of three-dimensional submicron structures using a continuous-wave laser at 532 nm. Appl Phys Lett 97:221102

57. Turunen S, Kapyla E, Terzaki K, Viitanen J, Fotakis C, Kellomaki M, Farsari M (2011) Pico- and femtosecond laser-induced crosslinking of protein microstructures: evaluation of processability and bioactivity. Biofabrication 3:045002

58. Wang I, Bouriau M, Baldeck PL, Martineau C, Andraud C (2002) Three-dimensional micro-fabrication by two-photon-initiated polymerization with a low-cost microlaser. Opt Lett 27:1348–1350

59. Sun HB, Takada K, Kim MS, Lee KS, Kawata S (2003) Scaling laws of voxels in two-photon photopolymerization nanofabrication. Appl Phys Lett 83:1104–1106

60. Sun HB, Maeda M, Takada K, Chon JWM, Gu M, Kawata S (2003) Experimental investigation of single voxels for laser nanofabrication via two-photon photopolymerization. Appl Phys Lett 83:819–821

61. Einstein A (1905) Über die von der molekularkinetischen Theorie der Wärme geforderte Bewegung von in ruhenden Flüssigkeiten suspendierten Teilchen. Annalen der Physik 322:549–560

62. von Smulochewski A (1906) Zur kinetischen Theorie der Brownschen Molekularbewegung und der Suspension. Annalen der Physik 326:756–780

63. Goodner MD, Bowman CN (2002) Development of a comprehensive free radical photopoly-merization model incorporating heat and mass transfer effects in thick films. Chem Eng Sci 57:887–900

64. Spichty M, Turro NJ, Rist G, Birbaum JL, Dietliker K, Wolf JP, Gescheidt G (2001) Bond cleavage in the excited state of acyl phosphene oxides. Insight on the role of conformation by model calculations: a concept. J Photochem Photobiol A Chem 142:209–213

65. Uppal N, Shiakolas PS (2008) Modeling of temperature-dependent diffusion and poly-merization kinetics and their effects on two-photon polymerization dynamics. J Micro-Nanolithography Mems Moems 7:043002

66. Uppal N, Shiakolas PS (2009) Process sensitivity analysis and resolution prediction for the two photon polymerization of micro/nano structures. J Manuf Sci Eng Trans ASME 131

67. Galajda P, Ormos P (2001) Complex micromachines produced and driven by light. Appl Phys Lett 78:249–251

68. von Freymann G, Ledermann A, Thiel M, Staude I, Essig S, Busch K, Wegener M (2010) Three-dimensional nanostructures for photonics. Adv Funct Mater 20:1038–1052

69. Fischer J, Wegener M (2012) Three-dimensional optical laser lithography beyond the diffrac-tion limit. Laser Photonics Rev 7:22–224

70. Sun HB, Tanaka T, Kawata S (2002) Three-dimensional focal spots related to two-photon excitation. Appl Phys Lett 80:3673–3675

71. DeVoe RJ, Kalweit H, Leatherdale CA, Williams TR (2003) Voxel shapes in two-photon microfabrication. Proc SPIE 4797:310–316. Ref Type: Journal (Full)

72. Hell S, Reiner G, Cremer C, Stelzer EHK (1993) Aberrations in confocal fluorescence microscopy induced by mismatches in refractive-index. J Microsc Oxford 169:391–405

73. Sun Q, Jiang HB, Liu Y, Zhou YH, Yang H, Gong QH (2005) Effect of spherical aberration on the propagation of a tightly focused femtosecond laser pulse inside fused silica. J Opt A Pure Appl Opt 7:655–659

74. Huot N, Stoian R, Mermillod-Blondin A, Mauclair C, Audouard E (2007) Analysis of the effects of spherical aberration on ultrafast laser-induced refractive index variation in glass. Opt Express 15:12395–12408

75. Torok P, Varga P, Booker GR (1995) Electromagnetic diffraction of light focused through a planar interface between materials of mismatched refractive-indexes-structure of the electromagnetic-field. J Opt Soc Am A Opt Image Sci Vis 12:2136–2144

76. Torok P, Varga P, Nemeth G (1995) Analytical solution of the diffraction integrals and inter-pretation of wave-front distortion when light is focused through a planar interface between materials of mismatched refractive-indexes. J Opt Soc Am A Opt Image Sci Vis 12:2660–2671

77. Egner A, Schrader M, Hell SW (1998) Refractive index mismatch induced intensity and phase variations in fluorescence confocal, multi-photon and 4Pi-microscopy. Opt Commun 153:211–217

78. Seet KK, Mizeikis V, Juodkazis S, Misawa H (2006) Three-dimensional horizontal circular spiral photonic crystals with stop gaps below 1 mu m. Appl Phys Lett 88

79. Sun HB, Kawakami T, Xu Y, Ye JY, Matuso S, Misawa H, Miwa M, Kaneko R (2000) Real three-dimensional microstructures fabricated by photopolymerization of resins through two-photon absorption. Opt Lett 25:1110–1112
80. Sun HB, Suwa T, Takada K, Zaccaria RP, Kim MS, Lee KS, Kawata S (2004) Shape precompensation in two-photon laser nanowriting of photonic lattices. Appl Phys Lett 85:3708–3710
81. Li Y, Qi FJ, Yang HH, Gong QZ, Dong XM, Duan X (2008) Nonuniform shrinkage and stretching of polymerized nanostructures fabricated by two-photon photopolymerization. Nanotechnology 19:055303
82. Wu DM, Fang N, Sun C, Zhang X (2002) Adhesion force of polymeric three-dimensional microstructures fabricated by microstereolithography. Appl Phys Lett 81:3963–3965
83. Wu DM, Fang N, Sun C, Zhang X (2006) Stiction problems in releasing of 3D microstructures and its solution. Sens Actuators A Phys 128:109–115
84. Ovsianikov A, Xiao SZ, Farsari M, Vamvakaki M, Fotakis C, Chichkov BN (2009) Shrinkage of microstructures produced by two-photon polymerization of Zr-based hybrid photosensitive materials. Opt Express 17:2143–2148
85. Vogel A, Noack J, Huttman G, Paltauf G (2005) Mechanisms of femtosecond laser nanosurgery of cells and tissues. Appl Phys B Lasers Opt 81:1015–1047
86. Schonle A, Hell SW (1998) Heating by absorption in the focus of an objective lens. Opt Lett 23:325–327
87. O'Brien AK, Bowman CN (2003) Modeling thermal and optical effects on photopolymerization systems. Macromolecules 36:7777–7782
88. Vogel A, Nahen K, Theisen D, Noack J (1996) Plasma formation in water by picosecond and nanosecond Nd:YAC laser pulses. 1. Optical breakdown at threshold and superthreshold irradiance. IEEE J Sel Top Quantum Electron 2:847–860
89. Vogel A, Busch S, Parlitz U (1996) Shock wave emission and cavitation bubble generation by picosecond and nanosecond optical breakdown in water. J Acoust Soc Am 100:148–165
90. Vogel A, Noack J, Nahen K, Theisen D, Busch S, Parlitz U, Hammer DX, Noojin GD, Rockwell BA, Birngruber R (1999) Energy balance of optical breakdown in water at nanosecond to femtosecond time scales. Appl Phys B Lasers Opt 68:271–280
91. Vogel A, Venugopalan V (2003) Mechanisms of pulsed laser ablation of biological tissues. Chem Rev 103:577–644
92. Williams F, Varma SP, Hillenius S (1976) Liquid water as a lone-pair amorphous-semiconductor. J Chem Phys 64:1549–1554
93. Sacchi CA (1991) Laser-induced electric breakdown in water. J Opt Soc Am B Opt Phys 8:337–345
94. Quinto-Su PA, Venugopalan V (2007) Mechanisms of laser cellular microsurgery. Laser Manipulation Cells Tissues 82:113–151
95. Oraevsky AA, DaSilva LB, Rubenchik AM, Feit MD, Glinsky ME, Perry MD, Mammini BM, Small W, Stuart BC (1996) Plasma mediated ablation of biological tissues with nanosecond-to-femtosecond laser pulses: relative role of linear and nonlinear absorption. IEEE J Sel Top Quantum Electron 2:801–809
96. Engler AJ, Sen S, Sweeney HL, Discher DE (2006) Matrix elasticity directs stem cell lineage specification. Cell 126:677–689
97. Tayalia P, Mendonca CR, Baldacchini T, Mooney DJ, Mazur E (2008) 3D cell-migration studies using two-photon engineered polymer scaffolds. Adv Mater 20:4494–4498
98. Nielsen LE (1969) Cross-linking-effect on physical properties of polymers. J Macromol Sci Rev Macromol Chem C 3:69–103
99. Meyer W, Engelhardt S, Novosel E, Elling B (2012) Soft polymers for building up small and smallest blood supplying systems by stereolithography. J Funct Biomater 3:257–268
100. Ligon SC, Baudis S, Nehl F, Wilke A, Bergmeister H, Bernhard D, Nigisch A, Stampfl J, Liska R, Husar B (2012) Improved elastomeric materials for CAD/CAM generation of vascular structures in soft tissue replacement therapies. J Tissue Eng Regen Med 6:301
101. Baudis S, Nehl F, Ligon SC, Nigisch A, Bergmeister H, Bernhard D, Stampfl J, Liska R (2011) Elastomeric degradable biomaterials by photopolymerization-based CAD-CAM for vascular tissue engineering. Biomed Mat 6

102. Baudis S, Schuster M, Turecek C, Bergmeister H, Weige G, Stampfl J, Varga F, Liska R (2007) Development of flexible biocompatible photopolymers as artificial vascular replacement materials. Int J Artif Organs 30:705

103. Burdick JA, Anseth KS (2002) Photoencapsulation of osteoblasts in injectable RGD-modified PEG hydrogels for bone tissue engineering. Biomaterials 23:4315–4323

104. Ovsianikov A, Malinauskas M, Schlie S, Chichkov B, Gittard S, Narayan R, Lobler M, Sternberg K, Schmitz KP, Haverich A (2011) Three-dimensional laser micro- and nanostructuring of acrylated poly(ethylene glycol) materials and evaluation of their cytoxicity for tissue engineering applications. Acta Biomaterialia 7:967–974

105. Kidoaki A, Matsuda T (2008) Microelastic gradient gelatinous gels to induce cellular mechanotaxis. J Biotechnol 133:225–230

106. Sant S, Hancock MJ, Donnelly JP, Iyer D, Khademhosseini A (2010) Biomimetic gradient hydrogels for tissue engineering. Can J Chem Eng 88:899–911

107. Stampfl J, Baudis S, Heller C, Liska R, Neumeister A, Kling R, Ostendorf A, Spitzbart M (2008) Photopolymers with tunable mechanical properties processed by laser-based high-resolution stereolithography. J Micromech Microeng 18:125014

108. Wong JY, Velasco A, Rajagopalan P, Pham Q (2003) Directed movement of vascular smooth muscle cells on gradient-compliant hydrogels. Langmuir 19:1908–1913

109. Araujo PHH, Sayer C, Poco JGR, Giudici R (2002) Techniques for reducing residual monomer content in polymers: a review. Polym Eng Sci 42:1442–1468

110. Khripin CY, Brinker CJ, Kaehr B (2010) Mechanically tunable multi-photon fabricated protein hydrogels investigated using atomic force microscopy. Soft Matter 6:2842–2848

111. Kaehr B, Shear JB (2008) Multi-photon fabrication of chemically responsive protein hydrogels for microactuation. Proc Natl Acad Sci USA 105:8850–8854

112. Liao CY, Bouriauand M, Baldeck PL, Leon JC, Masclet C, Chung TT (2007) Two-dimensional slicing method to speed up the fabrication of micro-objects based on two-photon polymerization. Appl Phys Lett 91:033108

113. Karageorgiou V, Kaplan D (2005) Porosity of 3D biornaterial scaffolds and osteogenesis. Biomaterials 26:5474–5491

114. Ovsianikov A, Deiwick A, Van Vlierberghe S, Pflaum M, Wilhelmi M, Dubruel P, Chichkov B (2011) Laser fabrication of 3D gelatin scaffolds for the generation of bioartificial tissues. Materials 4:288–299

115. Rumi M, Ehrlich JE, Heikal AA, Perry JW, Barlow S, Hu ZY, Cord-Maughon D, Parker TC, Rockel H, Thayumanavan S, Marder SR, Beljonne D, Bredas JL (2000) Structure-property relationships for two-photon absorbing chromophores: bis-donor diphenylpolyene and bis(styryl)benzene derivatives. J Am Chem Soc 122:9500–9510

116. Kato J, Takeyasu N, Adachi Y, Sun HB, Kawata S (2005), Multiple-spot parallel processing for laser micronanofabrication. Appl Phys Lett 86:044102

117. Formanek F, Takeyasu N, Tanaka T, Chiyoda K, Ishikawa A, Kawata S (2006) Three-dimensional fabrication of metallic nanostructures over large areas by two-photon polymerization. Opt Express 14:800–809

118. Dong XZ, Zhao ZS, Duan XM (2007) Micronanofabrication of assembled three-dimensional microstructures by designable multiple beams multi-photon processing. Appl Phys Lett 91:181109

119. Salter PS, Booth MJ (2011) Addressable microlens array for parallel laser microfabrication. Opt Lett 36:2302–2304

120. Ritschdorff ET, Nielson R, Shear JB (2012) Multi-focal multi-photon lithography. Lab A Chip 12:867–871

121. Obata K, Koch J, Hinze U, Chichkov BN (2010) Multi-focus two-photon polymerization technique based on individually controlled phase modulation. Opt Express 18:17193–17200

122. Haske W, Chen VW, Hales JM, Dong WT, Barlow S, Marder SR, Perry JW (2007) 65 nm feature sizes using visible wavelength 3-D multi-photon lithography. Opt Express 15:3426–3436

123. Staude I, Thiel M, Essig S, Wolff C, Busch K, von Freymann G, Wegener M (2010) Fabrication and characterization of silicon woodpile photonic crystals with a complete bandgap at telecom wavelengths. Opt Lett 35:1094–1096
124. Hell SW, Wichmann J (1994) Breaking the diffraction resolution limit by stimulated-emission—stimulated-emission-depletion fluorescence microscopy. Opt Lett 19:780–782
125. Hell SW (2003) Toward fluorescence nanoscopy. Nat Biotechnol 21:1347–1355
126. Willig KI, Kellner RR, Medda R, Hein B, Jakobs S, Hell SW (2006) Nanoscale resolution in GFP-based microscopy. Nat Methods 3:721–723
127. Willig KI, Keller J, Bossi M, Hell SW (2006) STED microscopy resolves nanoparticle assemblies. New J Phys 8:106
128. Willig KI, Rizzoli SO, Westphal V, Jahn R, Hell SW (2006) STED microscopy reveals that synaptotagmin remains clustered after synaptic vesicle exocytosis. Nature 440:935–939
129. Fischer J, von Freymann G, Wegener M (2010) The materials challenge in diffraction-unlimited direct-laser-writing optical lithography. Adv Mater 22:3578–3582
130. Fischer J, Wegener M (2011) Three-dimensional direct laser writing inspired by stimulated-emission-depletion microscopy [Invited]. Opt Mater Express 1:614–624
131. Fourkas JT (2010) Nanoscale photolithography with visible light. J Phys Chem Lett 1:1221–1227
132. Li LJ, Gattass RR, Gershgoren E, Hwang H, Fourkas JT (2009) Achieving lambda/20 resolution by one-color initiation and deactivation of polymerization. Science 324:910–913
133. Stocker MP, Li LJ, Gattass RR, Fourkas JT (2011) Multi-photon photoresists giving nanoscale resolution that is inversely dependent on exposure time. Nat Chem 3:223–227
134. Stichel T, Hecht B, Houbertz R, Sextl G (2010) Two-photon polymerization as method for the fabrication of large scale biomedical scaffold applications. J Laser Micro Nanoeng 5:209–212
135. Hsieh TM, Ng CWB, Narayanan K, Wan ACA, Ying JY (2010) Three-dimensional microstructured tissue scaffolds fabricated by two-photon laser scanning photolithography. Biomaterials 31:7648–7652
136. Gittard SD, Narayan R (2010) Laser direct writing of micro- and nano-scale medical devices. Expert Rev Med Devices 7:343–356
137. Melchels FPW, Feijen J, Grijpma DW (2010) A review on stereolithography and its applications in biomedical engineering. Biomaterials 31:6121–6130
138. Houbertz R, Steenhusen S, Stichel T, Sextl G (2010) Two-photon polymerization of inorganic-organic hybrid polymers as scalable technology using ultra-short laser pulses. In: Duarte FJ (ed) Coherence and ultrashort pulse laser emission. InTech, Rijeka
139. Peltola SM, Melchels FPW, Grijpma DW, Kellomaki M (2008) A review of rapid prototyping techniques for tissue engineering purposes. Ann Med 40:268–280
140. Engelhardt S, Refle O, Wehner M (2012) Method for the fabrication of macroscopic high resolution scaffolds by the combination of inkjet-printing and laser initiated polymerization. J Tissue Eng Regen Med 6:299–300
141. Refle O, Graf C, Engelhardt S, Visotschnig R (2012) New method for freeform fabrication for microstructured parts by combination of inkjet-printing and multi-photon polymerization. In: Direct digital manufacturing conference
142. Haycock JW (2011) 3D cell culture: a review of current approaches and techniques. Methods Mol Biol 695:1–15
143. Cukierman E, Pankov R, Stevens DR, Yamada KM (2001) Taking cell-matrix adhesions to the third dimension. Science 294:1708–1712
144. Petersen OW, Ronnovjessen L, Howlett AR, Bissell MJ (1992) Interaction with basement-membrane serves to rapidly distinguish growth and differentiation pattern of normal and malignant human breast epithelial-cells. Proc Natl Acad Sci USA 89:9064–9068
145. Berginski ME, Vitriol EA, Hahn KM, Gomez SM (2011) High-resolution quantification of focal adhesion spatiotemporal dynamics in living cells. Plos One 6:e22025
146. Koroleva A, Gittard S, Schlie S, Deiwick A, Jockenhoevel S, Chichkov B (2012) Fabrication of fibrin scaffolds with controlled microscale architecture by a two-photon polymerization-micromolding technique. Biofabrication 4:015001

147. Koroleva A, Schlie S, Fadeeva E, Gittard SD, Miller P, Ovsianikov A, Koch J, Narayan RJ, Chichkov BN (2010) Microreplication of laser-fabricated surface and three-dimensional structures. J Opt 12:124009
148. Allen R, Nielson R, Wise DD, Shear JB (2005) Catalytic three-dimensional protein architectures. Anal Chem 77:5089–5095
149. Kaehr B, Allen R, Javier DJ, Currie J, Shear JB (2004) Guiding neuronal development with in situ microfabrication. Proc Natl Acad Sci USA 101:16104–16108
150. Basu S, Cunningham LP, Pins GD, Bush KA, Taboada R, Howell AR, Wang J, Campagnola PJ (2005) Multi-photon excited fabrication of collagen matrixes cross-linked by a modified benzophenone dimer: bioactivity and enzymatic degradation. Biomacromolecules 6:1465–1474
151. Klein F, Striebel T, Fischer J, Jiang ZX, Franz CM, von Freymann G, Wegener M, Bastmeyer M (2010) Elastic fully three-dimensional microstructure scaffolds for cell force measurements. Adv Mater 22:868–871
152. Klein F, Richter B, Striebel T, Franz CM, von Freymann G, Wegener M, Bastmeyer M (2011) Two-component polymer scaffolds for controlled three-dimensional cell culture. Adv Mater 23:1341–1345
153. Dinca V, Kasotakis E, Catherine J, Mourka A, Ranella A, Ovsianikov A, Chichkov BN, Farsari M, Mitraki A, Fotakis C (2008) Directed three-dimensional patterning of self-assembled peptide fibrils. Nano Lett 8:538–543
154. Drakakis TS, Papadakis G, Sambani K, Filippidis G, Georgiou S, Gizeli E, Fotakis C, Farsari M (2006) Construction of three-dimensional biomolecule structures employing femtosecond lasers. Appl Phys Lett 89:144108
155. Claeyssens F, Hasan EA, Gaidukeviciute A, Achilleos DS, Ranella A, Reinhardt C, Ovsianikov A, Xiao S, Fotakis C, Vamvakaki M, Chichkov BN, Farsari M (2009) Three-dimensional biodegradable structures fabricated by two-photon polymerization. Langmuir 25:3219–3223
156. Hidai H, Hwang DJ, Grigoropoulos CP (2008) Self-grown fiber fabrication by two-photon photopolymerization. Appl Phys A Mater Sci Proces 93:443–445
157. Jeon H, Kim E, Grigoropoulos CP (2011) Measurement of contractile forces generated by individual fibroblasts on self-standing fiber scaffolds. Biomed Microdevices 13:107–115
158. Lee SH, Moon JJ, West JL (2008) Three-dimensional micropatterning of bioactive hydrogels via two-photon laser scanning photolithography for guided 3D cell migration. Biomaterials 29:2962–2968
159. Culver JC, Hoffmann JC, Poche RA, Slater JH, West JL, Dickinson ME (2012) Three-dimensional biomimetic patterning in hydrogels to guide cellular organization. Adv Mater 24:2344–2348
160. Hoffmann JC, West JL (2010) Three-dimensional photolithographic patterning of multiple bioactive ligands in poly(ethylene glycol) hydrogels. Soft Matter 6:5056–5063
161. Pitts JD, Campagnola PJ, Epling GA, Goodman SL (2000) Submicron multi-photon free-form fabrication of proteins and polymers: studies of reaction efficiencies and applications in sustained release. Macromolecules 33:1514–1523
162. Basu S, Campagnola PJ (2004) Properties of crosslinked protein matrices for tissue engineering applications synthesized by multi-photon excitation. J Biomed Mater Res Part A 71A:359–368
163. Basu S, Campagnola PJ (2004) Enzymatic activity of alkaline phosphatase inside protein and polymer structures fabricated via multi-photon excitation. Biomacromolecules 5:572–579
164. Kim DH, Provenzano PP, Smith CL, Levchenko A (2012) Matrix nanotopography as a regulator of cell function. J Cell Biol 197:351–360
165. Nikkhah M, Edalat F, Manoucheri S, Khademhosseini A (2012) Engineering microscale topographies to control the cell-substrate interface. Biomaterials 33:5230–5246
166. Rivron NC, Vrij EJ, Rouwkema J, Le GS, van den BA, Truckenmuller RK, van Blitterswijk CA (2012) Tissue deformation spatially modulates VEGF signaling and angiogenesis. Proc Natl Acad Sci USA 109:6886–6891

167. Unadkat HV, Hulsman M, Cornelissen K, Papenburg BJ, Truckenmuller RK, Carpenter AE, Wessling M, Post GF, Uetz M, Reinders MJ, Stamatialis D, van Blitterswijk CA, de Boer J (2011) An algorithm-based topographical biomaterials library to instruct cell fate. Proc Natl Acad Sci USA 108:16565–16570

168. Offenhausser A, Bocker-Meffert S, Decker T, Helpenstein R, Gasteier P, Groll J, Moller M, Reska A, Schafer S, Schulte P, Vogt-Eisele A (2007) Microcontact printing of proteins for neuronal cell guidance. Soft Matter 3:290–298

169. Costantino S, Heinze KG, Martinez OE, De KP, Wiseman PW (2005) Two-photon fluorescent microlithography for live-cell imaging. Microsc Res Technol 68:272–276

170. Cunningham LP, Veilleux MP, Campagnola PJ (2006) Freeform multi-photon excited microfabrication for biological applications using a rapid prototyping CAD-based approach. Opt Express 14:8613–8621

171. Liska R, Schuster M, Infuhr R, Tureeek C, Fritscher C, Seidl B, Schmidt V, Kuna L, Haase A, Varga F, Lichtenegger H, Stampfl J (2007) Photopolymers for rapid prototyping. J Coatings Technol Res 4:505–510

172. Pins GD, Bush KA, Cunningham LP, Carnpagnola PJ (2006) Multi-photon excited fabricated nano and micro patterned extracellular matrix proteins direct cellular morphology. J Biomed Mater Res Part A 78A:194–204

173. Chen XY, Brewer MA, Zou CP, Campagnola PJ (2009) Adhesion and migration of ovarian cancer cells on crosslinked laminin fibers nanofabricated by multi-photon excited photochemistry. Integrative Biol 1:469–476

174. Rowe RG, Weiss SJ (2008) Breaching the basement membrane: who, when and how? Trends Cell Biol 18:560–574

175. Sasaki T, Fassler R, Hohenester E (2004) Laminin: the crux of basement membrane assembly. J Cell Biol 164:959–963

176. Yurchenco PD, Cheng YS, Colognato H (1992) Laminin forms an independent network in basement-membranes. J Cell Biol 117:1119–1133

177. Ovsianikov A, Deiwick A, Van Vlierberghe S, Dubruel P, Moller L, Drager G, Chichkov B (2011) Laser fabrication of three-dimensional CAD scaffolds from photosensitive gelatin for applications in tissue engineering. Biomacromolecules 12:851–858

178. Schlie S, Ngezahayo A, Ovsianikov A, Fabian T, Kolb HA, Haferkamp H, Chichkov BN (2007) Three-dimensional cell growth on structures fabricated from ORMOCER® by two-photon polymerization technique. J Biomater Appl 22:275–287

179. Klein F, Striebel T, Fischer J, Jiang Z, Franz C, von Freymann G, Wegener M, Bastmeyer M (2010) Tailored three-dimensional microstructure scaffolds for cell culture. Eur J Cell Biol 89:57

180. Camci-Unal G, Nichol JW, Bae H, Tekin H, Bischoff J, Khademhosseini A (2012) Hydrogel surfaces to promote attachment and spreading of endothelial progenitor cells. J Tissue Eng Regen Med 7:337–347

181. Jha AK, Xu X, Duncan RL, Jia X (2011) Controlling the adhesion and differentiation of mesenchymal stem cells using hyaluronic acid-based, doubly crosslinked networks. Biomaterials 32:2466–2478

182. Skardal A, Sarker SF, Crabbe A, Nickerson CA, Prestwich GD (2010) The generation of 3-D tissue models based on hyaluronan hydrogel-coated microcarriers within a rotating wall vessel bioreactor. Biomaterials 31:8426–8435

183. Skardal A, Smith L, Bharadwaj S, Atala A, Soker S, Zhang Y (2012) Tissue specific synthetic ECM hydrogels for 3-D in vitro maintenance of hepatocyte function. Biomaterials 33:4565–4575

184. Yee D, Hanjaya-Putra D, Bose V, Luong E, Gerecht S (2011) Hyaluronic Acid hydrogels support cord-like structures from endothelial colony-forming cells. Tissue Eng Part A 17:1351–1361

185. Moon JJ, Saik JE, Poche RA, Leslie-Barbick JE, Lee SH, Smith AA, Dickinson ME, West JL (2010) Biomimetic hydrogels with pro-angiogenic properties. Biomaterials 31:3840–3847
186. Wylie RG, Ahsan S, Aizawa Y, Maxwell KL, Morshead CM, Shoichet MS (2011) Spatially controlled simultaneous patterning of multiple growth factors in three-dimensional hydrogels. Nat Mater 10:799–806

# Chapter 3
# Biomimetic Photonic Materials by Direct Laser Writing

Mark D. Turner, Gerd E. Schröder-Turk and Min Gu

**Abstract** Direct laser writing is a nanofabrication method used to develop three-dimensional nanostructures with almost arbitrary geometry. The fabrication of nanophotonic devices has been the major application of this technology to date. However, recent growth in the adoption of this technology and even commercialization of direct laser writing systems has extended the access of this nanofabrication method to a broader range of researchers including those in the fields of biology and biomimetics. In this review chapter the direct laser writing method and its recent application in developing biomimetic photonic materials are introduced.

## 3.1 Introduction

Biomimetics is both, the adaptation and the translation of underlying working principles of structures that have been found to exhibit certain properties in living systems for the design of synthetic systems with the same or similar functionality. Biomimetic designs, often motivated by resource-efficiency of the natural system, have been employed to achieve a range of properties, including stiff but light-weight structure adapted from the bees honeycomb, strong water repellence by mimicry of the Lotus leaf [1], the strong adhesion of gecko paws on dry surfaces [2]. Biomimetic design of photonic nanostructures that exploit the ingenious photonic geometries employed by living organisms–in particular in insects, beetles and crustaceans–is a further current field of research [3–5]. The focus of this chapter is the mimicry of three-dimensional

M. D. Turner (✉) · M. Gu
Centre for Micro-Photonics and CUDOS, Faculty of Engineering and Industrial Sciences,
Swinburne University of Technology, Hawthorn, VIC 3122, Australia
e-mail: mdturner32@gmail.com

G. E. Schröder-Turk
Theoretische Physik, Friedrich-Alexander Universität Erlangen-Nürnberg, Staudtstrasse 7B,
Erlangen, Germany

V. Schmidt and M. R. Belegratis (eds.), *Laser Technology in Biomimetics*,
Biological and Medical Physics, Biomedical Engineering,
DOI: 10.1007/978-3-642-41341-4_3, © Springer-Verlag Berlin Heidelberg 2013

(3D) biophotonic designs. Modern 3D electron microscopy methods (including electron tomography [6]) allow us to decipher increasingly complex spatial structures formed by nature. Advances in nanofabrication methods such as electron beam lithography (EBL) with resolutions below 10 nm [7] and direct laser writing (DLW) with resolutions below 70 nm [8] allow for the truthful/accurate replication of these structures for custom-designed photonic applications.

Many photonic properties of biological nanostructures have been reported, and indeed often their biological functionality and purpose identified. The pigmentless generation of colour by a biological photonic crystals is well documented for various organisms including butterfly wing scales [9, 10], insect cuticles [11], weevils [12], bird feathers [13] and marine life forms [14], and documented in reviews and books [15–18]. Iridescence, the dependence of the reflection rate on incident angle of the light, is a common biological feature [14, 19–21]. Natural antireflection coatings based on gratings have been identified in the visual systems of moths and have been mimicked to demonstrate their potential in industrial applications such as, wide-angle antireflection coatings and antiglare coatings for glass or plastic panels [22]. Similarly, photonic designs are in place in the eyes of various organisms to allow polarization sensitive vision, in particular linear-polarization [23, 24] but also, less well studied, to circular polarization [24, 25].

Many of the biophotonic structures are two-dimensional (such as the hexagonal cylinder pattern in the sea mouse [14]) or variations of a two-dimensional pattern (such as the tree-like lamellae grating responsible for the strong iridescence in the *Morpho* butterfly [20]). However, intricate and inherently three-dimensional nanostructures based on ordered or disordered spatial networks are also observed in several organisms. Amongst these are the chiral ordered porous structure known as the gyroid or **srs**-network [27] (a cubic chiral network named after the $Sr\,Si_2$ crystal [28]) found in several green butterfly species [9, 29, 30] (Fig. 3.1). The chiral structure of this photonic crystal [29, 30] causes circular dichroism [31]. A further ordered structure is the achiral ordered diamond structures, observed both in beetles [32] and in weevils [12]. Disordered porous geometries have been identified in bird feathers [33] and in the cuticles of white beetles [5]. Their structures have been suggested for adaptation of efficient paper coating [5].

Amongst the current topics in biophotonics is the response of the biological nanostructures to circularly polarized light. A structure, biological or synthetic, that discriminates between left-handed (LHD) and right-handed (RHD) circularly polarized light must have a chiral, or handed, spatial structure (chirality is the geometric property of an object that a structure and its mirror image cannot be transformed into one another by rotations and translations alone, excluding mirror operations and point inversion). It is by now well-established that some organisms have visual sensitivity to circular polarization, including mantis shrimps [25] and beetles [34]. Birefringent effects have been reported for the cuticles of crustaceans [35] and beetles [36]. In these examples, the circular polarization effects are achieved through biological quarter-wave plates (in the cuticles or eyes), i.e. by utilising the linear birefringence of the structure. In green butterflies, the inherently three-dimensional chiral gyroid or **srs**-network structure, realized in the porous chitin phase of the wing scales [9, 29, 30],

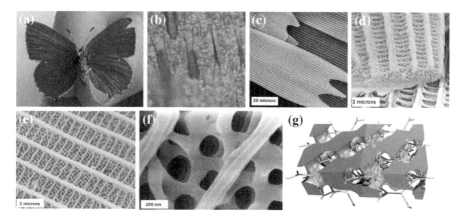

**Fig. 3.1** Photograph (**a**) and optical (**b**) and SEM microscopy images (**c–f**) of the wing and wing-scales of *Callophrys rubi* butterfly (courtesy of M. Thiel, Karlsruhe Institute of Technology). **g** Spatial structure of the chitin phase of *Callophrys rubi*. The *right* fraction of the 3D body represents a subset of the tomographic data; the *left* side represents a solid body bounded by a parallel surface to Schoen's triply-periodic gyroid minimal surface [26]. Also shown is a single **srs**-network tracing the centers of the void phase (*orange*)

has been shown to lead to a difference in the reflection of LHD and RHD circularly polarized light [31]. It appears that the improved understanding of circular polarization effects in biophotonics can open up numerous possibilities for biomimetic replication of these designs and effects. The adaptation of the butterflies' gyroid structures for nanophotonic applications, described in literature [37] and below, is certainly a step in that direction.

Nanophotonics, the structuring of materials at the nanoscale is an emerging field of research that utilises the unique interactions between light and nanoscale structures to engineer novel materials with superior optical properties. Recently, chiral structures have been the focus of many nanophotonics designs due to the strong chiral optical phenomena [31, 37–45].

The engineering of nanoscale chiral metallic nanoparticles has been used to develop nanoscale plasmonic motors that can be rotated via application of an optical source [44]. This phenomenon was attributed to the chiral asymmetry of the nanoparticle leading to a net rotational force acting on the particle. In another report [38], the same chiral metallic nanoparticle was also shown to have extremely high local field chirality leading to a great enhancement in the sensitivity of chiral biomolecules detection via circular dichroism spectroscopy [38]. Thus the design of nanostructures with strong geometrical chirality can lead to greatly enhanced chiral light-matter interactions.

Metamaterials (artificial materials whose optical properties are governed by their highly subwavelength structure) have also recently been developed with chiral geometries, demonstrating huge optical activity (i.e. optical rotary power) [45–47] and strong circular dichroism (difference in the transmission, reflection or absorption of light) [39, 48]. These nanostructured metamaterials were constructed from metallic components leading to their greatly enhanced chiral-optical properties compared to that found in natural biophotonic structures which are typically made of dielectric materials [29]. Amazingly, the chirality of these chiral metamaterials can be larger than the refractive index of these materials, leading to unnatural phenomena such as negative refractive indices [40, 49–51].

All of these chiral-optical phenomena arise from the interaction between light and the 3D chiral asymmetries of the highly subwavelength nanostructures. Fabrication technologies such as electron beam lithography can be used to make planar nanostructures with 2D chirality [38, 44]. By applying a multi-step process of lithography involving multiple layers of planar nanostructures carefully aligned on top of each other [46, 47, 52], this technique can be used to extend these planar geometries to 3D chiral nanostructures. However, this process is very time consuming, hence expensive and designs are typically limited to just a few layers.

Another fabrication method applied commonly in the fabrication of 3D structures with nanoscale resolution is the DLW method, which can be used to trace out nearly arbitrary 3D nanostructures, with resolutions now down to 68 nm [8]. This is a very suitable technology for mimicking and fabricating of biophotonic designs such as chiral photonic microstructures [37, 41], 3D cell scaffolds [53–55] and biomimetic microchannels [56]. Unlike self-assembling biological nanostructures that typically have a narrow range of structural parameters such as size and porosity, DLW is a flexible technology that can freely change these geometrical parameters (including chirality), limited only by the fabrication resolution and overall mechanical stability of the 3D nanostructure.

The next section discusses DLW in more detail and how it can be used for the development of 3D biomimetic photonic materials. The following section, discusses biologically self-assembled nanostructures which have inspired the DLW of novel 3D photonic devices. The final section reviews state-of-the-art fabrication results in the development of 3D biomimetic microstructures. This chapter specifically reviews the recent work on the chiral **srs**-networks found in the *Callophrys rubi*, as chiral nanostructures are an excellent example of a structural property only found in truly 3D geometries and is an asymmetry that is less common in nature, but easily achievable using modern technologies such as DLW.

## 3.2 Three-Dimensional Direct Laser Writing

Direct laser writing is a 3D nanofabrication technique developed over the past two decades [57–60] to what is now a well-established technology, with several commercially available systems on the market. Being able to create arbitrary-shaped 3D nano/micro-structures with resolutions as small as 68 nm [8], DLW is a powerful technology for many fields of research, including the development of novel biomimetic materials.

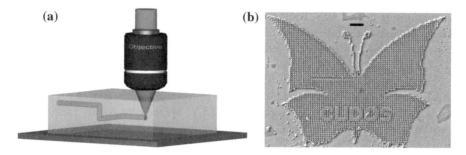

**Fig. 3.2 a** The DLW method. A pulsed *green* laser is focused using an objective lens into a transparent photoresist. The laser focal spot initiates a nonlinear photoreaction at the *center* and is then traced out in an arbitrary 3D pattern as designed by the user. **b** SEM image of a *butterfly-shaped* microstructure inspired by the 3D nanostructures found within the wing scales of the *Callophrys rubi*. The scale bar is $10\,\mu$m

The DLW method typically uses an ultrafast (i.e. short pulsed) laser (typically femtosecond or picoseconds lasers). However recent work has shown sub-micron resolution from DLW with continuous wave lasers in certain materials [61]. The laser source is tightly focussed to a diffraction limited focal spot, using a high numerical aperture objective lens, forming an ultrahigh intensity of light (see Fig. 3.2a). The sample, which contains a photoresist that is transparent at the wavelength of operation is placed at the focal spot. The transparency is critical for the ability to write 3D structures, as it allows the laser to pass through the entire photoresist without attenuation. However, due to the very high intensity supplied by the tight focusing condition, nonlinear processes such as two-photon absorption occur [57], which trigger reactions such as photopolymerisation [57, 59], micro-explosion [58, 62] or even photoreduction of metals [63]. Due to the nonlinearity of these photoreactions, these reaction can only take place at the centre of the focal spot where the intensity is highest, leaving the surrounding material relatively unmodified. The sample is then moved using a 3D nanotranslation stage and the focal spot drawn through the sample, tracing out the desired 3D design (as shown in Fig. 3.2a). In the case of a negative polymer photoresist, after the DLW has completed, the sample is then rinsed with a solvent, to remove the unwritten material, leaving behind only the regions where the focal spot has traced through and caused photopolymerisation to occur.

An example of the flexibility of DLW is shown in Fig. 3.2b, which contains a scanning electron microscope (SEM) image of a butterfly-shaped microstructure inspired by the 3D nanostructures formed within the wings of the *Callophrys rubi* butterfly. The microstructure consists of a 3D chiral **srs**-network, with cubic unit cell $2\,\mu$m in size and an overall size of $100 \times 100 \times 4\,\mu$m$^3$. The logo *CUDOS* is formed by fabricating a second **srs**-network that intertwines with the first. More details on this structure are given in section Direct laser writing of 3D biomimetic microstructures.

Much effort has gone into the development of DLW in a range of different photoresists including: photopolymers [57, 64], photoreduction of silver [63], chalcogenide glass [8, 65], bio-compatible materials [53] and even quantum dot nanocomposites [66, 67]. Post processing techniques can also be applied to modify the material properties of the 3D structure, including inversion techniques with silicon [68], infiltration of quantum dots [67, 69], electroless metal coating [70–73] and electrodeposition with conducting metals such as gold [48].

The maximum height of these 3D microstructures fabricated by DLW is typically tens of microns, limited by the working distance restrictions of the objective lens. However, the recently developed dip-in DLW optical lithography method has demonstrated the fabrication of 3D microstructures with hundreds of microns in height [74].

Whilst in theory arbitrarily shaped microstructures can be written using DLW, there are two practical limitations of this method. Firstly, the 3D voxel (or pixel) that is traced out to form the microstructure has a limitation on its size. The exact maximal resolution that is achievable is dependent on the photoresist and wavelength of operation. For example, when using commercially available polymer photoresists such as IP-L [68], Ormocer® [75] and SU-8 [41], the smallest voxel size that can be produced is typically around 100 nm in the lateral direction and 300 nm in the vertical direction. The reason for the unequal sizes in the lateral and vertical axes of the voxel is due to the unavoidable elongation of the diffraction limited focal spot [76]. In higher index materials such as chalcogenide glass, the diffraction limited focal spot is even more elongated, and can even become worse when spherical or birefringent aberrations are introduced [77].

Due to the ever increasing demand for higher resolution recent efforts have been made to further improve the resolution well beyond the diffraction limit [78–82]. These methods achieve their high resolution features by using two laser beams to illuminate the sample, inspired by the work on stimulated emission depletion microscopy [83]. The first laser beam is focused as in standard DLW to excite the laser writing photoreaction (e.g. photopolymerisation). The second beam is then applied to stop this photoreaction, either via stimulated emission depletion [79, 81] or photoinhibition [78, 82] (the photoexcitation of a chemical reaction that inhibits the writing process). By carefully aligning the position and shape these two laser focal spots on top of each other, the resulting fabrication voxel volume can be significantly decreased, and in theory has no limit. Whilst, in its early stages of development this technology is promising for the ability for 3D laser writing to achieve resolutions comparable to that of nature's nanostructure self-assembling methods, allowing scientists to truly replicate 3D biological nanostructures.

The second practical limitation as with any 3D structure is structural integrity. An important feature of any photoresists is their mechanical strength and lack of distortions such as shrinkage [64, 84]. As well as the material dependence, the design of the 3D structure, must take into account these mechanical requirements and it is quite common for microstructures to be fabricated with square [59, 85] or circular [86] frames to support the structure.

## 3.3  Chiral Structures in Self-Assembly and Circular Dichroism in Biology

Biomimetic designs are naturally mechanical robust, as evolution rarely leads to the development of mechanically unstable structures. Along with their superb mechanical properties, many biological specimens contain interesting geometrical features such as cubic symmetry and chirality, thus are great inspirations for the designs of 3D nanophotonic devices fabricated by DLW. In this section the biologically inspired **srs**-network and its exotic geometrical properties useful for photonic devices are discussed.

The **srs**-network [27] named after the $Si$ network in the poly-cationic $SrSi_2$ crystal [28] (also known as (10, 3)a or Laves' graph [87–89]) is a cubic chiral network with space group $I4_132$ (see Fig. 3.3). The **srs**-network has a four-fold screw helix along [100] and a three-fold screw helix, of opposite handedness along [111]. A RHD **srs**-network is one whose four-fold helices along [100] are RHD. Related to the **srs**-network, is Schoen's gyroid minimal surface. The gyroid surface bisects space into two domains that are mirror images of each other. The topology of these two domains is represented by the LHD and a RHD **srs**-networks. The gyroid surface geometry is ubiquitous in self-assembled materials, including lipid systems [90–92], copolymer systems [93–96]), mesoporous silicas [97, 98], germanium oxides [99] and also in cubic inner-cellular membranes [100, 101]. Materials structured according to just a single chiral **srs**-network are less common, but have been reported in zeolites [102], terblock-copolymers [96], butterfly wing scales [9, 29, 30, 103], mesoporous silica [104] and mesoporous germanium oxides [99].

Materials structured according to the **srs**-network can be obtained by chemical removal of two of the three components of the $I4_132$ phase in linear tri-block copolymers, e.g. in gold [105], at a length scale of $\sim$50 nm. A self-assembly for larger length scale is currently not available, but using the butterfly structure as template for inorganic replica has been demonstrated (e.g. in silica [106] and titania [107]), at $\sim$300 nm lattice parameter. On the contrary, 3D nanofabrication techniques such as DLW as discussed in the next section have the ability to create arbitrarily shaped 3D structures with sizes from hundreds of nanometres to hundreds of microns.

## 3.4  Direct Laser Writing of 3D Biomimetic Microstructures

The development of novel photonic materials remains one of the major applications of DLW today. Chiral microstructures are an excellent example of the development of photonic devices with complex 3D nanoscale features, yet easily achievable via DLW. An example of a simple 3D chiral microstructure is the spiral photonic crystal (consisting of a square array of helices), which has been fabricated using DLW [86, 108]. Due to their chiral asymmetry strong circular dichroism regions are formed, manifesting in the existence of polarization stop bands [109]. However, these

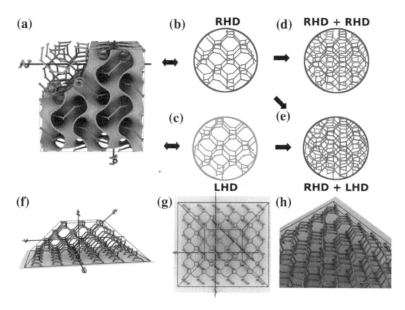

Fig. 3.3 **a** The gyroid minimal surface and its two complementary LHD and RHD chiral srs-networks. **b** RHD srs-network. **c** LHD srs-network. **d** Chiral composite consisting of two RHD srs-networks. **e** Achiral composite consisting of RHD and LHD srs-networks. **f–h** Illustrations of the *pyramid* shaped srs-networks viewed from the side (**f**), *top* (**g**) and at an oblique angle (**h**) [37]

spiral photonic crystals (PCs) have only uniaxial chirality and are highly anisotropic, greatly limiting their potential applications in photonics. The development of novel photonic structures providing complete 3D control of chirality is important for the advancement of photonic devices in a broad range of applications where polarization manipulation is important.

Recently, the bi-chiral PC [41] shown in Fig. 3.4, was developed which consisted of helices orientated along all three Cartesian axes forming an interconnected network with both chirality and cubic symmetry. The bi-chiral PC in [41] was inspired by blue phase cholesteric liquid crystals [110–112] and consisted of a fully interconnected network of helices. By choosing the handedness of the helices as well as the corners of these helical arrangements, the strength of the circular dichroism was controlled [41]. Specifically, when the handedness of the helices was opposite to that of the handedness of the corners as in the naturally occurring blue phase cholesteric liquid crystals the circular dichroism was weak. On the other hand, when the handed helices and corners were equal, the PC showed stronger circular dichroism. Thus use of DLW to develop 3D biomimetic photonic microstructures, allows for a broader range of geometries including those with superior optical properties. The application of cubic chiral designs such as the bi-chiral network may have applications in metamaterials, where the chiral geometry can be utilised to create phenomena such as negative refraction. Recently, a metallic version of the bi-chiral structure has also

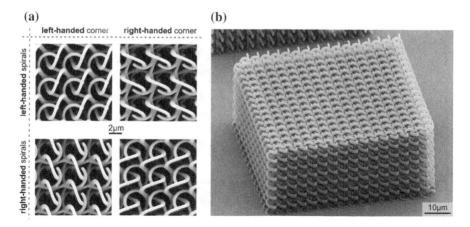

**Fig. 3.4 a, b** SEM images of the bi-chiral PCs from [41] fabricated using a commercial DLW system (Nanoscribe Gmbh.), inspired by the *blue* phase cholesteric liquid crystals. The scale bars are 2 μm in (**a**) and 10 μm in (**b**)

been demonstrated [72] that possesses broadband strong circular dichroism with reduced angular dependence, than uniaxial or planar designs.

Another 3D cubic chiral network found in nature is the **srs**-network found in the *Callophrys rubi* butterfly discussed in the previous section. The **srs**-network has recently been shown to possess unique photonic properties such as circular dichroism [31]. It was also predicted that a composite material consisting of four identical **srs**-networks that intertwine with each other would lead to superior circular dichroism properties [31]. Inspired by these theoretical findings, recent efforts have been made to replicate of these **srs**-networks at the micro-scale via DLW [37].

An illustration of the gyroid surface and the **srs**-networks contained within the surface is given in Fig. 3.3. **Srs**-networks that have RHD and LHD chirality are shown in Fig. 3.3b, c respectively. A unique feature of the **srs**-networks that is a result of its simplicity is the ability to intertwine multiple networks to form more chiral composites structures [37] such as the chiral 2-**srs** composite (see Fig. 3.4) and the achiral 2-**srs** composite (see Fig. 3.3e). Even more networks can be intertwined to form what are known as the 3, 4 and 8-**srs** nets.

Figure 3.3f–h show the design of the **srs**-network used in the DLW fabrication of [37]. The DLW was performed using a custom built DLW setup. A beam of femtosecond pulses (∼150 fs) operating at a wavelength of 580 nm was focused by an oil immersion objective (Olympus, N.A. 1.4, 100X) in the commercial photoresist IP-L (Nanoscribe Gmbh). In order to maintain a uniform writing speed with high precision, a slow writing speed of 10 μm/s was used for the 3D nanotranslation stage.

The overall shape of the **srs**-network shown in Fig. 3.3f–h has the shape of a flat top pyramid. The reason for this is to ensure that there are no free standing rods, by imitating cleaving planes in crystallography. This leads to boundaries with good structural integrity, leading to minimal distortions of the overall structure. Note this

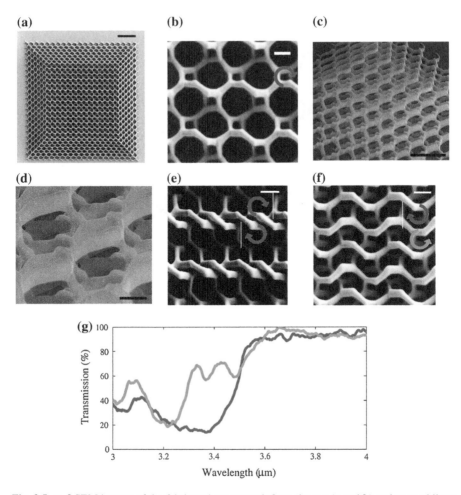

**Fig. 3.5 a–f** SEM images of the fabricated srs-network from the *top* (**a** and **b**) and at an oblique angle (**c** and **d**). **e, f** SEM images of the chiral (**e**) and achiral (**f**) composites consisting of two intertwined srs-networks. **g** Transmission spectra of RCP (*blue*) and LCP (*red*) waves for normal incidence, showing the formation of a circular dichroism band. The scale bars are are $10\,\mu$m (**a** and **c**) and $1\,\mu$m (**b, d–f**)

is a powerful advantage of DLW microstructures over natural self-assembly, which has limited (if none at all) control over the boundaries of these crystals.

The DLW of this **srs**-network design is given in Fig. 3.5, which contains scanning electron microscope (SEM) images of the **srs**-network with a $3\,\mu$m cubic unit cell size. This RHD chiral microstructure has excellent uniformity, demonstrating the combination of the engineering quality of DLW and the structural integrity of the **srs**-network. From close examination of the topology in Fig. 3.5a, b one can see the RHD 4-screw axes, as highlighted by the blue arrow.

Figure 3.5g contains the transmission spectra of infrared waves passing through the **srs**-network at normal incidence (i.e. along [1]), measured using an FTIR microscope in conjunction with a broadband linear polariser, and quarter-wave plate to achieve RHD (blue) and LHD (red) polarization. A deep bandgap at $3.4\,\mu m$ is observed only for the RHD polarization, which matches with the handedness of the **srs**-network. This demonstrates the existence of a circular dichroism band as theoretically investigated in [31]. These results imply that the chiral networks found within the wing scales of the *Callophrys rubi*, may have chiral-optical features such as circular dichroism [31], but at the UV and visible wavelength regime due to the much smaller unit cell sizes.

The great flexibility of the DLW method also allows one to fabricate not only single 3D networks, but multiple intertwining networks. In Fig. 3.4d, e this is illustrated with two different chiral (Fig. 3.4d) and achiral (Fig. 3.4e) composites, each consisting of two **srs**-networks. These multi-network designs can also be realized via DLW and the fabrication of the chiral and achiral composites are shown in Fig. 3.5e, f respectively. Thus, the **srs**-network makes an excellent building block for the design of chiral composites whose chirality can be controlled, a desirable feature for many applications whose functionality relies on the chiral light-matter interactions.

Other designs that involve the intertwining of multiple 3D networks inspired by biology and self-assembly may lead to photonic devices with novel photonic properties realisable by DLW. These chiral PCs could be used for the development of compact circularly polarized filters, beamsplitters, cavities and waveguides. Such devices are of great interest for integrated photonic circuits requiring the manipulation of polarization at microscopic sizes.

## 3.5 Conclusion and Outlook

DLW is a flexible technology for the development of 3D microstructures for applications in photonics. The 3D network designs such as the **srs**-networks found in the *Callophrys rubi* are practical geometrical designs, with useful physical phenomena such as photonic bandgaps and circular polarization stop bands. With a huge variety of biological nanostructures that have been discovered (and are yet to be discovered), other biologically inspired designs may lead to interesting 3D blueprints for the design of biomimetic materials for photonics applications. With the recent developments of super-resolution writing techniques, we may observe true 1-to-1 replication of these biological nanostructures in the near future.

**Acknowledgments** We would like to thank Michael Thiel for the photographs and electron microscopy images of the *Callophrys rubi*. This work was conducted by the Australian Research Council Centre of Excellence for Ultrahigh Bandwidth Devices for Optics Systems (project CE110001018).

# References

1. Yan YY, Gao N, Barthlott W (2011) Mimicking natural superhydrophobic surfaces and grasping the wetting process: a review on recent progress in preparing superhydrophobic surfaces. Adv Colloid Interfac 169:80–105
2. Huber G, Mantz H, Spolenak R, Mecke K, Jacobs K, Gorb SN, Arzt E (2005) Evidence for capillarity contributions to gecko adhesion from single spatula nanomechanical measurements. Proc Natl Acad Sci USA 102:16293–16296
3. Parker AR, Townley HE (2007) Biomimetics of photonic nanostructures. Nat Nanotechnol 2:347–353
4. Kolle M, Salgard-Cunha PM, Scherer MRJ, Huang F, Vukusic P, Mahajan S, Baumberg JJ, Steiner U (2010) Mimicking the colourful wing scale structure of the Papilio blumei butterfly. Nat Nanotechnol 5:511–515
5. Hallam BT, Hiorns AG, Vukusic P (2009) Developing optical efficiency through optimized coating structure: biomimetic inspiration from white beetles. Appl Opt 48:3243–3249
6. Midgley PA, Dunin-Borkowski RE (2009) Electron tomography and holography in materials science. Nat Mater 8:271–280
7. Broers AN, Hoole ACF, Ryan JM (1996) Electron beam lithography–resolution limits. Microelectron Eng 32:131–142
8. Nicoletti E, Bulla D, Luther-Davies B, Gu M (2011) Generation of lambda/12 nanowires in chalcogenide glasses. Nano Lett 11:4218–4221
9. Michielsen K, Stavenga DG (2008) Gyroid cuticular structures in butterfly wing scales: biological photonic crystals. J R Soc Interface 5:85–94
10. Prum RO, Quinn T, Torres RH (2006) Anatomically diverse butterfly scales all produce structural colours by coherent scattering. J Exp Biol 209:748–765
11. Seago AE, Brady P, Vigneron JP, Schultz TD (2009) Gold bugs and beyond: a review of iridescence and structural colour mechanisms in beetles (Coleoptera). J R Soc Interface 6:S165–S184
12. Wilts BD, Michielsen K, Kuipers J, De Raedt H, Stavenga DG (2012) Brilliant camouflage: photonic crystals in the diamond weevil, Entimus imperialis. Proc R Soc B 279:2524–2530
13. Stavenga DG, Leertouwer HL, Marshall NJ, Osorio D (2011) Dramatic colour changes in a bird of paradise caused by uniquely structured breast feather barbules. Proc R Soc B 278:2098–2104
14. Parker AR, McPhedran RC, McKenzie DR, Botten LC, Nicorovici N-AP (2001) Photonic engineering: Aphrodite's iridescence. Nature 409:36–37
15. Parker AR (2000) 515 million years of structural colour. J Opt A: Pure Appl Opt 2:R15–R28
16. Vukusic P, Sambles JR (2003) Photonic structures in biology. Nature 424:852–855
17. Srinivasarao M (1999) Nanooptics in the biological world: beetles, butterflies, birds, and moths. Chem Rev 99:1935–1962
18. Kinoshita S, Yoshioka S (2005) Structural colors in biological systems: principles and applications. Osaka University Press, Osaka
19. Bosi SG, Hayes J, Large MCJ, Poladian L (2008) Color, iridescence, and thermoregulation in Lepidoptera. Appl Opt 47:5235–5241
20. Vukusic P, Sambles JR, Lawrence CR, Wootton RJ (1999) Quantified interference and diffraction in single morpho butterfly scales. Proc R Soc Lond B 266:1403–1411
21. Poladian L, Wickham S, Lee K, Large MCJ (2009) Iridescence from photonic crystals and its suppression in butterfly scales. J R Soc Interface 6:S233–S242
22. Wilson SJ, Hutley MC (1982) The optical properties of "moth eye" antireflection surfaces. Opt Acta 29:993–1009
23. Horváth G, Varjú D (2003) Polarized light in animal vision. Springer, Berlin
24. Marshall J, Cronin TW (2011) Polarization vision. Curr Biol 21:R101–R105
25. Kleinlogel S, White AG (2008) The secret world of shrimps: polarization vision at its best. PLoS ONE 3:e2190

26. Schoen AH (1970) Infinite periodic minimal surfaces without self-intersections. NASA
27. Delgado Friedrichs O, O'Keeffe M, Yaghi OM (2003) Three-periodic nets and tilings: semi-regular nets. Acta Crystallogr A 59:515–525
28. Pringle GE (1972) The structure of $SrSi_2$: a crystal of class O(432). Acta Crystallogr B 28:2326–2328
29. Schröder-Turk GE, Wickham S, Averdunk H, Brink F, Fitz Gerald JD, Poladian L, Large MCJ, Hyde ST (2011) The chiral structure of porous chitin within the wing-scales of Callophrys rubi. J Struct Biol 174:290–295
30. Saranathan V, Osuji CO, Mochrie SGJ, Noh H, Narayanan S, Sandy A, Dufresne ER, Prum RO (2010) Structure, function, and self-assembly of single network gyroid (I4132) photonic crystals in butterfly wing scales. P Natl Acad Sci USA 107:11676–11681
31. Saba M, Thiel M, Turner MD, Hyde ST, Gu M, Grosse-Brauckmann K, Neshev DN, Mecke K, Schröder-Turk GE (2011) Circular dichroism in biological photonic crystals and cubic chiral nets. Phys Rev Lett 106:103902
32. Galusha JW, Richey LR, Gardner JS, Cha JN, Bartl MH (2008) Discovery of a diamond-based photonic crystal structure in beetle scales. Phys Rev E 77:050904
33. Dufresne ER, Noh H, Saranathan V, Mochrie SGJ, Cao H, Prum RO (2009) Self-assembly of amorphous biophotonic nanostructures by phase separation. Soft Matter 5:1792–1795
34. Brady P, Cummings M (2010) Differential response to circularly polarized light by the jewel scarab beetle Chrysina gloriosa. Am Nat 175:614–620
35. Neville AC, Luke BM (1971) Form optical activity in crustacean cuticle. J Insect Physiol 17:519–526
36. Hegedüs R, Szél G, Horváth G (2006) Imaging polarimetry of the circularly polarizing cuticle of scarab beetles (Coleoptera: Rutelidae, Cetoniidae). Vision Res 46:2786–2797
37. Turner MD, Schröder-Turk GE, Gu M (2011) Fabrication and characterization of three-dimensional biomimetic chiral composites. Opt Express 19:10001–10008
38. Hendry E, Cornelius T, Johnston J, Popland M, Mikhaylovskiy RV, Lapthorn AJ, Kelly SM, Barron LD, Gadegaard N, Kadodwala M (2010) Ultrasensitive detection and characterization of biomolecules using superchiral fields. Nat Nanotechnol 5:783–787
39. Decker M, Klein MW, Wegener M, Linden S (2007) Circular dichroism of planar chiral magnetic metamaterials. Opt Lett 32:856–858
40. Plum E, Zhou J, Dong J, Fedotov VA, Koschny T, Soukoulis CM, Zheludev NI (2009) Metamaterial with negative index due to chirality. Phys Rev B 79:035407
41. Thiel M, Rill MS, Freymann G, Wegener M (2009) Three-dimensional bi-chiral photonic crystals. Adv Mater 21:4680–4682
42. Andryieuski A, Menzel C, Rockstuhl C, Malureanu R, Lederer F, Lavrinenko A (2010) Homogenization of resonant chiral metamaterials. Phys Rev B 82:235107
43. Hur K, Francescato Y, Giannini V, Maier SA, Hennig RG, Wiesner U (2011) Three-dimensionally isotropic negative refractive index materials from block copolymer self-assembled chiral gyroid networks. Angew Chem 123:12191–12195
44. Liu M, Zentgraf T, Liu Y, Bartal G, Zhang X (2010) Light-driven nanoscale plasmonic motors. Nat Nanotechnol 5:570–573
45. Kuwata-Gonokami M, Saito N, Ino Y, Kauranen M, Jefimovs K, Vallius T, Turunen J, Svirko Y (2005) Giant optical activity in quasi-two-dimensional planar nanostructures. Phys Rev Lett 95:227401
46. Decker M, Ruther M, Kriegler CE, Zhou J, Soukoulis CM, Linden S, Wegener M (2009) Strong optical activity from twisted-cross photonic metamaterials. Opt Lett 34:2501–2503
47. Decker M, Zhao R, Soukoulis CM, Linden S, Wegener M (2010) Twisted split-ring-resonator photonic metamaterial with huge optical activity. Opt Lett 35:1593–1595
48. Gansel JK, Thiel M, Rill MS, Decker M, Bade K, Saile V, von Freymann G, Linden S, Wegener M (2009) Gold helix photonic metamaterial as broadband circular polarizer. Science 325:1513–1515
49. Zhou J, Dong J, Wang B, Koschny T, Kafesaki M, Soukoulis CM (2009) Negative refractive index due to chirality. Phys Rev B 79:121104

50. Pendry JB (2004) A chiral route to negative refraction. Science 306:1353–1355
51. Tretyakov S, Nefedov I, Sihvola A, Maslovski S, Simovski C (2003) Waves and energy in chiral nihility. J Electromagnet Wave 17:695–706
52. Helgert C, Pshenay-Severin E, Falkner M, Menzel C, Rockstuhl C, Kley E-B, Tünnermann A, Lederer F, Pertsch T (2011) Chiral metamaterial composed of three-dimensional plasmonic nanostructures. Nano Lett 11:4400–4404
53. Melissinaki V, Gill AA, Ortega I, Vamvakaki M, Ranella A, Haycock JW, Fotakis C, Farsari M, Claeyssens F (2011) Direct laser writing of 3D scaffolds for neural tissue engineering applications. Biofabrication 3:045005
54. Ovsianikov A, Deiwick A, Van Vlierberghe S, Dubruel P, Möller L, Dräger G, Chichkov B (2011) Laser fabrication of three-dimensional CAD scaffolds from photosensitive gelatin for applications in tissue engineering. Biomacromolecules 12:851–858
55. Klein F, Striebel T, Fischer J, Jiang Z, Franz CM, von Freymann G, Wegener M, Bastmeyer M (2010) Elastic fully three-dimensional microstructure scaffolds for cell force measurements. Adv Mater 22:868–871
56. Kam DH, Mazumder J (2008) Three-dimensional biomimetic microchannel network by laser direct writing. J Laser Appl 20:185–191
57. Maruo S, Nakamura O, Kawata S (1997) Three-dimensional microfabrication with two-photon-absorbed photopolymerization. Opt Lett 22:132–134
58. Straub M, Ventura M, Gu M (2003) Multiple higher-order stop gaps in infrared polymer photonic crystals. Phys Rev Lett 91:043901
59. Straub M, Gu M (2002) Near-infrared photonic crystals with higher-order bandgaps generated by two-photon photopolymerization. Opt Lett 27:1824–1826
60. Gu M, Jia B, Li J, Ventura MJ (2010) Fabrication of three-dimensional photonic crystals in quantum-dot-based materials. Laser Photonics Rev 4:414–431
61. Thiel M, Fischer J, von Freymann G, Wegener M (2010) Direct laser writing of three-dimensional submicron structures using a continuous-wave laser at 532 nm. Appl Phys Lett 97:221102
62. Sun H-B, Xu Y, Matsuo S, Misawa H (1999) Microfabrication and characteristics of two-dimensional photonic crystal structures in vitreous silica. Opt Rev 6:396–398
63. Cao Y, Takeyasu N, Tanaka T, Duan X, Kawata S (2009) 3D metallic nanostructure fabrication by surfactant-assisted multiphoton-induced reduction. Small 5:1144–1148
64. Ovsianikov A, Viertl J, Chichkov B et al (2008) Ultra-low shrinkage hybrid photosensitive material for two-photon polymerization microfabrication. ACS Nano 2:2257–2262
65. Nicoletti E, Zhou G, Jia B, Ventura MJ, Bulla D, Luther-Davies B, Gu M (2008) Observation of multiple higher-order stopgaps from three-dimensional chalcogenide glass photonic crystals. Opt Lett 33:2311–2313
66. Jia B, Buso D, van Embden J, Li J, Gu M (2010) Highly non-linear quantum dot doped nanocomposites for functional three-dimensional structures generated by two-photon poly-merization. Adv Mater 22:2463–2467
67. Ventura MJ, Gu M (2008) Engineering spontaneous emission in a quantum-dot-doped polymer nanocomposite with three-dimensional photonic crystals. Adv Mater 20:1329–1332
68. Staude I, Thiel M, Essig S, Wolff C, Busch K, von Freymann G, Wegener M (2010) Fabrication and characterization of silicon woodpile photonic crystals with a complete bandgap at telecom wavelengths. Opt Lett 35:1094–1096
69. Li J, Jia B, Zhou G, Gu M (2006) Fabrication of three-dimensional woodpile photonic crystals in a PbSe quantum dot composite material. Opt Express 14:10740–10745
70. Formanek F, Takeyasu N, Tanaka T, Chiyoda K, Ishikawa A, Kawata S (2006) Selective elec-troless plating to fabricate complex three-dimensional metallic micro/nanostructures. Appl Phys Lett 88:083110
71. Malureanu R, Zalkovskij M, Andryieuski A, Lavrinenko AV (2010) Controlled Ag electroless deposition in bulk structures with complex three-dimensional profiles. J Electrochem Soc 157:K284–K288

72. Radke A, Gissibl T, Klotzbücher T, Braun PV, Giessen H (2011) Three-dimensional bichiral plasmonic crystals fabricated by direct laser writing and electroless silver plating. Adv Mater 23:3018–3021
73. Chen Y-S, Tal A, Kuebler SM (2007) Route to three-dimensional metallized microstructures using cross-linkable epoxide SU-8. Chem Mater 19:3858–3860
74. Bückmann T, Stenger N, Kadic M, Kaschke J, Frölich A, Kennerknecht T, Eberl C, Thiel M, Wegener M (2012) Tailored 3D mechanical metamaterials made by dip-in direct-laser-writing optical lithography. Adv Mater 24:2710–2714
75. Serbin J, Gu M (2006) Experimental evidence for superprism effects in three-dimensional polymer photonic crystals. Adv Mater 18:221–224
76. Sun H-B, Tanaka T, Kawata S (2002) Three-dimensional focal spots related to two-photon excitation. Appl Phys Lett 80:3673–3675
77. Cumming BP, Jesacher A, Booth MJ, Wilson T, Gu M (2011) Adaptive aberration compensation for three-dimensional micro-fabrication of photonic crystals in lithium niobate. Opt Express 19:9419–9425
78. Scott TF, Kowalski BA, Sullivan AC, Bowman CN, McLeod RR (2009) Two-color single-photon photoinitiation and photoinhibition for subdiffraction photolithography. Science 324:913–917
79. Li L, Gattass RR, Gershgoren E, Hwang H, Fourkas JT (2009) Achieving lambda/20 resolution by one-color initiation and deactivation of polymerization. Science 324:910–913
80. Fischer J, von Freymann G, Wegener M (2010) The materials challenge in diffraction-unlimited direct-laser-writing optical lithography. Adv Mater 22:3578–3582
81. Fischer J, Wegener M (2011) Three-dimensional direct laser writing inspired by stimulated-emission-depletion microscopy [Invited]. Opt Mater Express 1:614–624
82. Cao Y, Gan Z, Jia B, Evans RA, Gu M (2011) High-photosensitive resin for super-resolution direct-laser-writing based on photoinhibited polymerization. Opt Express 19:19486–19494
83. Hell SW, Wichmann J (1994) Breaking the diffraction resolution limit by stimulated emission: stimulated-emission-depletion fluorescence microscopy. Opt Lett 19:780–782
84. Ovsianikov A, Shizhou X, Farsari M, Vamvakaki M, Fotakis C, Chichkov BN (2009) Shrinkage of microstructures produced by two-photon polymerization of Zr-based hybrid photosensitive materials. Opt Express 17:2143–2148
85. Deubel M, von Freymann G, Wegener M, Pereira S, Busch K, Soukoulis CM (2004) Direct laser writing of three-dimensional photonic-crystal templates for telecommunications. Nat Mater 3:444–447
86. Thiel M, Decker M, Deubel M, Wegener M, Linden S, von Freymann G (2007) Polarization stop bands in chiral polymeric three-dimensional photonic crystals. Adv Mater 19:207–210
87. Heesch H, Laves F (1933) Über dünne kugelpackungen. Z Kristallogr 85:443–453
88. Wells AF (1954) The geometrical basis of crystal chemistry. Part 1. Acta Crystallographica 7:535–544
89. Wells AF (1977) Three-dimensional nets and polyhedra. Wiley, New York
90. Luzzati V, Spegt PA (1967) Polymorphism of lipids. Nature 215:701–704
91. Luzzati V, Taredieu A, Gulik-Krzywicki T, Rivas E, Reiss-Husson F (1968) Structure of the cubic phases of lipid-water systems. Nature 220:485–488
92. Alexandridis P, Olsson U, Lindman B (1998) A record nine different phases (four cubic, two hexagonal, and one lamellar lyotropic liquid crystalline and two micellar solutions) in a ternary isothermal system of an amphiphilic block copolymer and selective solvents (water and oil). Langmuir 14:2627–2638
93. Hajduk DA, Harper PE, Gruner SM, Honeker CC, Kim G, Thomas EL, Fetters LJ (1994) The gyroid: a new equilibrium morphology in weakly segregated diblock copolymers. Macromolecules 27:4063–4075
94. Laurer JH, Hajduk DA, Fung JC, Sedat JW, Smith SD, Gruner SM, Agard DA, Spontak RJ (1997) Microstructural analysis of a cubic bicontinuous morphology in a neat SIS triblock copolymer. Macromolecules 30:3938–3941

95. Avgeropoulos A, Dair BJ, Hadjichristidis N, Thomas EL (1997) Tricontinuous double gyroid cubic phase in triblock copolymers of the ABA type. Macromolecules 30:5634–5642
96. Epps TH, Cochran EW, Bailey TS, Waletzko RS, Hardy CM, Bates FS (2004) Ordered network phases in linear poly(isoprene-b-styrene-b-ethylene oxide) triblock copolymers. Macromolecules 37:8325–8341
97. Beck JS, Vartuli JC, Roth WJ, Leonowicz ME, Kresge CT, Schmitt KD, Chu CTW, Olson DH, Sheppard EW (1992) A new family of mesoporous molecular sieves prepared with liquid crystal templates. J Am Chem Soc 114:10834–10843
98. Kresge CT, Leonowicz ME, Roth WJ, Vartuli JC, Beck JS (1992) Ordered mesoporous molecular sieves synthesized by a liquid-crystal template mechanism. Nature 359:710–712
99. Zou X, Conradsson T, Klingstedt M, Dadachov MS, O'Keeffe M (2005) A mesoporous germanium oxide with crystalline pore walls and its chiral derivative. Nature 437:716–719
100. Deng Y, Landh T (1995) The cubic gyroid-based membrane structure of the chloroplast in zygnema (chlorophyceae zygnematales). Zool Stud 34:175–177
101. Almsherqi ZA, Landh T, Kohlwein SD, Deng Y (2009) Chapter 6 cubic membranes: the missing dimension of cell membrane organization. Int Rev Cel Mol Biol 274:275–342
102. Sun J, Bonneau C, Cantin A, Corma A, Diaz-Cabanas MJ, Moliner M, Zhang D, Li M, Zou X (2009) The ITQ-37 mesoporous chiral zeolite. Nature 458:1154–1157
103. Hyde ST, O'Keeffe M, Proserpio DM (2008) A short history of an elusive yet ubiquitous structure in chemistry, materials, and mathematics. Angew Chem 47:7996–8000
104. Terasaki O, Liu Z, Ohsuna T, Shin HJ, Ryoo R (2002) Electron microscopy study of novel Pt nanowires synthesized in the spaces of silica mesoporous materials. Microsc Microanal 8:35–39
105. Vignolini S, Yufa NA, Cunha PS, Guldin S, Rushkin I, Stefik M, Hur K, Wiesner U, Baumberg JJ, Steiner U (2012) A 3D optical metamaterial made by self-assembly. Adv Mater 24:OP23–OP27
106. Mille C, Tyrode EC, Corkery RW (2011) Inorganic chiral 3D photonic crystals with bicontinuous gyroid structure replicated from butterfly wing scales. Chem Commun 47:9873–9875
107. Mille C, Tyrode EC, Corkery RW (2013) 3D titania photonic crystals replicated from gyroid structures in butterfly wing scales: approaching full band gaps at visible wavelengths. RSC Adv 3:3109–3117
108. Thiel M, Fischer H, von Freymann G, Wegener M (2010) Three-dimensional chiral photonic superlattices. Opt Lett 35:166–168
109. Lee J, Chan C (2005) Polarization gaps in spiral photonic crystals. Opt Express 13:8083–8088
110. Cao W, Munoz A, Palffy-Muhoray P, Taheri B (2002) Lasing in a three-dimensional photonic crystal of the liquid crystal blue phase II. Nat Mater 1:111–113
111. Coles HJ, Pivnenko MN (2005) Liquid crystal blue phases with a wide temperature range. Nature 436:997–1000
112. Kikuchi H, Yokota M, Hisakado Y, Yang H, Kajiyama T (2002) Polymer-stabilized liquid crystal blue phases. Nat Mater 1:64–68

# Chapter 4
# Selective Laser Sintering and Its Biomedical Applications

**Bin Duan and Min Wang**

**Abstract** Selective laser sintering (SLS), a mature and versatile rapid prototyping (RP) technology, uses a laser beam to selectively sinter powdered materials to form three-dimensional objects, porous or non-porous, according to the computer-aided design which can be based on data obtained from advanced medical imaging technologies such as magnetic resonance imaging (MRI) and computer tomography (CT). In this chapter, major RP technologies suitable for biomedical applications are briefly introduced first. A review is made on SLS, including its working principle, modification of commercial SLS machines for fabricating biomedical products, biomedical SLS materials, and optimization of SLS parameters. Finally, a detailed presentation is given on the biomedical application of SLS, focusing on the fabrication of tissue engineering scaffolds and drug or biomolecule delivery vehicles. It is shown that SLS has great potential for many biomimetic and biomedical applications.

## 4.1 Introduction to Rapid Prototyping Technologies

With the advances in computer technology, imaging techniques, laser techniques and control technology, rapid prototyping (RP) technologies, or alternatively termed "solid free-form fabrication" (SFF), which can automatically construct solid objects through additive manufacturing, have been developed over the past few decades. RP comprises a group of techniques that can generate physical models directly from computer-aided design (CAD) data, computer-based medical imaging techniques or

B. Duan · M. Wang (✉)
Department of Mechanical Engineering, Faculty of Engineering,
The University of Hong Kong, Pokfulam Road, Pokfulam, Hong Kong
e-mail: memwang@hku.hk

B. Duan
Department of Biomedical Engineering, College of Engineering, Cornell University,
Ithaca, NY 14853-7202, USA

V. Schmidt and M. R. Belegratis (eds.), *Laser Technology in Biomimetics*,
Biological and Medical Physics, Biomedical Engineering,
DOI: 10.1007/978-3-642-41341-4_4, © Springer-Verlag Berlin Heidelberg 2013

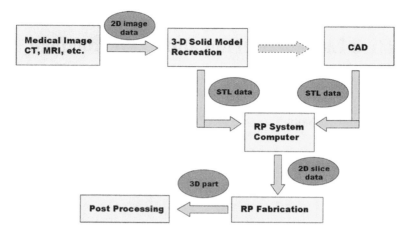

**Fig. 4.1** The rapid prototyping (RP) manufacturing process

other computer-based technologies in a layer-by-layer manner and each layer is in the shape of the cross-section of the physical model at a specific level [1]. In RP, all computer-based information or data is in general firstly converted to the STL-type file format, STL being derived from "stereolithography" which is the oldest RP technology [2, 3]. The two-dimensional (2D) layers generated by the STL file are then created in the RP machine so as to construct a solid three-dimensional (3D) physical model, starting from the bottom of the physical model and proceeding upwards. Each layer is bonded to the previous layer, thus forming a solid object based on the design. The typical process for RP is illustrated in Fig. 4.1.

According to the working principle, there are three main categories of RP techniques: (1) laser polymerization-based techniques, including stereolithography apparatus (SLA) and two-photon polymerization (TPP); (2) nozzle deposition-based techniques, such as fused deposition modeling (FDM) and 3D plotting; (3) powder-based techniques, such as 3D printing and selective laser sintering (SLS) [4, 5]. Each type of these techniques requires a specific form of materials for constructing 3D objects and has its pros and cons. For the last decade, RP techniques have been widely investigated for biomedical applications, such as fabricating tissue engineering scaffolds with or without human cells, making drug delivery vehicles, producing medical devices, and constructing physical models for surgical planning. In this section, commonly used RP techniques are briefly introduced and some of these techniques, which are suitable for tissue engineering and drug delivery applications, are described and illustrated.

## 4.1.1 Stereolithography Apparatus

Stereolithography apparatus (SLA) is based on the use of an electromagnetic radiation source (e.g. a laser or a UV light source) to initiate photopolymerization of resins [6]. The photopolymerizable resins are usually mixtures of simple, low-molecular-weight monomers capable of forming polymers when activated by the radiation energy within specific wavelength ranges. Typically, the controlled laser beam or digital light projector is directed onto preprogrammed regions of a layer of liquid resin, initiating polymerization and causing the radiation-exposed region to solidify. The first solid layer on the machine platform is then lowered into the liquid resin such that a new layer of resin is solidified by the radiation on the surface of the first solid layer at a defined layer thickness. The polymerization process is repeated, layer by layer, until a 3D object is built. The object produced may be mechanically weak and therefore is subjected to post-SLA treatment after being removed from the platform.

With the developments in polymer science and engineering, more and more photopolymerizable biomaterials, including hydrogels and multifunctional monomers, are investigated for the SLA process for biomedical applications. Some photoreactive and crosslinkable groups such as acrylates or methacrylates can be easily attached to poly(ethylene glycol) (PEG) and then crosslinked into PEG hydrogel to be used for tissue engineering and even for the encapsulation of cells in the presence of a cytocompatible photoinitiator [7, 8]. The major technological challenges for the SLA technique are the removal of uncured resin from the objects constructed and the change of resin reservoirs for using multiple resins to fabricate composite porous or non-porous structures for tissue engineering applications.

## 4.1.2 Two-Photon Polymerization

Two-photon polymerization (TPP) is based on the simultaneous absorption of two photons, which induces chemical reactions between starter molecules and monomers within a transparent matrix [9]. An ultrashort-pulse laser is needed to provide the high intensity. The current capability of TPP allows the generation of 3D structures with a resolution down to 100 nm [10]. For tissue engineering applications, TPP has been used to manufacture 3D scaffolds for hepatocyte culture by triggering free-radical polymerization processes within polymerizable precursors and to fabricate micro-patterns of biomolecules for guiding cell migration within 3D PEG-based hydrogels [11, 12]. The hydrogel network produced with well-defined chemical and physical properties has been further functionalized and modified for elucidating more complex relationships in cell behavior and tissue development and also for introducing pathways to engineer complex tissues [13]. With high resolution and high fidelity, TPP is capable to fabricate several types of small-scale medical devices, including micro-needles, microfluidic devices, etc. [14, 15]. In contrast to other RP techniques, TPP is not restricted to a layer-by-layer build-up of a 3D structure.

### *4.1.3 Fused Deposition Modeling*

In an a fused deposition modeling (FDM) process, a spool of thermoplastic filament is fed into a heated FDM extrusion head and the filament is melted into liquid polymer (normally 1 °C above the melting temperature) by a heater. As the extrusion head moves horizontally in the x and y axes following a programmed path, the filament material is extruded from a nozzle directly onto the FDM machine platform by a precision pump. Once a layer is finished, the extrusion head moves up for a programmed distance in the z direction for depositing the next layer. Each layer is bonded to the previous layer through thermal heating [16]. Solid objects are made layer by layer, with the layer thickness depending on the nozzle diameter [17]. The FDM technique is constrained by the use of thermoplastic materials with good melt viscosity and human cells cannot be encapsulated into FDM-built structures during the fabrication process because of the high temperature involved. It was shown that poly($\varepsilon$-caprolactone) (PCL) and poly(D,L-lactide-co-glycolide) (PLGA) filaments could be fabricated from polymer pellets, which had constant diameters. These filaments were then fit for FDM systems and 3D scaffolds of these polymers were made [18, 19]. Furthermore, for bone tissue engineering, biodegradable polymer-bioceramic composite scaffolds such as calcium phosphate (Ca-P)/PCL scaffolds were fabricated using the FDM technique [20, 21].

### *4.1.4 3D Plotting*

3D plotting, also termed "bioplotting" when used in the biomedical field, is based on an automatic dispenser for a hydrogel material which is forced to go through the tip of a syringe and then laid down on a platform in air or into a liquid medium [22]. Hydrogel formation can be achieved by further chemical reaction and crosslinking. The main advantage of 3D plotting is the mild processing environment, which is a prerequisite for the incorporation of biomolecules such as growth factors or even human cells. Many biodegradable materials can be used for 3D plotting, including gelatin, PEG, chitosan, and composites such as gelatin/hyaluronan and starch/PCL [23–25]. However, its limited resolution and low mechanical properties of the products formed have restricted its applications in tissue engineering.

With appropriate modification, 3D plotting can also be employed to lay down fibers to construct 3D solid scaffolds in a process similar to that of FDM. In this process, a highly viscous polymer in the molten state is deposited from an xyz motor-driven syringe onto a platform by pressure [26]. Woodfield et al. used this technique to produce 3D scaffolds containing deposited polyethyleneoxide-terephtalate (PEOT) or poly(ethylene glycol) (PEGT)/polybutylene-terephtelate (PBT) block copolymer. They investigated the effects of various parameters such as fiber diameter, fiber spacing and layer thickness in the internal structure of scaffolds on the mechanical properties of scaffolds [27, 28]. The PEGT/PBT scaffolds produced by 3D fiber deposition could create an in vivo environment that enhanced cartilaginous matrix deposition and therefore have the potential for treating articular cartilage defects [29, 30].

## 4.1.5 3D Printing

3D Printing (3DP) was first developed at Massachusetts Institute of Technology, USA, and is probably one of the most widely investigated RP techniques for fabricating tissue engineering scaffolds. During the 3DP process, a thin layer of powder is first spread over the building platform and then an "inkjet" print head prints or deposits a binder solution onto the powder bed. After the 2D layer profile is printed, the piston of the building platform lowers the platform so that the next layer of powder can be spread by a roller and laid down. This layer-by-layer process repeats until the whole 3D object is built. After the binder has dried in the object composed of the powder, the solid object is retrieved and unbound powder is removed. Many biodegradable polymers in the fine powder form may be processed into 3D scaffolds through the 3DP process. For synthetic poly($\alpha$-hydroxy esters) such as poly(L-lactic acid) (PLLA), PLGA, PCL, organic solvents (such as chloroform) are chosen as the binder [31–33]. The conventional particulate leaching technique for scaffold fabrication could be combined with the 3DP technique to create porous scaffolds [34]. (The particulate leaching technique is a very common method used for producing tissue engineering scaffolds. In this process, porogens, i.e. dissolvable particles (such as sodium chloride), are firstly mixed with a polymer solution. The solvent in the polymer solution is then evaporated and hence a polymeric object (film, bar, rod, etc.) which contains dispersed porogens is formed. This object is subsequently immersed in a liquid (normally distilled water) to dissolve the porogens. After the total dissolution of porogens while avoiding possible damages to the polymer structure in the immersion process, a porous scaffold is eventually formed). The use of an organic solvent as the binder can cause problems to the incorporation of cells into scaffolds. Therefore, a starch-based biomaterial consisting of cornstarch, dextran and gelatin was used for 3DP, with water being used as the binder [35].

## 4.1.6 Selective Laser Sintering

Selective laser sintering (SLS) uses a laser beam, such as a $CO_2$ laser, to make solid 3D objects by selectively sintering thin layers of suitable and powdered materials (mainly polymers and polymer-based composites). This technique has many advantages, including high part accuracy, material availability and easy post-SLS processing [36], which make SLS a very attractive technique for producing physical models for surgical planning and for prototyping medical devices. In addition, the capacity of SLS to generate 3D constructs with irregular shapes and also structures such as channels and overhanging features enables this technique to fabricate tissue engineering scaffolds with controlled pore shape, pore size and customized scaffold architecture [1]. In this chapter, the working principle of SLS is firstly introduced and the modification of commercial SLS machines for fabricating tissue engineering scaffolds is then presented. The materials used in SLS are reviewed and the optimization of SLS parameters using various methods for fabricating porous objects of good quality is discussed. Biomimetic and biomedical applications of the SLS technique, particularly for tissue engineering and drug delivery, are demonstrated.

## 4.2 Selective Laser Sintering

SLS was first developed and patented by Deckard and Beaman of the University of Texas at Austin, USA, in the mid-1980s. It was commercialized by DTM Corporation, USA, in 1987.

SLS is an additive manufacturing process and CAD files can be used for SLS machines to fabricate 3D objects, porous or non-porous. The external and internal structures of the objects to be formed can be designed using professional computer software for 3D drawing, be described using mathematical equations, or be derived from computerized medical imaging techniques such as magnetic resonance imaging (MRI) and computer tomography (CT). The CAD file or the reconstructed imaging file with the information of geometry and size of the object is converted into an STL file which can be used by the computer of an SLS machine. According to the design of the 3D object, the laser beam of the SLS machine selectively scans the layer of the fine powder material to heat the powder and fuse the particles together, forming a thin layer (a slice) of the 3D object. Layer-by-layer, a 3D object is constructed.

### 4.2.1 Principle of Selective Laser Sintering and Modification of Commercial SLS Machines

Figure 4.2 displays a schematic diagram for SLS. Before the SLS process is started, the entire part bed of the SLS machine is heated to just below the melting temperature of the material or near the glass transition temperature of the material (if it is an amorphous polymer) in order to minimize thermal distortion of the sintered layer (and hence the sintered object) and facilitate fusion of the layer being sintered to the previous layer [37]. During SLS, following the cross-sectional profiles from the slice data, the laser beam scans the surface of powders to heat up the powders, causing the particles in the powder to fuse together to form a solid layer. The powders that have not been scanned by the laser beam remain in place to serve as the support for the next layer of powder and will be removed and recycled after the whole object is formed via SLS. After one layer is finished, the part bed is lowered and the powder

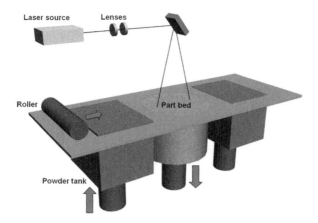

**Fig. 4.2** A schematic diagram showing the principle of SLS

tank containing fine powders is raised. A new layer of powder is then spread on the sintered part by a roller and the selective sintering process is repeated. After SLS, the non-fused powder is removed from the 3D object formed by shaking the object manually or using a compressed air jet for blowing away loose powder.

With the application of powerful, high-quality lasers, complete melting of powder materials can occur, giving rise to a new technique, selective laser melting (SLM). The SLM technique is very effective to produce, from metallic powders, objects with complex geometries and good mechanical properties [38]. SLM has a sintering mechanism similar to that of SLS, which includes the following phenomena: absorption and scattering of laser radiation, heat transfer, phase transformation, fluid flow within the molten pool caused by surface-tension gradient, evaporation and emission of material, and chemical reaction [38]. Many metallic materials in the powder form, such as stainless steel, titanium and Ti-6Al-4V which are widely used as materials for medical implants, can be fabricated into either non-porous or porous objects [39, 40]. Therefore, SLM for biomimetic and biomedical applications is also reviewed in this chapter.

Commercial SLS systems normally produce relatively large objects–this is in comparison with the size of most of the individualized medical devices for patients– and hence require large amounts of powder materials which are needed to fill the powder tanks and to build the object. (In the Sinterstation® 2,000 SLS system, the size of the powder tank is $325 \times 250 \times 370 \, mm^3$; and in the Sinterstation® 2,500 SLS system, the size of the powder tank is $330 \times 280 \times 380 \, mm^3$) A typical SLS operation requires a powder supply of at least one-third of the power tank volume. Therefore, a commercial SLS machine is not material-efficient for producing objects of small sizes or for constructing tissue engineering scaffolds using expensive biomaterials. In order to reduce the consumption of biopolymer powders in the fabrication of tissue engineering scaffold using the SLS technique, modifications of commercial SLS machines need to be made. For example, for producing bone tissue engineering

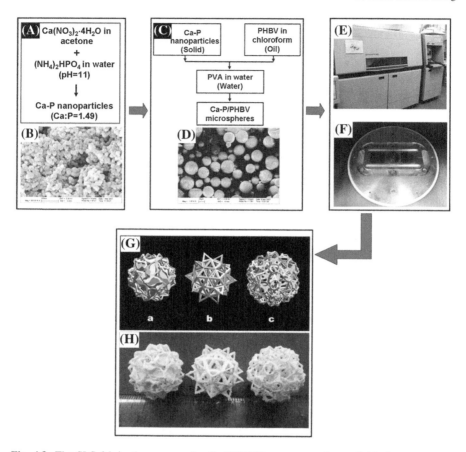

**Fig. 4.3** The SLS fabrication process for Ca-P/PHBV nanocomposite scaffolds for bone tissue engineering or as a demonstration, highly porous structures: **A** Wet-synthesis of Ca-P nanoparticles; **B** Morphology of Ca-P nanoparticles synthesized; **C** Fabrication of Ca-P/PHBV microspheres using the S/O/W emulsion–solvent evaporation method; **D** Morphology of Ca-P/PHBV nanocomposite microspheres produced; **E** A commercial Sinterstation® 2,000 SLS system; **F** A miniature sintering platform for modifying the Sinterstation® 2,000 SLS system; **G** Complex models designs for SLS: *a* salamanders, *b* elevated icosidodecahedron, *c* snarl; **H** Sintered Ca-P/PHBV nanocomposite porous structures based on the models

scaffolds, a miniature sintering platform was designed, fabricated and installed in the build tank of a Sinterstation® 2,000 system (Fig. 4.3E) [41]. The miniature sintering platform consisted primarily of a miniature build part and two powder tanks similar to those in the commercial machine but with a much smaller size. The movement of miniature build part was synchronized with the existing build part of Sinterstation® 2,000 and the two miniature powder tanks were driven by two additional stepping motors fixed within the miniature platform (Fig. 4.3F). After the machine modification, only small amounts of powders were needed to fill the miniature powder tank, with the original powder tanks being kept empty. Two sensors were installed to sense the roller positions and the signals were fed back to a control panel which could control the movement of miniature powder tanks. In another investigation, a compact adaptation device was developed for a Sinterstation® 2,500 system, transferring the motion of the SLS part bed of the Sinterstation® 2,500 system to the part bed of the compact adaptation device [42]. This device was an integrated attachment that was fixed onto the building platform of the SLS machine. When the compact adaptation device was used, the amount of powders required was only 15 % of that used in the full build version of Sinterstation® 2,500 machine. With the careful modification of commercial SLS machines, small-size tissue engineering scaffolds and medical devices can be fabricated economically, using much less biomaterial and also much shorter time.

For some biomedical applications, surface selective laser sintering (SSLS) appears very attractive. Different from conventional SLS in which a polymeric material absorbs infrared radiation (e.g. $CO_2$ laser at $\lambda = 10.6\,\mu m$), resulting in a volumetric absorption by the whole polymer particle, SSLS uses near-infrared laser radiation ($\lambda = 0.97\,\mu m$) to melt the surface of polymer particles [43]. It is therefore possible to sinter polymer microparticles into solid 3D structures by melting only the near-surface layer of microparticles instead of the whole microparticles. This is advantageous for sintering thermally unstable polymers (e.g. PLLA or PLGA) and for producing scaffolds which contain biomolecules. Bukharova and co-workers thus used SSLS to fabricate PLLA scaffolds onto which bone marrow-derived mesenchymal stromal cells were seeded [44]. The cell–scaffold constructs were subsequently implanted subcutaneously on the back of rats. Neoangiogenesis and invasion of constructs by the surrounding tissue were studied. Results showed that the cell–scaffold constructs did not evoke inflammatory response and could provide the conditions for organotypic regeneration (a high degree of blood supply and considerable amount of immature precursor cells).

## 4.2.2 Materials for SLS

For constructing porous or non-porous objects, SLS requires materials in the powder form. Many powdered materials, from polymers to ceramics to metals, can be processed into solid objects by SLS in the general engineering fields as well as in biomedical engineering. Usually, $CO_2$ lasers with the wavelength of 10.6 µm are selected for sintering polymers or materials with low melting temperatures. These materials, which are commercially available, include wax, polycarbonate, nylons and their composites, and acrylics [45]. However, they lack biocompatibility or bioactivity and therefore their biomedical applications are limited. With the development of biomaterials and emergence of regenerative medicine, biopolymers, including biodegradable and non-biodegradable polymers, have been used to fabricate tissue engineering scaffolds or porous implants via SLS.

Neodymium: yttrium aluminium garnet (Nd:YAG) lasers with a short wavelength of 1.06 µm can also be used for SLS. They may outperform $CO_2$ lasers for sintering metallic and ceramic materials, which absorb much better at short wavelengths [46]. Consequently, pure titanium, Ti-6Al-4V and NiTi shape memory alloy, which are known for their biocompatibility and good corrosion resistance, have been successfully sintered into 3D porous structures for medical implantation using Nd:YAG laser [47, 48]. Bioceramics such as hydroxyapatite (HA) could also be sintered to form customized implants for bone substitution [49]. The type of laser used in SLS can affect the properties (mechanical properties, density, and surface texture) of SLS-formed objects.

In theory, any material that can be processed into the powder form and can be sintered by heat may be used in SLS to form solid objects. And the powdered materials for SLS should be of appropriate particle sizes for sintering and for having good flowability for spreading on the part bed of an SLS machine. For SLS, powdered materials with particle sizes in the range of 10–150 $\mu$m are preferred [50]. It has been shown that the polymer particle size has significant effects on the properties of sintered scaffolds. Salmoria et al. fabricated starch–cellulose and cellulose acetate scaffolds using SLS and found that the scaffolds fabricated from small polymer particles had a higher degree of sintering and a significant level of closed pores [51]. Owing to lower degrees of sintering and low density of unions, larger polymer particles caused decreases in mechanical properties of sintered scaffolds (lower elastic modulus and tensile strength). Commercial powdered materials can be either used directly for SLS or mixed with other particles of suitable size to form composite powders to produce 3D objects via SLS. For fabricating nanocomposite scaffolds for bone tissue engineering, in order to homogeneously distribute bioactive bioceramic particles in the polymer matrix of scaffolds, bioceramics such as HA particles could be blended with polymer granules and the mixtures could be then compounded in a twin-screw extruder to form composites [52]. The extruded composites were subsequently pelletized and powdered for SLS, leading to the formation of implants for bone replacement. Another strategy for using nanocomposites in SLS is to fabricate polymer microspheres or composite microspheres first and then use the microspheres as raw materials for 3D scaffold fabrication [41]. Adopting such a strategy, as shown in Fig. 4.3A, B, bioactive and biodegradable Ca-P nanoparticles having sizes in the range of 10–30 nm were firstly synthesized. The Ca-P nanoparticles were amorphous and had a Ca:P molar ratio of about 1.49, which is close to that of tricalcium phosphate (TCP), a bioactive and biodegradable bioceramic for bone tissue repair. Ca-P/poly(hydroxybutyrate-co-hydroxyvalerate) (PHBV) nanocomposite microspheres consisting of Ca-P nanoparticles and biodegradable PHBV matrix were then fabricated using the solid-in-oil-in-water (S/O/W) emulsion–solvent evaporation method, as shown in Fig. 4.3C. The nanocomposite microspheres were of the suitable particle size for SLS and were successfully fabricated into bone tissue engineering scaffolds via SLS. As a demonstration of the capability of SLS when proper powdered materials are used and optimized SLS parameters are employed, porous structures with very complex shapes and architectures could be made by SLS using Ca-P/PHBV nanocomposite microspheres (Fig. 4.3H), according to the complex designs by Hart and shared through internet [53] (Fig. 4.3G).

**Table 4.1** Definition of important SLS parameters

| SLS parameter | Unit | Definition |
|---|---|---|
| Laser power | W | The power of the applied $CO_2$ laser |
| Scan spacing | mm | On the part bed, the distance between adjacent scan lines |
| Layer thickness | mm | The distance that the build part piston moves after each layer is sintered |
| Part bed temperature | °C | The temperature of the area in the build part that contains both the powder being sintered and the sintered part |
| Scan speed | mm/s | The speed at which the laser travels in a scan line across the part bed |
| Roller speed | mm/s | The speed of the roller that moves across the build part |

## *4.2.3 SLS Parameters*

During SLS, a number of parameters, including laser power, scan spacing, layer thickness, part bed temperature, scan speed and roller speed need to be controlled. The definitions of these parameters can be found in Table 4.1. Most of these parameters determine the energy that the part being sintered will absorb and therefore they can significantly affect the quality of sintered products. The laser power and scan spacing, along with scan speed, determine the laser energy input into the part bed in the SLS machine. Without sufficient laser energy, it is impossible to produce objects with designed structure, sufficient mechanical properties and good structural stability. A lack of input laser energy will result in products that are too fragile to handle. However, excessive input laser energy (due to excessive laser power and/or small scan spacing) will lead to inaccurate product dimensions owing to oversintering, i.e. undesirable/unintentional melting due to excessive heating. The part bed temperature should be maintained just below the melting temperature of crystalline polymers or just below the glass-transition temperature of amorphous polymers. Overheating the powder materials with a high part bed temperature will reduce the flowability of the powders or even cause the unsintered powders, which should not be melted by the part bed temperature, to partially fuse with the sintered structure, making them difficult to be removed after SLS. On the other hand, if the part bed temperature is not sufficiently high, the strength of the sintered products will be low and hence they tend to fall apart during post-SLS handling.

## *4.2.4 Optimization of SLS Parameters*

In order to obtain high-quality 3D objects, viz., 3D objects with accurate dimensions according to design, good structural stability and handability, and desired mechanical properties, SLS parameters must be optimized, and by using appropriate SLS parameters, the objects will be properly sintered and constructed. For the optimization of SLS parameters, the one-factor-at-a-time method of experimentation is often used, in which one SLS parameter is varied at a time while the other parameters are kept constant [54]. This method is useful for initial explorations of any subjects of interest in R & D and is adopted by many researchers (even for in-depth studies beyond the initial investigations). But the results obtained through studies using this method do not provide information about interactions among two or more influencing factors and thus do not reflect the real situation. To be closer to the actual process during SLS, the factorial design methodology can be employed to evaluate the main effects and interactions of joint factor effects on the response variables. The determination of factors, levels and responses is important in the factorial design approach for the optimization of SLS parameters. Partee et al. employed the two-level factorial design of experiments (DOE) technique to determine optimal SLS parameters for processing PCL [55]. Five SLS parameters (laser power, scan speed, scan spacing, part bed temperature, and powder layer delay time) were investigated and a mathematical model was developed to relate their influences on the PCL part quality (dimensional accuracy, structural integrity, mechanical strength, etc.). It was demonstrated that after optimization of these parameters, non-porous PCL objects and porous PCL scaffolds with 1D, 2D and 3D orthogonal, periodic porous architectures could be produced using SLS [56]. The mechanical properties of their sintered products were experimentally determined and then modeled using the finite element analysis method. In investigations of fabricating Ca-P/PHBV nanocomposite scaffolds for bone tissue engineering, three aspects of the scaffolds, namely, the dimensional accuracy, structure and handling stability and compressive properties, were considered and DOE with three factors and three levels was used to optimize the SLS parameters [57]. In order to demonstrate the usefulness of optimized SLS parameters, a porous Ca-P/PHBV nanocomposite scaffold in the shape of the proximal end of a human femur but with a reduced size was successfully produced via SLS [58]. It is believed that SLS parameters should be optimized for the materials that are used for constructing the 3D objects, be they porous or non-porous. It has also been found that the mechanical behavior of scaffolds fabricated by SLS was not isotropic and was strongly dependent on the manufacturing direction [59]. The sintered objects were the weakest in the x-direction, where scan lines were parallel to the loading direction. The highest elastic modulus values were recorded for struts in the y-direction, where scan lines were perpendicular to the loading direction. Further investigations need to be conducted for elucidating the mechanisms.

## 4.2.5 Applications of SLS in the General Manufacturing Industry

Except for biomedical applications, SLS is most widely used for making prototypes in the manufacturing industry. By carefully choosing suitable commercial SLS materials such as DuraForm™ (a nylon-based powder) and CastForm™ (a polystyrene material coated with wax), prototypes with properties very close to those of injection moulded parts can be fabricated via SLS. These thermoplastic raw materials provide users with a choice of producing durable and flexible objects or stiff and rigid objects. The SLS technique is highly capable of fabricating prototypes with complex geometries and irregular shapes, including structures containing channels and overhanging features. In addition, the finished parts and/or assemblies have a good surface finish and contain detailed features. Apart from making functional prototypes, SLS can also be used to produce presentation models for marketing purposes or showpieces for display. These SLS-formed objects usually require master finish, which is a post-production process for the objects to be sanded, primed, plated or painted according to different purposes. Another regular application that utilizes the SLS technique is creating casting patterns. TrueForm (an acrylic-based powder) and CastForm™ are the commercial SLS materials for creating patterns for investment casting (an industrial process based on one of the metal-forming techniques) [45]. These SLS materials are easily burned off in the foundry process and are capable of making patterns with moderate strength, high accuracy and intricate details. Compared to other techniques, SLS is attractive and well suited to fabricate complex and customized parts with shorter time and lower cost.

## 4.3  Biomedical Applications of SLS

SLS has found many applications in the biomedical field, including the fabrication of tissue engineering scaffolds, drug delivery vehicles, medical device prototypes, biomedical devices themselves, and physical models for surgical planning and simulation. This section highlights these biomedical applications and pays special attention to the creation of tissue engineering scaffold and drug delivery system using SLS.

### 4.3.1  Physical Models for Surgical Planning

The first medical applications of SLS from the general manufacturing industry were the production of accurate models for surgical planning and simulation, particularly for craniofacial and maxillofacial surgeries and for neurosurgery. Surgical planning is needed for accurately and effectively performing surgeries, for minimizing the duration of complex and time-consuming surgeries and for reducing the risk of complications [60]. Using high-resolution, multiplanar medical imaging techniques such as MRI and CT, 3D medical images can be obtained from patients, whose computer data are subsequently used for 3D model fabrication using the SLS technique. Physical models of the brain and skull were produced via SLS for the purpose of surgical training in the procedures of deep microvascular anastomosis [61]. Similarly, neurosurgeons could use SLS-formed models to practice with real sense and judgment when manipulating operative instruments. Therefore, the accuracy of medical models formed by RP techniques is very important. The shape, dimensions and anatomic details of models produced can be affected by errors at any stage of the SLS process. Silva et al. compared the accuracy of a craniomaxillary model made by SLS to that of the model fabricated via 3DP [62]. The model generated by SLS showed better dimensional precision and reproduced craniomaxillary anatomy more accurately than the 3DP-formed model (dimensional errors of 2.10 and 2.67 % for SLS-formed and 3DP-formed models, respectively). SLS-formed models could reproduce better anatomic details, except for thin bones, small foramina and acute bone projections. When SLS and 3DP were compared for reproducing mandibular anatomy, it was found that the SLS-formed model had a greater dimensional accuracy than the 3DP-formed model [63].

## 4.3.2 Medical Device Prototypes

A major advantage of using RP technologies to produce medical device prototypes is the rapid fabrication of customized products which are tailored to the needs of individual patients. Using the SLS technique, Wu et al. fabricated a wax pattern for a partial nasal prosthesis [64]. The precision of the computerized model allowed for the satisfactory restoration of facial contours and was advantageous for both the patient and the maxillofacial prosthodontist. Montgomery et al. designed an actively actuated prosthetic socket (which is the portion of the prosthesis that attaches to the residual limb of an amputee and serves a variety of functions, from stationary support to the transfer of forces necessary for making movements) and using nylon, manufactured the prosthetic socket according to the design [65]. Tests with this socket showed great potential for developing new sockets that would provide greater comfort and fit for patients to avoid movement problems and rehabilitation difficulties. These studies have provided good examples of using the SLS technique to produce the so-called "bionic man", i.e. to make functioning extremities (arms, legs, etc.) for patients who unfortunately have lost or do not have these limbs due to accidents, diseases or congenital anomalies. Also using SLS, Faustini et al. manufactured ankle-foot orthoses (AFOs) which are used as assistive or therapeutic devices to improve gait performance for persons with impaired lower limb function [66]. SLS allowed the fabrication of AFOs directly from digital shape information of a patient's limb and was capable of optimizing the characteristics of AFOs (e.g. weight and stiffness) by changing the materials and by using different reinforcements such as carbon fibers. Similarly, customized foot orthoses, which are used to lessen the symptoms of rheumatoid arthritis and to improve function and mobility, were fabricated using SLS [67]. The trials by patients indicated that these tailor-made orthoses performed well in terms of patients' gait and evaluation for fit and comfort by the patients.

## 4.3.3 Medical Implants and Prostheses

One of the disadvantages of polymers or polymer-based composites for orthopedic applications is their low mechanical properties. They may only serve to repair or regenerate cortical and/or cancellous bone in human bodies in non-load-bearing areas or in load-bearing parts but without the load-bearing requirement. In contrast, implantable metals such as titanium and titanium alloys can be used to construct 3D objects as bioactive and load-bearing implants. Hao et al. fabricated HA/316L stainless steel (SS) composite using the SLM technique and studied the effects of SLM parameters and HA particles on the properties of HA/SS composite [68]. The highest tensile strength of HA/SS parts produced using optimal SLM parameters was close to the tensile strength of human bone and was sufficient for making load-bearing bone implants. In another study, porous Ti scaffolds characterized by high porosity ($\sim$70%), connected Ti walls and open porous structures with macropores (in the range of 200–500 μm) were successfully produced using a laser power of 1000 W and a scan speed of 0.02 m/s [69]. Porous Ti structures were developed primarily to circumvent the problem of mechanical mismatch between the implant and bone and provide a suitable environment for bone ingrowth, thus achieving a strong fixation between the implant and surrounding bone. Mullen et al. used a regular unit cell to design and produce porous Ti structures with a large range of physical and mechanical properties [70]. These properties could be tailored to suit specific requirements. Particularly, functionally graded structures with surfaces that could enhance bone ingrowth were manufactured. They showed porosity dependent compressive strength (porosity: 10–95%; strength: 0.5–350 MPa) comparable to those of human bone. For investigating biocompatibility of sintered products, human osteoblasts were cultured on SLM-formed Ti-6Al-4V meshes and their behavior was studied [71]. SEM examinations revealed osteoblasts with well-spread morphology and multiple contact points. Other in vitro experiments confirmed osteoblast vitality and proliferation on the Ti-6Al-4V meshes. SLM proves to be capable of producing metallic or ceramic porous structures which are biocompatible and have sufficient mechanical properties for load-bearing bone implants.

## 4.3.4 Tissue Engineering Scaffolds

Traditionally, autograft, allograft and synthetic biomaterials are used in the treatment of diseased or traumatized tissues or organs in human bodies. Even though the treatments can be successful, however, major limitations, such as shortage of donors, graft rejections and lack of bioactivity for integration with the host tissue, do exist. As an alternative for human tissue repair, tissue engineering emerged in the late-1980s, which involves the use of biodegradable scaffolds and growth factors with or without specific populations of living cells [72]. There are several strategies in tissue engineering, which are generally divided into cell-based tissue engineering, factor-based tissue engineering and scaffold-based engineering, as schematically illustrated in Fig. 4.4. For scaffold-based tissue engineering, controlling the macro- and micro-architecture of the scaffold and fulfilling a customized design with a complex anatomic shape are of significant importance for the clinical application of the scaffold [73]. Unlike non-designed manufacture techniques for scaffolds such

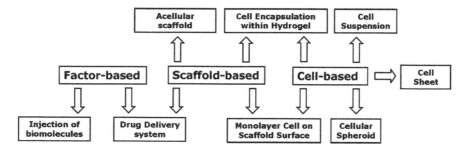

**Fig. 4.4** Tissue engineering strategies

as solvent casting/particulate leaching, phase separation, gas foaming/high pressure processing, etc. SLS and other RP techniques are capable of achieving extensive and detailed control over scaffold architecture using CAD designs and computer-based medical imaging technologies [17, 74].

Scaffolds as bone graft substitutes or for bone tissue engineering can be made via SLS using biopolymers (including both biodegradable and non-biodegradable polymers), bioceramics and biomedical composites. In one investigation, Rimell and Marquis used a simplified SLS apparatus to sinter an ultrahigh molecular weight polyethylene (UHMWPE) [75]. They found that solid, linear and continuous objects could be formed but the high degree of shrinkage caused problems in the formation of sheet-like structures. Other powdered non-biodegradable polymers such as polyetheretherketone (PEEK) could also be used in SLS [76]. PEEK is a high temperature, semi-crystalline thermoplastic polymer and is often chosen for manufacturing medical instruments or implants due to its good mechanical properties and good biocompatibility. These non-biodegradable polymers (UHMWPE, PEEK, etc.) can form some useful implants for bone tissue repair. On the other hand, it is often shown that only one single biomaterial cannot fully fulfill the requirements for bone graft substitutes. Therefore, composite scaffolds based on non-degradable polymer and HA, including HA/PEEK and HA/high density polyethylene (HDPE), were produced using SLS [77, 78]. HA/HDPE scaffolds were made and investigated as potential customized maxillofacial implants. It was observed that HA/HDPE composite scaffolds with high HA contents enhanced cell proliferation, increased alkaline phosphatase (ALP) activity and produced more osteocalcin (OCN) [79]. In the study of HA/PEEK scaffolds, different amounts of HA were incorporated into PEEK to produce porous scaffolds with well-defined pore interconnectivity. By varying the HA content, the porosity of the HA/PEEK composite scaffolds changed gradually and the bioactivity and mechanical properties were also altered [80].

For scaffold-based tissue engineering, with conventional notion, the scaffold should be biodegradable and the degradation rate should be synchronized with the rate of neo-tissue formation. Therefore, for tissue engineering scaffolds, several well established biodegradable polymers as well as their composites have often been used. One widely used biodegradable polymer is PCL which is a semi-crystalline polymer with high thermal stability, good biocompatibility and low melting temperature [81]. It is a US Food and Drug Administration-approved biodegradable polymer. Williams el al. used PCL to fabricate tissue engineering scaffolds via SLS [82]. The sintered PCL scaffolds had a porous structure with orthogonal interconnecting pores and were seeded with human gingival fibroblastic cells. The cell–scaffold constructs were subcutaneously implanted in 5–8 week old immuno-compromised mice. Subsequent studies using μCT and histological staining showed that bone formed on or inside the orthogonal pore scaffolds. The sintered scaffolds possessed mechanical properties in the lower range of trabecular bone. In order to demonstrate the potential for clinical application of SLS-formed scaffolds, a mandibular condyle scaffold based on an actual pig condyle was designed, fabricated and evaluated. Ciardelli et al. applied SLS to sinter blends of microparticles of PCL and a polysaccharide (starch, dextran or gelatin) and systematically studied the morphology, thermal behavior and cell compatibility of the scaffolds produced [83]. Some other biodegradable polymers, such as poly(vinyl alcohol) (PVA), PLLA and PLGA (LA:GA $= 95:5$), were also successfully processed into scaffolds via SLS [84, 85]. Particulate bioceramics such as HA can be incorporated into biodegradable polymers to form biodegradable composite scaffolds and the composite scaffolds are expected to combine the characteristics of constituent materials. For example, dry blends of PCL and HA particles were used to fabricate composite scaffolds via SLS [54]. The SLS parameters for sintering HA/PCL scaffolds were varied for achieving good-quality scaffolds. The HA/PCL scaffolds produced exhibited good bioactivity in a simulated body fluid (SBF) and good cell compatibility in cell culture experiments. Other composite scaffolds such as HA/PVA and HA/PLGA (LA:GA $= 95:5$) were made and investigated by other researchers [85, 86].

For fabricating novel and/or improved biomaterials, the research has been moving from conventional technologies to microtechnology and further to nanotechnology. Consequently, the mechanical properties and the biological responses of materials are improved. With large surface-area-to-volume ratios, nano-bioceramics are reported to exhibit good ductility before fracture owing to the grain-boundary phase and lower sintering temperature [87]. Nanostructured Ca-P and nanocomposites consisting of nano-sized Ca-P can provide better biocompatibility and osseointegration than their larger-particle-size counterparts [88, 89]. In reported studies, for achieving a homogeneous distribution of Ca-P nanoparticles in the composite, nanocomposite microspheres consisting of nano-sized Ca-P and a PLLA or PHBV matrix were fabricated first using the S/O/W emulsion solvent evaporation method [90]. The nanocomposite microspheres subsequently served as raw materials which would be subjected to SLS for the fabrication of nanocomposite scaffolds for bone tissue engineering, as illustrated in Fig. 4.5A. It has been shown that the incorporation of nano-sized Ca-P in PHBV-matrix scaffolds significantly promoted cell growth, with higher metabolic activity as compared to PHBV polymer scaffolds. Moreover, the ALP activity of osteoblastic SaOS-2 cells on Ca-P/PHBV nanocomposite scaffolds was remarkably higher than that of SaOS-2 cells on PHBV scaffolds.

Besides bone tissue engineering, SLS can also be used to fabricate 3D porous scaffolds for the repair or regeneration of other tissues. Liu et al. physically

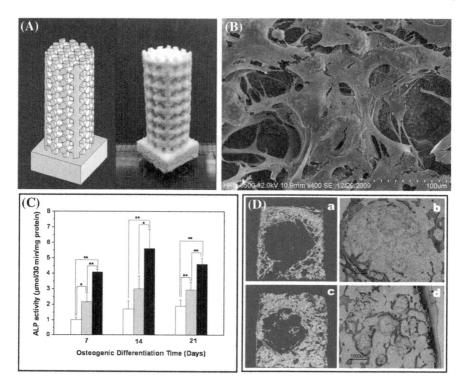

**Fig. 4.5** Biological performance of SLS-formed Ca-P/PHBV nanocomposite scaffolds: **A** A Ca-P/PHBV scaffold fabricated via SLS (*Left* design model; *Right* sintered scaffold); **B** Morphology of C3H10T1/2 cells cultured on surface modified Ca-P/PHBV scaffold for 21 days in an osteogenic medium; **C** ALP activity of C3H10T1/2 cells cultured on different scaffolds in an osteogenic medium (*White column* Ca-P/PHBV scaffolds; *Gray column* Ca-P/PHBV scaffolds with simply adsorbed rhBMP-2; *Black column* surface modified Ca-P/PHBV scaffolds loaded with rhBMP-2; for the statistical analysis of results, * indicates $p < 0.05$ and ** indicates $p < 0.01$); **D** In vivo evaluation of bone regeneration: *a, c* μCT 3D images, and *b, d* histological images of different scaffolds after their implantation in the ilium of rabbits for 6 weeks. *a, b* for Ca-P/PHBV scaffolds, and *c, d* for surface modified Ca-P/PHBV scaffolds loaded with rhBMP-2

blended epoxy resin E-12 (a polymer, acting as a binder) with $K_2O$-$Al_2O_3$-$SiO_2$ series of dental glass-ceramics, forming fine composite powders. Subsequently, they used the powders to fabricate dental restoration devices through SLS [91]. After pre-densification and sintering, glass-ceramic teeth could be made under the optimized SLS condition. PCL scaffolds with designed architectural and mechanical characteristics were also made using SLS to accommodate C2C12 myoblast cells for cardiac tissue engineering [92]. The SLS-formed PCL scaffolds with a relatively low stiffness (300–400 kPa) could support myoblast cells of a high initial density and with a spatially uniform distribution and maintain their viability and function. To engineer implantable liver tissue, Huang et al. designed a PCL scaffold with a 3D branching and joining flow-channel network comprising multiple tetrahedral units and fabricated the scaffolds using SLS [93]. Human hepatoma Hep G2 cells were then seeded in the scaffold using avidin–biotin (AB) binding and cultured in a perfusion system. The 3D flow channels were shown to be essential to the cell growth and function and the AB binding-based seeding improved remarkably the overall performance of the cell-loaded scaffolds.

## 4.3.5 *Drug or Biomolecule Delivery Systems*

SLS can also be used to produce delivery vehicles for drugs or biomolecules. Leong et al. fabricated drug delivery devices (DDDs) by selectively sintering mixtures of nylon powder and methylene blue dye which acted as a model drug [94]. In order to improve the release behaviour, two features, viz., porous microstructure and dense wall formation, were introduced for DDD and then studied for their roles in drug storage and in controlling drug release through the diffusion process [95]. In vitro release experiments showed that SLS-formed DDDs were capable of controlling the release of the model drug into a simulated body environment. Although non-biodegradable polymer matrix, reservoir-type DDDs can provide zero-order drug release by the diffusion of drug molecules from the core through the shell, these drug delivery systems are not suitable for tissue engineering. Using biodegradable polymers as device matrices, DDDs are potentially much more useful for drug or biomolecule delivery. Therefore, a biodegradable DDD was made via SLS using PCL-methylene blue mixtures [96]. In order to reduce the level of initial burst release, a number of barrier rings were designed and made in the cylindrical DDD. The drug release profiles could therefore be modified by changing the design and fabrication of DDD structure for polymeric delivery systems. Generally, drug molecules are loaded in a DDD either by blending them with polymer powders before SLS or by drug adsorption on a DDD after the formation of DDD via SLS. A PCL-progesterone (PG, a steroid hormone used for endometrial cancer treatment) drug delivery system was fabricated by sintering mixed powders of PCL and PG [97]. The introduction of PG particles provided reinforcement for the DDD, resulting in a higher strength than that of sintered PCL samples. In one investigation, as a demonstration of SLS capability, bovine serum albumin (BSA)-containing nanocomposite scaffolds were produced [98]. BSA was used as a model protein (biomolecule) and it was firstly encapsulated in Ca-P/PHBV nanocomposite microspheres using the double emulsion solvent evaporation method. BSA-loaded Ca-P/PHBV microspheres were then processed into 3D porous scaffolds with good dimensional accuracy using the SLS technique. The nanocomposite microspheres served as protective carriers for BSA and maintained the bioactivity of BSA during SLS. In the subsequent in vitro BSA release study, an initial burst release was observed, which was followed by a slow release. The BSA encapsulation efficiency in both BSA-loaded microspheres and SLS-formed scaffolds was relatively low owing to the materials used, the microsphere fabrication method and the SLS process. However, this manufacturing route could be effective for incorporating drugs which are not sensitive to high temperatures. On the other hand, SSLS might be an alternative technique for making delivery vehicles for bio-molecules. It was investigated for producing poly(D, L-lactic acid) (PDLLA)-based scaffolds containing ribonuclease A (a model enzyme) [99]. Composite powders of PDLLA and ribonuclease A were made into 3D porous scaffolds via SSLS. Ribonuclease A in scaffolds sintered at various laser intensities retained substantial activity after the SSLS process.

For tissue regeneration, scaffolds alone may not be adequate due to the lack of biochemical stimulation for promoting cell proliferation and differentiation. The stimulants include hormones, proteins such as cytokines and growth factors, and they are responsible for providing biosignals that will prompt specific cell behavior and functions. Ideally, an adequate amount of growth factor(s) should be incorporated in the scaffold according to the specific tissue targeted for regeneration and be released in a temporal and spatial way for a desired period of time. Although biomolecule-containing scaffolds could be made by sintering mixtures of polymer powder and biomolecules or by sintering biomolecule-loaded microspheres, the SLS process may damage the biomolecules due to high heat. One solution to this problem is to bind biomolecules to the scaffolds after SLS and the bioactive biomolecules could be released later in a controlled manner in vitro and in vivo. One example was the loading of recombinant human bone morphogenetic protein-2 (rhBMP-2) onto surface modified Ca-P/PHBV nanocomposite scaffolds [58]. The surface modification of Ca-P/PHBV scaffolds was achieved through the physically attachment of gelatin on scaffolds and then the immobilization of heparin on attached gelatin. rhBMP-2 was loaded onto surface modified scaffolds (and later released from the scaffolds in vitro or in vivo in a sustained manner) due to the specific affinity between heparin and growth factors. This strategy not only provided a means to protect the loaded rhBMP-2 but also improved the sustained release profile for rhBMP-2. To investigate the effect of rhBMP-2 release from scaffolds and also the osteogenic differentiation of mesenchymal stem cells (MSCs), pluripotent mesenchymal cell line C3H10T1/2 was seeded on bare Ca-P/PHBV scaffolds and on surface modified Ca-P/PHBV scaffolds loaded with rhBMP-2. Figure 4.5B shows an SEM micrograph of C3H10T1/2 cells after 21-day culture on a surface modified and rhBMP-2-loaded Ca-P/PHBV scaffold using an osteogenic medium. The proliferating cells were observed to cover the whole surface of the scaffold. The ALP activity assay and mRNA expression results showed that as compared with scaffolds with simple adsorption of rhBMP-2, the ALP levels were significantly up-regulated for cells cultured on surface modified Ca-P/PHBV scaffolds loaded with rhBMP-2 (Fig. 4.5C). Consistent with the up-regulation of ALP and gene expression, C3H10T1/2 cells cultured on surface modified Ca-P/PHBV scaffolds loaded with rhBMP-2 also displayed much higher OCN expression. In in vivo experiments, Ca-P/PHBV nanocomposite scaffolds with or without rhBMP-2 loading were implanted in a drilled hole in the ilium of New Zealand white rabbits. 3D images of explanted Ca-P/PHBV scaffolds were reconstructed using μCT, as shown in Fig. 4.5D(a) and D(c). After 6 week implantation, there was very limited bone formation in Ca-P/PHBV scaffolds without rhBMP-2 loading (Fig. 4.5D(a)), with the drilled hole being occupied by the scaffold. In contrast, using surface modified scaffolds loaded with rhBMP-2, new bone was formed and it penetrated into the pores (Fig. 4.5D(c)). The histological images showed that for the surface modified Ca-P/PHBV scaffolds loaded with rhBMP-2, the center of the defect (the drilled hole) was filled with newly formed bone which bridged with the host bone (Fig. 4.5D(d)), whereas in the bone defect treated with a Ca-P/PHBV scaffold without rhBMP-2 loading, new bone formation was observed only at the periphery of the defect (Fig. 4.5D(b)). For the surface modified Ca-P/PHBV scaffolds loaded with rhBMP-2, the bony tissue grew from the host bone and the original site of the drilled hole exhibited enhanced osteogenesis. All these results indicated that the surface modification could provide binding sites for rhBMP-2 and control its sustained release behavior. The sustained local release of rhBMP-2 could attract MSCs in the bone marrow, stimulate their differentiation into osteoblasts and promote the ingrowth of new bone in the scaffold.

## 4.4 Summary

SLS is a mature and versatile RP technique that can find many biomedical applications. It has already been intensively investigated for producing high-quality tissue engineering scaffolds for the regeneration of different human body tissues. Unlike the conventional use of SLS in industries such as microelectronics, telecommunication and automobiles where the cost of SLS material is not an issue for much consideration, in most biomedical applications, the availability and cost of materials can significantly affect the outcome of the strategy based on the use of SLS. It has been shown that with certain modifications, commercial SLS machines can accommodate the requirement of using small amounts of powdered materials for producing medical devices and tissue engineering scaffolds. For medical devices that will be implanted in human bodies and tissue engineering scaffolds, current general purpose SLS materials are not suitable and hence new approaches of using existing biomaterials and/or new biomaterials have been and will be investigated and developed. On using SLS in the biomedical field itself, it has been demonstrated that the optimization of SLS parameters is of great importance for the fabrication of good-quality products. Using data from computerized medical imaging techniques such as MRI and CT, customized biomedical devices and tissue engineering scaffolds can be made through SLS. SLS can also be employed to produce delivery systems for the controlled release of drugs or biomolecules. But the way a drug or a type of biomolecules is incorporated in the delivery vehicle must be carefully considered and developed.

Within a relatively short period of time, many RP technologies, including SLS, have been explored for fabricating implantable medical devices and tissue engineering scaffolds owing to their many advantages over conventional manufacturing techniques. Still, new explorations need to be conducted and new efforts made to fully take the advantages of SLS in the biomedical field. Issues such as making a device using different materials for different parts and effectively incorporating desired amounts of delicate biomolecules without denaturation must be tackled. Some future biomedical application of SLS will require higher resolutions for the products (and hence the SLS machine) and higher dimensional accuracy of sintered products. But as we have witnessed over the past decade, with researchers of diverse disciplines (mechanical engineering, materials science and engineering, biological science, and clinical science) working together, SLS can have many biomedical applications. And SLS certainly has great potential for further developments in the biomedical field.

**Acknowledgments** Our research on applying SLS in the biomedical field was supported by Hong Kong Research Grants Council through GRF grants and by The University of Hong Kong (HKU) through a research grant in its Basic Research Programme. We are grateful to staff and students at HKU for their assistance in our SLS research.

# References

1. Leong KF, Cheah CM, Chua CK (2003) Solid freeform fabrication of three-dimensional scaffolds for engineering replacement tissues and organs. Biomaterials 24:2363–2378
2. Yang SF, Leong KF, Du ZH, Chua CK (2002) The design of scaffolds for use in tissue engineering. Part II. Rapid prototyping techniques. Tissue Eng 8:1–11
3. Borah B, Gross GJ, Dufresne TE, Smith TS, Cockman MD, Chmielewski PA, Lundy MW, Hartke JR, Sod EW (2001) Three-dimensional microimaging (MR mu I and mu CT), finite element modeling, and rapid prototyping provide unique insights into bone architecture in osteoporosis. Anat Rec 265:101–110
4. Peltola SM, Melchels FPW, Grijpma DW, Kellomaki M (2008) A review of rapid prototyping techniques for tissue engineering purposes. Ann Med 40:268–280
5. Hollister SJ (2009) Scaffold design and manufacturing: from concept to clinic. Adv Mater 21:3330–3342
6. Melchels FPW, Feijen J, Grijpma DW (2010) A review on stereolithography and its applications in biomedical engineering. Biomaterials 31:6121–6130
7. Arcaute K, Mann BK, Wicker RB (2006) Stereolithography of three-dimensional bioactive poly(ethylene glycol) constructs with encapsulated cells. Ann Biomed Eng 34:1429–1441
8. Chan V, Zorlutuna P, Jeong JH, Kong H, Bashir R (2010) Three-dimensional photopatterning of hydrogels using stereolithography for long-term cell encapsulation. Lab Chip 10:2062–2070
9. Cumpston BH, Ananthavel SP, Barlow S, Dyer DL, Ehrlich JE, Erskine LL, Heikal AA, Kuebler SM, Lee IYS, McCord-Maughon D, Qin JQ, Rockel H, Rumi M, Wu XL, Marder SR, Perry JW (1999) Two-photon polymerization initiators for three-dimensional optical data storage and microfabrication. Nature 398:51–54
10. Weiss T, Hildebrand G, Schade R, Liefeith K (2009) Two-photon polymerization for microfabrication of three-dimensional scaffolds for tissue engineering application. Eng Life Sci 9:384–390
11. Hsieh TM, Ng CWB, Narayanan K, Wan ACA, Ying JY (2010) Three-dimensional microstructured tissue scaffolds fabricated by two-photon laser scanning photolithography. Biomaterials 31:7648–7652
12. Hoffmann JC, West JL (2010) Three-dimensional photolithographic patterning of multiple bioactive ligands in poly(ethylene glycol) hydrogels. Soft Matter 6:5056–5063
13. Kasko AM, Wong DY (2010) Two-photon lithography in the future of cell-based therapeutics and regenerative medicine: a review of techniques for hydrogel patterning and controlled release. Future Med Chem 2:1669–1680
14. Schade R, Weiss T, Berg A, Schnabelrauch M, Liefeith K (2010) Two-photon techniques in tissue engineering. J Artif Organ 33:219–227
15. Narayan RJ, Doraiswamy A, Chrisey DB, Chichkov BN (2010) Medical prototyping using two photon polymerization. Mater Today 13:42–48

16. Yan X, Gu P (1996) A review of rapid prototyping technologies and systems. Comput Aided Des 28:307–318
17. Hutmacher DW, Sittinger M, Risbud MV (2004) Scaffold-based tissue engineering: rationale for computer-aided design and solid free-form fabrication systems. Trends Biotechnol 22:354–362
18. Hutmacher DW (2001) Scaffold design and fabrication technologies for engineering tissues—state of the art and future perspectives. J Biomat Sci-Poly E 12:107–124
19. Yen HJ, Tseng CS, Hsu SH, Tsai CL (2009) Evaluation of chondrocyte growth in the highly porous scaffolds made by fused deposition manufacturing (FDM) filled with type II collagen. Biomed Microdevices 11:615–624
20. Schantz JT, Brandwood A, Hutmacher DW, Khor HL, Bittner K (2005) Osteogenic differentiation of mesenchymal progenitor cells in computer designed fibrin-polymer-ceramic scaffolds manufactured by fused deposition modeling. J Mater Sci-Mater M 16:807–819
21. Sawyer AA, Song SJ, Susanto E, Chuan P, Lam CXF, Woodruff MA, Hutmacher DW, Cool SM (2009) The stimulation of healing within a rat calvarial defect by mPCL-TCP/collagen scaffolds loaded with rhBMP-2. Biomaterials 30:2479–2488
22. Landers R, Mulhaupt R (2000) Desktop manufacturing of complex objects, prototypes and biomedical scaffolds by means of computer-assisted design combined with computer-guided 3D plotting of polymers and reactive oligomers. Macromol Mater Eng 282:17–21
23. Zhang T, Yan YN, Wang XH, Xiong Z, Lin F, Wu RD, Zhang RJ (2007) Three-dimensional gelatin and gelatin/hyaluronan hydrogel structures for traumatic brain injury. J Bioact Compat Pol 22:19–29
24. Ang TH, Sultana FSA, Hutmacher DW, Wong YS, Fuh JYH, Mo XM, Loh HT, Burdet E, Teoh SH (2002) Fabrication of 3D chitosan-hydroxyapatite scaffolds using a robotic dispensing system. Mat Sci Eng C-Mater 20:35–42
25. Oliveira AL, Malafaya PB, Costa SA, Sousa RA, Reis RL (2007) Micro-computed tomography (micro-CT) as a potential tool to assess the effect of dynamic coating routes on the formation of biomimetic apatite layers on 3D-plotted biodegradable polymeric scaffolds. J Mater Sci-Mater M 18:211–223
26. Woodfield TBF, Malda J, de Wijn J, Peters F, Riesle J, van Blitterswijk CA (2004) Design of porous scaffolds for cartilage tissue engineering using a three-dimensional fiber-deposition technique. Biomaterials 25:4149–4161
27. Moroni L, Poort G, Van Keulen F, de Wijn JR, van Blitterswijk CA (2006) Dynamic mechanical properties of 3D fiber-deposited PEOT/PBT scaffolds: An experimental and numerical analysis. J Biomed Mater Res A 78A:605–614
28. Woodfield TBF, Van Blitterswijk CA, De Wijn J, Sims TJ, Hollander AP, Riesle J (2005) Polymer scaffolds fabricated with pore-size gradients as a model for studying the zonal organization within tissue-engineered cartilage constructs. Tissue Eng 11:1297–1311
29. Malda J, Woodfield TBF, van der Vloodt F, Wilson C, Martens DE, Tramper J, van Blitterswijk CA, Riesle J (2005) The effect of PEGT/PBT scaffold architecture on the composition of tissue engineered cartilage. Biomaterials 26:63–72
30. Wang HJ, van Blitterswijk CA (2010) The role of three-dimensional polymeric scaffold configuration on the uniformity of connective tissue formation by adipose stromal cells. Biomaterials 31:4322–4329
31. Giordano RA, Wu BM, Borland SW, Cima LG, Sachs EM, Cima MJ (1996) Mechanical properties of dense polylactic acid structures fabricated by three dimensional printing. J Biomat Sci-Poly E 8:63–75
32. Koegler WS, Griffith LG (2004) Osteoblast response to PLGA tissue engineering scaffolds with PEO modified surface chemistries and demonstration of patterned cell response. Biomateirals 25:2819–2830
33. Tay BY, Zhang SX, Myint MH, Ng FL, Chandrasekaran M, Tan LKA (2007) Processing of polycaprolactone porous structure for scaffold development. J Mater Process Tech 182:117–121

34. Kim SS, Utsunomiya H, Koski JA, Wu BM, Cima MJ, Sohn J, Mukai K, Griffith LG, Vacanti JP (1998) Survival and function of hepatocytes on a novel three-dimensional synthetic biodegradable polymer scaffold with an intrinsic network of channels. Ann Surg 228:8–13
35. Lam CXF, Mo XM, Teoh SH, Hutmacher DW (2002) Scaffold development using 3D printing with a starch-based polymer. Mat Sci Eng C-Mater 20:49–56
36. Liu QB, Leu MC, Schmitt SM (2006) Rapid prototyping in dentistry: technology and application. Int J Adv Manuf Tech 29:317–335
37. Pham DT, Gault RS (1998) A comparison of rapid prototyping technologies. Int J Mach Tool Manu 38:1257–1287
38. Yadroitsev I, Gusarov A, Yadroitsava I, Smurov I (2010) Single track formation in selective laser melting of metal powders. J Mater Process Tech 210:1624–1631
39. Thijs L, Verhaeghe F, Craeghs T, Van Humbeeck J, Kruth JP (2010) A study of the micro structural evolution during selective laser melting of Ti-6Al-4V. Acta Mater 58:3303–3312
40. Shen YF, Gu DD, Wu P (2008) Development of porous 316L stainless steel with controllable microcellular features using selective laser melting. Mater Sci Tech-Lond 24:1501–1505
41. Zhou WY, Lee SH, Wang M, Cheung WL, Ip WY (2008) Selective laser sintering of porous tissue engineering scaffolds from poly(L)/carbonated hydroxyapatite nanocomposite microspheres. J Mater Sci-Mater M 19:2535–2540
42. Wiria FE, Sudarmadji N, Leong KF, Chua CK, Chng EW, Chan CC (2010) Selective laser sintering adaptation tools for cost effective fabrication of biomedical prototypes. Rapid Prototyping J 16:90–99
43. Antonov EN, Bagratashvili VN, Howdle SM, Konovalov AN, Popov VK, Panchenko VY (2006) Fabrication of polymer scaffolds for tissue engineering using surface selective laser sintering. Laser Phys 16:774–787
44. Bukharova TB, Antonov EN, Popov VK, Fatkhudinov TK, Popova AV, Volkov AV, Bochkova SA, Bagratashvili VN, Gol'dshtein D (2010) Biocompatibility of tissue engineering constructions from porous polylactide carriers obtained by the method of selective laser sintering and bone marrow-derived multipotent stromal cells. Exp Biol Med 149:148–153
45. Pham DT, Dimov S, Lacan F (1999) Selective laser sintering: applications and technological capabilities. Proc Inst Mech Eng Part B-J Eng Manuf 213:435–449
46. Savalani MM, Hao L, Harris RA (2006) Evaluation of CO$_2$ and Nd : YAG lasers for the selective laser sintering of HAPEX (R). Proc Inst Mech Eng Part B-J Eng Manuf 220:171–182
47. Tolochko NK, Savich VV, Laoui T, Froyen L, Onofrio G, Signorelli E, Titov VI (2002) Dental root implants produced by the combined selective laser sintering/melting of titanium powders. Proc Inst Mech Eng Pt L-J Mater-Design Appl 216:267–270
48. Shishkovsky IV, Volova LT, Kuznetsov MV, Morozov YG, Parkin IP (2008) Porous biocompatible implants and tissue scaffolds synthesized by selective laser sintering from Ti and NiT. J Mater Chem 18:1309–1317
49. Comesana R, Lusquinos F, del Val J, Malot T, Lopez-Alvarez M, Riveiro A, Quintero F, Boutinguiza M, Aubry P, De Carlos A, Pou J (2011) Calcium phosphate grafts produced by rapid prototyping based on laser cladding. J Eur Ceram Soc 31:29–41
50. Chung H, Das S (2006) Processing and properties of glass bead particulate-filled functionally graded Nylon-11 composites produced by selective laser sintering. Mater Sci Eng A-Struct Mater Prop Microstruct Process 437:226–234
51. Salmoria G, Klauss P, Paggi RA, Kanis LA, Lago A (2009) Structure and mechanical properties of cellulose based scaffolds fabricated by selective laser sintering. Polym Testing 28:648–652
52. Hao L, Savalani MM, Zhang Y, Tanner KE, Harris RA (2006) Selective laser sintering of hydroxyapatite reinforced polyethylene composites for bioactive implants and tissue scaffold development. Proc Inst Mech Eng Part H-J Eng Med 220:521–531
53. Hart GW (2009) The State University of New York in Stony Brook. http://www.georgehart.com/rp/rp.html. New York, USA. Accessed 1st Dec 2009
54. Wiria FE, Leong KF, Chua CK, Liu Y (2007) Poly-epsilon-caprolactone/hydroxyapatite for tissue engineering scaffold fabrication via selective laser sintering. Acta Biomater 3:1–12

55. Partee B, Hollister SJ, Das S (2006) Selective laser sintering process optimization for layered manufacturing of CAPA (R) 6501 polycaprolactone bone tissue engineering scaffolds. J Manuf Sci Eng-Trans ASME 128:531–540
56. Eshraghi S, Das S (2010) Mechanical and microstructural properties of polycaprolactone scaffolds with one-dimensional, two-dimensional, and three-dimensional orthogonally oriented porous architectures produced by selective laser sintering. Acta Biomater 6:2467–2476
57. Duan B, Wang M, Cheung WL (2011) Optimized fabrication of Ca-P/PHBV nanocomposite scaffolds via selective laser sintering for bone tissue engineering. Biofabrication 3:015001
58. Duan B, Wang M (2010) Customized Ca-P/PHBV nanocomposite scaffolds for bone tissue engineering: design, fabrication, surface modification and sustained release of growth factor. J R Soc Interface 7:S615–S629
59. Eosoly S, Brabazon D, Lohfeld S, Looney L (2010) Selective laser sintering of hydroxyapatite/poly-epsilon-caprolactone scaffolds. Acta Biomater 6:2511–2517
60. Petzold R, Zeilhofer HF, Kalender WA (1999) Rapid prototyping technology in medicine-basics and applications. Comput Med Imaging Graph 23:277–284
61. Wanibuchi M, Ohtaki M, Fukushima T, Friedman AH, Houkin K (2010) Skull base training and education using an artificial skull model created by selective laser sintering. Acta Neurochir 152:1055–1060
62. Silva DN, De Oliveira MG, Meurer E, Meurer MI, Da Silva JVL, Santa-Barbara A (2008) Dimensional error in selective laser sintering and 3D-printing of models for craniomaxillary anatomy reconstruction. J Cranio-Maxillo Fac Surg 36:443–449
63. Ibrahim D, Broilo TL, Heitz C, de Oliveira MG, de Oliveira HW, Nobre SMW, Dos Santos JHG, Silva DN (2009) Dimensional error of selective laser sintering, three-dimensional printing and PolyJet $^{(TM)}$ models in the reproduction of mandibular anatomy. J Cranio-Maxillo Fac Surg 37:167–173
64. Wu GF, Zhou B, Bi YP, Zhao YM (2008) Selective laser sintering technology for customized fabrication of facial prostheses. J Prosthet Dent 100:56–60
65. Montgomery JT, Vaughan MR, Crawford RH (2010) Design of an actively actuated prosthetic socket. Rapid Prototyping J 16:194–201
66. Faustini MC, Neptune RR, Crawford RH, Stanhope SJ (2008) Manufacture of passive dynamic ankle-foot orthoses using selective laser sintering. IEEE Trans Biomed Eng 55:784–790
67. Pallari JHP, Dalgarno KW, Woodburn J (2010) Mass customization of foot orthoses for rheumatoid arthritis using selective laser sintering. IEEE Trans Biomed Eng 57:1750–1756
68. Hao L, Dadbakhsh S, Seaman O, Felstead M (2009) Selective laser melting of a stainless steel and hydroxyapatite composite for load-bearing implant development. J Mater Process Tech 209:5793–5801
69. Wang Y, Shen YF, Wang ZY, Yang JL, Liu N, Huang WR (2010) Development of highly porous titanium scaffolds by selective laser melting. Mater Lett 64:674–676
70. Mullen L, Stamp RC, Brooks WK, Jones E, Sutcliffe CJ (2009) Selective laser melting: a regular unit cell approach for the manufacture of porous, titanium, bone in-growth constructs, suitable for orthopedic applications. J. Biomed Mater Res Part B 89B:325–334
71. Warnke PH, Douglas T, Wollny P, Sherry E, Steiner M, Galonska S, Becker ST, Springer IN, Wiltfang J, Sivananthan S (2009) Rapid prototyping: porous titanium alloy scaffolds produced by selective laser melting for bone tissue engineering. Tissue Eng Part C 15:115–124
72. Salgado AJ, Coutinho OP, Reis RL (2004) Bone tissue engineering: state of art and future trends. Macromol Biosci 4:743–765
73. Hutmacher DW, Cool S (2007) Concepts of scaffold-based tissue engineering-the rationale to use solid free-form fabrication techniques. J Cell Mol Med 11:654–669
74. Yeong WY, Chua CK, Leong KF, Chandrasekaran M (2004) Rapid prototyping in tissue engineering: challenges and potential. Trends Biotechnol 22:643–652
75. Rimell JT, Marquis PM (2000) Selective laser sintering of ultra high molecular weight polyethylene for clinical applications. J Biomed Mater Res 53:414–420
76. Schmidt M, Pohle D, Rechtenwald T (2007) Selective laser sintering of PEEK. CIRP Ann-Manuf Technol 56:205–208

77. Hao L, Savalani MM, Zhang Y, Tanner KE, Harris RA (2006) Effects of material morphology and processing conditions on the characteristics of hydroxyapatite and high-density polyethylene biocomposites by selective laser sintering. Proc Inst Mech Eng Pt L-J Mater-Design Appl 220:125–137

78. Tan KH, Chua CK, Leong KF, Cheah CM, Cheang P, Abu Bakar MS, Cha SW (2003) Scaffold development using selective laser sintering of polyetheretherketone-hydroxyapatite biocomposite blends. Biomateirals 24:3115–3123

79. Zhang Y, Hao L, Savalani MM, Harris RA, Di Silvio L, Tanner KE (2009) In vitro biocompatibility of hydroxyapatite-reinforced polymeric composites manufactured by selective laser sintering. J Biomed Mater Res Part A 91A:1018–1027

80. Tan KH, Chua CK, Leong KF, Naing MW, Cheah CM (2005) Fabrication and characterization of three-dimensional poly(ether-ether-ketone)/-ydroxyapatite biocomposite scaffolds using laser sintering. Proc Inst Mech Eng Part H-J Eng Med 219:183–194

81. Woodruff MA, Hutmacher DW (2010) The return of a forgotten polymer- Polycaprolactone in the 21st century. Prog Polym Sci 35:1217–1256

82. Williams JM, Adewunmi A, Schek RM, Flanagan CL, Krebsbach PH, Feinberg SE, Hollister SJ, Das S (2005) Bone tissue engineering using polycaprolactone scaffolds fabricated via selective laser sintering. Biomaterials 26:4817–4827

83. Ciardelli G, Chiono V, Vozzi G, Pracella M, Ahluwalia A, Barbani N, Cristallini C, Giusti P (2005) Blends of poly-(epsilon-caprolactone) and polysaccharides in tissue engineering applications. Biomacromolecules 6:1961–1976

84. Tan KH, Chua CK, Leong KF, Cheah CM, Gui WS, Tan WS, Wiria FE (2005) Selective laser sintering of biocompatible polymers for applications in tissue engineering. Bio-Med Mater Eng 5:113–124

85. Simpson RL, Wiria FE, Amis AA, Chua CK, Leong KF, Hansen UN, Chandraselkaran M, Lee MW (2008) Development of a 95/5 poly(L-lactide-co-glycolide)/hydroxylapatite and beta-tricalcium phosphate scaffold as bone replacement material via selective laser sintering. J Biomed Mater Res Part B 84B:17–25

86. Wiria FE, Chua CK, Leong KF, Quah ZY, Chandrasekaran M, Lee W (2008) Improved biocomposite development of poly(vinyl alcohol) and hydroxyapatite for tissue engineering scaffold fabrication using selective laser sintering. J Mater Sci-Mater Med 19:989–996

87. Kalita SJ, Bhardwaj A, Bhatt HA (2007) Nanocrystalline calcium phosphate ceramics in biomedical engineering. Mater Sci Eng C-Mater Biol Appl 27:441–449

88. Li JJ, Dou Y, Yang J, Yin YJ, Zhang H, Yao F, Wang HB, Yao KD (2009) Surface characterization and biocompatibility of micro- and nano-hydroxyapatite / chitosan-gelatin network films. Mater Sci Eng C-Mater Biol Appl 29:1207–1215

89. Heo SJ, Kim SE, Wei J, Hyun YT, Yun HS, Kim DH, Shin JW, Shin JW (2009) Fabrication and characterization of novel nano- and micro-HA/PCL composite scaffolds using a modified rapid prototyping process. J Biomed Mater Res Part A 89A:108–116

90. Duan B, Wang M, Zhou WY, Cheung WL, Li ZY, Lu WW (2010) Three-dimensional nanocomposite scaffolds fabricated via selective laser sintering for bone tissue engineering. Acta Biomater 6:4495–4505

91. Liu J, Zhang B, Yan CZ, Shi YS (2010) The effect of processing parameters on characteristics of selective laser sintering dental glass-ceramic powder. Rapid Prototyping J 16:138–145

92. Yeong WY, Sudarmadji N, Yu HY, Chua CK, Leong KF, Venkatraman SS, Boey YCF, Tan LP (2010) Porous polycaprolactone scaffold for cardiac tissue engineering fabricated by selective laser sintering. Acta Biomater 6:2028–2034

93. Huang H, Oizumi S, Kojima N, Niino T, Sakai Y (2007) Avidin-biotin binding-based cell seeding and perfusion culture of liver-derived cells in a porous scaffold with a three-dimensional interconnected flow-channel network. Biomaterials 28:3815–3823

94. Leong KF, Phua KKS, Chua CK, Du ZH, Teo KOM (2001) Fabrication of porous polymeric matrix drug delivery devices using the selective laser sintering technique. Proc Inst Mech Eng Part H-J Eng Med 215:191–201

95. Cheah CM, Leong KF, Chua CK, Low KH, Quek HS (2002) Characterization of microfeatures in selective laser sintered drug delivery devices. Proc Inst Mech Eng Part H-J Eng Med 216:369–383

96. Leong KF, Wiria FE, Chua CK, Li SH (2007) Characterization of a poly-epsilon-caprolactone polymeric drug delivery device built by selective laser sintering. Bio-Med Mater Eng 17:147–157

97. Salmoria CV, Klauss P, Paggi RA, Souza M, Kanis LA, Zepon KM (2010) Rapid manufacturing of Polycaprolactone/Progesterone drug delivery device by SLS. In: da Silva (et al) Proceedings of VR@P4. Innov Dev Des Manuf-Adv Res Virtural Rapid Prototyping. CRC Press, Leiria, Portugal, 229–232

98. Duan B, Wang M (2010) Encapsulation and release of biomolecules from Ca-P/PHBV nanocomposite microspheres and three-dimensional scaffolds fabricated by selective laser sintering. Polym Degrad Stab 95:1655–1664

99. Antonov EN, Bagratashvili VN, Whitaker MJ, Barry JJA, Shakesheff KM, Konovalov AN, Popov VK, Howdle SM (2005) Three-dimensional bioactive and biodegradable scaffolds fabricated by surface-selective laser sintering. Advd Mater 17:327–330

# Chapter 5
# Biomimetic Assemblies by Matrix-Assisted Pulsed Laser Evaporation

**Felix Sima and Ion N. Mihailescu**

**Abstract** The Matrix-Assisted Pulsed Laser Evaporation (MAPLE) technique emerged more than one decade ago as an alternative and complementary method to Pulsed Laser Deposition (PLD) in view of transferring organic materials onto solid substrates. In contrast to PLD, MAPLE proved to be a less harmful approach for transporting and depositing delicate, heat sensitive molecules. Since origin, MAPLE developed fast and was generally applied for organic biomaterials. It turned recently to inorganic compounds and has become a competitor to PLD. An important benefit of MAPLE is the capability of transferring films of nanoparticles with largely extended active areas. Such films can play an essential role in biology, pharmaceutics or sensing applications. This chapter reviews the mechanisms and recent progresses of MAPLE in thin film assembling for biomimetic applications in drug delivery systems, biosensors and advanced implant coatings.

## 5.1 Introduction

MAPLE was developed as a derivation of PLD and introduced [1, 2] for depositing thin films of organic and polymeric materials with a minimal thermal and chemical decomposition. In PLD, solid and compact inorganic materials are laser ablated and transferred via plasma onto a parallel substrate. Nevertheless, even for low laser energies, organic materials would be damaged by this approach. In MAPLE, organic biomolecules are dissolved in a laser wavelength absorbent solvent which is next frozen to form a solid target and exposed to laser irradiation. This way, the violent

F. Sima · I. N. Mihailescu (✉)
Lasers Department, National Institute for Lasers, Plasma and Radiation Physics,
409 Atomistilor Street, 77125 Magurele, Ilfov, Romania
e-mail: felix.sima@inflpr.ro

I. N. Mihailescu
e-mail: ion.mihailescu@inflpr.ro

V. Schmidt and M. R. Belegratis (eds.), *Laser Technology in Biomimetics*,
Biological and Medical Physics, Biomedical Engineering,
DOI: 10.1007/978-3-642-41341-4_5, © Springer-Verlag Berlin Heidelberg 2013

interaction of photons with the active material is diminished since the main fraction of energy is absorbed by the solvent. The energy is converted to thermal energy which helps to vaporize the solvent molecules entraining the organic molecules. The volatile solvent molecules are eliminated by vacuum system while organic molecules are reaching a facing substrate. There is a similarity of MAPLE with Matrix-Assisted Laser Desorption/Ionization (MALDI) [1, 3] but in the latter case the matrix is more complex from the chemical point of view and not appropriate for deposition.

With respect to solvent processing techniques, MAPLE preserves all the advantages of PLD and, in particular, it allows a better control of the film thickness and surface morphology, enhanced film/substrate adhesion, multi-layer deposition and patterning. Furthermore, being a non-contact procedure, eliminates a major source of contamination and can be integrated with other sterile processes. A more detailed discussion is found in Sect. 5.5.

Biomimetics emerged as a novel technological approach based on the biodiversity of the natural environment in order to reproduce the structure, physiology and function of biological entities. By mimicking the organization and the mechanisms in human body, the aim is to find solutions of replacing or repairing affected parts by the creation and design of new bioinspired materials. Human structural assembly is composed of complex configurations from nanometric to macroscopic scale. Although there were many attempts to develop it, the biomimetic nanotechnology is still in formative years, with no applications on the market. Nonetheless, exploring and imitating the biology at the nanometer scale is challenging and could bring new ideas and solutions in different domains such as tissue engineering, drug delivery systems or biosensors. During the last decade, one new approach of nanomaterials and of nanotechnologies focused on tissue engineering and regenerative medicine fields with the view of developing new tissue substitutes with superior biological properties [4].

Research in biomimetic tissue engineering and regenerative medicine takes advantage of biology inspired structures and mechanism and follows generally three directions: the development of bio-inspired functional surfaces for repair and regeneration of damaged tissues, the use of biomimetic drug delivery systems integrated into engineering materials in view of a controllable local release to prevent e.g. post-surgical infections in orthopedic implants, and to utilize engineered tissues with biologically implantable biosensor microdevices in order to monitor physiology and disease [5–8].

*Development of Bio-Inspired Functional Surfaces For Repair and Regeneration of Damaged Tissue*

Nowadays, metallic prostheses or bone grafts are used to repair bone defects. Titanium (Ti) and its alloys are the materials currently chosen in orthopedical and stomatological implants due to their high corrosion resistance in biological aggressive media, low mass density, high specific strength and biocompatibility [9]. However, in some cases implant failures due to encapsulation by a fibrous tissue or mismatch in elasticity modulus between bone and implant lead to revision surgeries [9]. To improve the regenerative capacity of bone tissue, bone associated materials are used for two main reasons: mimic the environment and speed up the healing processes at interfaces [10–12]. Thin bioactive coatings were developed on implant surfaces to form solid chemical bonds between implant surfaces and bone tissue in respect with long-term stability [12, 13]. Inorganic materials such as calcium phosphates (CaP) and in particular, hydroxyapatite (HA), which represents the major mineral phase of native bone, bioglasses (BG) or CaP- and BG- based composite materials are the most widely studied in the literature [14, 15] due to their biological properties for bone bonding and protein adsorption. On the other hand, organic structures are interesting beside their biocompatibility, because of their biodegradability and mechanical properties. Collagen-based matrices with complex pore orientation, pore size and alignment anisotropy resembling the extracellular matrix (ECM) scaffolds [16], porous crosslinked chitosan hydrogels [17], silk fibroin scaffolds with good stability [18], fibronectin (FN) crosslinked within hyaluronic acid hydrogels [19] were suggested for damaged soft tissues regeneration and repair. In view of mimicking the mineral-organic composition of bone and interaction processes, inorganic-organic composite materials were focusing recently a great attention. HA-gelatin (HA-GEL) composites were extensively studied since gelatin is a protein obtained by hydrolysis of collagen, the main organic phase of bone [20–22]. CaP introduction into biopolymer matrices such as polylactic acid (PLA) or poly (lactic acid-co-glycolic acid) (PLGA) were found to improve mechanical performance as compared to CaP alone [23]. Poly(lactide-co-caprolactone) (PLCL) with good mechanical properties and a bone-mimicking gelatin-apatite system were combined into a functional composite membrane with improved biological functions for hard tissues regeneration [24]. In addition to composites, which are intended to interfacial tissue reconstruction between soft (e.g. cartilage) and hard (e.g. bone) tissues, gradient biomaterials of multilayer coatings were synthesized for repairing or regenerating the functions of damaged parts at the interface of different tissue types [25].

*Use of Biomimetic Drug Delivery Systems*

The second approach in biomimetic tissue engineering resorts to biomimetic principles that have been extended to drug delivery systems. Such systems are integrated into engineering materials in view of constructing cellular microenvironments for different biomedical applications. Delivery of a drug into a precise place via bioactive molecules is of a great significance for tumor or damaged tissues treatment. Based upon a biomimetic approach antibiotics with calcium phosphate coatings were obtained on titanium alloy (Ti6Al4V) substrates capable to release the drug with a pH-dependent rate [26]. This approach is considered helpful to prevent post-surgical infections in orthopedic implants. Calcium apatite has also been used for the retention and local delivery of osteogenic factors such as recombinant human bone morphogenetic protein-2 (rhBMP-2) to locally induce the osteogenic transdifferentiation, [27], which means that mature cells could differentiate into bone forming osteoblasts. Thermosensitive polymers connected with peptides act as a dual stimulus-sensitive polymer capable of both, forming a gel at body temperature and allowing degradation in the presence of peptides. The controlled biodegradation has potential application in delivery systems, where the polymer gels can release the incorporated drug in a bioresponsive manner [28], sensitive to variations of physiological environment.

*Engineered Tissues With Biologically Implantable Biosensor Microdevices*

Bioreceptor entities such as proteins, peptides, enzymes or antibodies have been extensively studied for biosensing applications. In this respect, natural (biomimetic) recognition elements introduced in biomaterials extend the range of application of biosensors. Immobilization of biomaterials on artificial devices are usually attained by self-assembled monolayers (SAM), Langmuir-Blodgett (LB) films, or layer-by-layer (LBL) assembling [29]. Genetically manipulated proteins capable to fabricate 2D or 3D structures via bottom-up processes to control mineralization in biological systems were synthesized [30]. Engineered tissues containing biologically implantable biosensor microdevices which are able to observe tissue functions are expected to improve the feedback loop of implementation by the biosensor recordings [8]. Enzymes were stabilized via encapsulation in liposomes, polymers or gels to maintain their activity for longer time and be used as biosensors [31].

## 5.2 Biomimetic Design-Mimicking Aspects of a Natural Organism

Research studies demonstrated that all living systems are governed by molecular processes at nanometric scale. In particular, cellular organization and tissue characteristics depend on the extracellular matrix (ECM) composition. ECM is a non-cellular part present in tissues and organs and has significant roles in tissue morphogenesis, cellular differentiation or homeostasis [32, 33]. It is also involved in the physical arrangement of the cellular constituents. ECM is composed of water, proteins and polysaccharides acting as mediators between cellular components and the growing microenvironment [32]. It is a hierarchical complex structure with nanometer to the centimeter spatial order. Tissue-derived cells require strong attachment to a solid surface to ensure viability and growth. In contact with solid surfaces, the cells are adhering and communicating via integrins, trans-membrane receptors that mediate interactions between the cytoskeleton and the ECM. Studies confirmed the biological productive interaction of a material surface organized at nanometric scale [4]. In fact, human cells are in permanent contact with nanostructured surfaces [34]. In case of an implant, surface features (such as morphology, composition and structure) noticeably influence the adsorption of proteins which consequently mediate the cellular adhesion, proliferation and differentiation [35, 36].

Following a biomimetic approach, the future biomaterials will benefit from the understanding of the biological processes such as wound healing or inflammation and will develop reactions with high accuracy. The natural materials are generally composites based on polymers and minerals exhibiting a diversity of properties depending on their structure at various length scales [37]. To avoid an uncontrolled cellular response to surfaces, both, nanoscale and microscale characteristics are therefore desired. In a recent review, porous structures were proposed for tissue replacement and regeneration that could mimic hierarchically structured porous natural materials [38]. It was stressed upon the different scales of porosity in natural materials (which ranges from nanometers to millimeters): pores with small dimensions (nanometers) are in charge with the bioactivity and protein interaction, micrometer sized pores are involved in cellular attachment, while larger ones are responsible for cellular growing, blood flux, mechanical resistance and implant functionality. In this view, the use of inorganic materials (such as ceramics) is not always convenient for the structural or functional restoration of a tissue, particularly in large bone defects. It exists therefore a growing interest to deliver and release in a controllable manner biologically active proteins, such as growth factors, adhesion proteins, or antibodies via inorganic layer dissolution and to stimulate the cellular response toward mineralization [39]. Deposition of composites, patterns or multilayer structures were found clinically significant and innovative designs were proposed [40].

A material-inspired strategy of nano-assembling in biomedical applications for enamel repair was recently reported [41]. The combined action of a glutamic acid and nano apatite particles was capable to regenerate an enamel-like structure under physiological conditions.

A novel design based on graded materials was proposed. The structure consists of layers of biomineralized collagen, hyaluronic acid-charged collagen, and an intermediate layer of the same nature as the biomineralized collagen, but with a lower content of mineral. The aim was to develop composite osteochondral scaffolds organized in different integrated layers with biomimetic features for articular cartilage and subchondral bone [42].

## 5.3 Scaffold Fabrication and Deposition Methods

One novelty in the field of biomimetics is the hybrid deposition methods with the view of obtaining multilayer or nanocomposite structures for new applications and improved performances. Many attempts were mentioned to combine magnetron sputtering with cathodic arc deposition, plasma-enhanced chemical vapor and electron beam evaporation, or flash evaporation with physical vapor deposition to synthesize innovative inorganic films, which are reviewed comprehensively in [43]. In case of organic thin films it is mandatory to avoid thermal phenomena. Techniques such as sol-gel, dip-coating or spin-coating which are in present employed to obtain thin films of polymers or proteins are difficult to extend to multistructure generation because of solvent implication and combination problems.

In recent years, there was a major interest in thin coatings technology to fabricate multilayer coatings or high spatial resolution patterns on solid substrates with the view of developing appropriate deposition techniques for available biomaterials in suitable shape for specific applications [44, 45].

Table 5.1 summarizes widely used materials and deposition methods along with specific advantages, drawbacks and relevant examples.

Multi-nozzle Deposition Manufacturing (MDM) was introduced as a non-thermal process for manufacturing porous scaffolds for tissue engineering [14]. To obtain inorganic-organic composites, MDM was applied in a single-nozzle deposition process where the material was prepared by dissolving poly (L-lactic acid) into dioxane and mixing with tricalcium phosphate (TCP) particles. Thus, a macroporous scaffold structure was obtained by phase separation and solvent sublimation while consequently a growth factor such as BMP (bone morphogenetic protein) could be loaded into the scaffold by vacuum suction.

Organ-printing technology is a rapid prototyping computer-aided 3D method based on layer by layer deposition of different types of hydrogels and cells in order to fabricate 3D constructs for perfused, vascularized human tissues or structural and functional units of human organs [65].

Direct assembling of cells and extracellular matrices for the construction of functional 3D tissue/organ substitutes was achieved by an optimized cell-assembly printing technique [66].

However, rapid prototyping techniques working with hydrogels, laser-based, nozzle-based or printer-based systems, suffer from poor mechanical strength [67].

Another method to manufacture 3D hierarchical structures is the layer-by-layer microfluidics process which involves immobilization of a cell-matrix assembly, cell-matrix contraction, and pressure-driven microfluidic delivery to fabricate hybrid biopolymer structures for tissue engineering [68]. Electrospinning enables the fabrication of scaffolds with micro and nanoscale topography and high porosity [69]. A new generation of scaffolds comprising living cells was developed by electrospinning technology. Antibiotics, proteins as well as living cells were incorporated into the advanced scaffolds and electrospun [70].

Direct laser writing techniques such as Laser induced forward transfer (LIFT), allow for the deposition of biomolecule patterns without degradation and with high spatial resolution. Under the action of a laser pulse focused through a transparent layer on a thin metallic film, a small fraction of an organic material coating is transferred to a receptor substrate, which is placed closely and parallel. In this configuration, biomaterials such as polyethylene glycol and eukaryotic cells, were deposited with a spatial resolution of $\sim 10\,\mu m$ without damage of structures or genotype [40]. In other studies, microarrays of DNA have been spotted and were found capable to maintain gene discrimination capacity [60].

Biological laser printing (BioLP) has been proposed as an alternative to the above mentioned techniques for assembling and micropatterning biomaterials and cells. High-throughput laser printing of a biopolymer, hydroxyapatite and human endothelial cells was achieved demonstrating the capability of the method for three-dimensional tissue construction [59]. LIFT-derived cell seeding pattern was shown to modify the growth characteristics of cell co-cultures resulting in vessel formation and in an efficient regeneration of infarcted hearts after transplantation of a LIFT-tissue engineered cardiac patch [71].

**Table 5.1** Common deposition methods of biomimetic thin films

| Method | Advantage | Disadvantage | Materials |
|---|---|---|---|
| Plasma spray (PS) | Simple, not expensive | Poor coating adhesion to substrate/ weak ceramic-metal interface, limited to inorganic coatings | Hydroxyapatite (HA)/ Bioglass (BG) [46] |
| Radio frequency magnetron sputtering (RF-MS) | Uniform dense coating | Amorphous coatings, expensive, limited to inorganic coatings | Hydroxyapatite (HA)/ Bioglass (BG) [47, 48] |
| Pulsed laser deposition (PLD) | Film stoichiometry (especially for doped materials) and good adhesion | Small covering areas, limited to inorganic coatings | Hydroxyapatite (HA)/ Bioglass (BG) [13] |
| Sol-gel (SG) | Application to both organic and inorganic coatings | Poor coating adhesion; Involvement of liquid media limit multi-layer assembling | Hydroxyapatite (HA)/ Bioglass (BG) [49] |
| Electrophoretic deposition (EPD) | Simple processing equipment, application to both organic and inorganic coatings | Poor coating adhesion; Involvement of liquid media limit multi-layer assembling (affect interfaces) | Hydroxyapatite (HA)/ Bioglass (BG) [50] |
| Electrostatic deposition (ED) | Solvent-free coating process | Limited to single coating | Poly (lactic acid-co-glycolic acid) (PLGA) [51] |
| Sol-gel (SG) | Macroporous bioactive scaffold | Poor coating adhesion; Involvement of liquid media limit multi-layer assembling (affect interfaces) | Chitosan/BG composite [52] |
| Layer-by-layer (LBL) | Viscoelasticity/bioactivity | Involvement of liquid media limit multi-layer assembling (affect interfaces) | Chitosan/BG composite [53] |
| Adsorption on surface | Simple, rapid | Poor adhesion and uniformity, Involvement of liquid media limit multi-layer assembling (affect interfaces) | Fibronectin (FN), vitronectin (VN), collagen I [54] |
| Spin coating | Simple, uniform coatings | Solvent issue during multilayers, adherence | PLGA/polycaprolactone (PCL) composite [55] |

(continued)

**Table 5.1** (continued)

| Method | Advantage | Disadvantage | Materials |
|---|---|---|---|
| Langmuir-Blodgett dip coating | Monolayers | Limited to very thin films | Glucose [56] |
| Ink-jet printing | Thicker films | Possible nozzle blockage for composites | PCL/chitosan [57] |
| Multi-nozzle Deposition Manufacturing (MDM) | Porous structures | Possible nozzle blockage for composites | Poly (L-lactic acid) (PLA)/tricalcium phosphate (TCP) [14] |
| Laser-induced forward transfer (LIFT) | Patterns with high spatial resolution | Limited to patterns; difficulties for large area thin coatings | Hydroxyapatite (HA) [58, 59], proteins [40, 60, 61] |
| Matrix assisted pulsed laser evaporation (MAPLE) | Nanoparticulate films, application to both organic and inorganic coatings multilayers and multistructures | Small covering areas | Fibronectin (FN) [62], vitronectin (VN) [63], Octocalcium phosphate (OCP) [64] |

2D structures of hybrid polymers were produced by Two-Photon-Polymerization (2PP). The scaffolds were tested in vitro for applications in tissue engineering with the aim of developing a dermal graft [72, 73].

One alternative single step fabrication method of complex constructions of multilayers and multistructures is MAPLE due to the potential of transferring and depositing both organic and inorganic materials. Because the transfer is supposed to be dry (the solvent molecules are eliminated during the transfer by vacuum pumping) the solvent implication is circumvented while the synthesis of multilayer structures keeps rather simple. The method is a non-contact technique which proved very versatile and challenging in respect with other laser based techniques [1, 74, 75]. One can produce by MAPLE coatings with adhesion better than by other methods whilst film uniformity and thickness on either rough or flat substrates can be well controlled. Moreover, using appropriate masks in MAPLE one can manufacture microsized samples (single or multilayered) for microarray chip applications [40].

## 5.4 Basics of MAPLE

Since its invention in late 1990s [2] as an alternative to spray coating of thin films for chemical vapor sensors [76], MAPLE was successfully applied to a large class of organic compounds for various applications [1, 77]. Thin films of pullulan [78] or triacetate-pullulan [79] polysaccharides for drug delivery, polyfluorene and polythiophene copolymers for metal-insulator-semiconductor and field-effect transistor (FET) structures [80], chemoselective polymers for microsensors [81] or proteins with applications in tissue engineering [62, 82] or biosensing [83, 84] were obtained. A recent review on MAPLE deposition of organic, biological and composite thin films summarized several potential applications of thin coatings obtained by this method [85]. MAPLE was recently used to deposit uniform, ultra stable and nanostructured glassy polymer films with superior thermal and kinetic stability [86].

### 5.4.1 Experimental Conditions and Mechanisms of MAPLE

According to the introductory remarks about the principle and prerequisites of MAPLE, a typical MAPLE experiment starts with the preparation of a homogenous solution consisting of a small amount of solute (0.5–5 %wt) dissolved in a solvent (matrix). Next, the solution is frozen with liquid nitrogen to obtain a solid target (Fig. 5.1), which is cooled and kept frozen during the laser irradiation and deposition process.

Adequate deposition substrates are chosen according to specific applications or analyses to be carried out. Before use, they are purged with acetone in an ultrasonic cleaner, ethanol and deionized water and dried with high purity $N_2$ gas. Subsequently, the substrates are placed parallel at an optimized distance in front of the

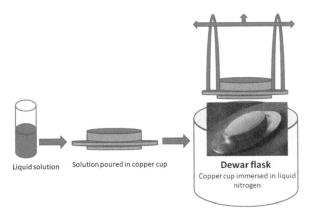

**Fig. 5.1** Preparation stages of a solid frozen target starting from a liquid solution

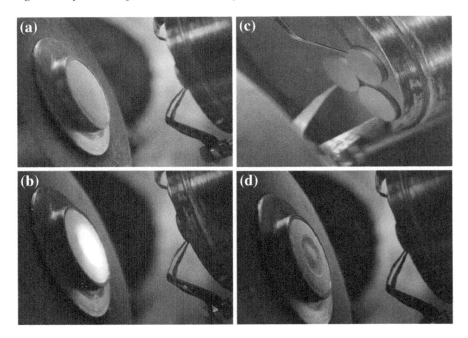

**Fig. 5.2** Photos of a MAPLE configuration inside the reaction chamber: non-irradiated target mounted on cooler (**a**), cryogenic target illuminated during laser irradiation (**b**), deposition substrates dimly lit by irradiated target (**c**) and eroded target after multi-pulse laser irradiation (**d**)

rotating frozen target in the reaction chamber and gently heated below the degradation temperature of the solute (Fig. 5.2).

The solvent plays an important role in the MAPLE process, since it is the carrier of the solute molecules. Therefore it must be chemically inert and must not interact with the solute during laser irradiation. The frozen solvent must efficiently absorb

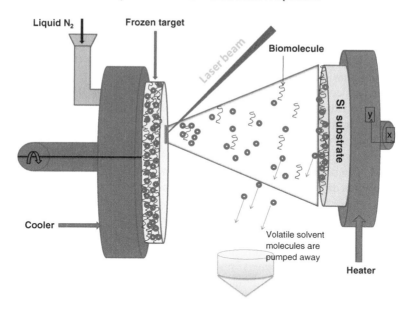

**Fig. 5.3** MAPLE scheme. The solution is frozen and forms the target (*left*) which is cooled by a liquid nitrogen flow and maintained in constant rotation and translation. The laser beam (*center top*) irradiates the target, disrupting material from the surface which is deposited on the heated substrate (*right*). A warm substrate encourages solvent evaporation and adherence of film

the incident laser power and be easily evaporated. The volatile solvent molecules are vaporized and evacuated from the reaction chamber by the pumping system, guiding the solute molecules onto the substrates where they are deposited without degradation (Fig. 5.3). The most important MAPLE parameters are laser fluence, laser pulse repetition rate, substrate temperature, and target-collector separation distance. A dynamic pressure ($<10^{-1}$ mbar) is maintained inside the deposition chamber during the process. Typically, several hundred up to few thousand laser pulses are applied in order to achieve the desired thickness of the growing film.

The preferred laser systems in MAPLE experiments are pulsed UV lasers, such as excimer lasers (ArF* (wavelength: 193 nm), KrF* (wavelength: 248 nm), XeCl* (wavelength: 308 nm)), which generate pulses with a duration in the range of 20–30 ns or solid state Nd:YAG lasers (wavelength: 266 nm (fourth harmonic) or 355 nm (third harmonic)) with durations in the range of 5–10 ns. A typical applied laser fluence varies from 0.05 to 1 J/cm$^2$ [87]. In addition, a uniform intensity over the laser spot is generally desired (top hat profile) for a defined and stable material evaporation. At first sight, it seems difficult to apply MAPLE for the coating of large substrates due to the concentrated and localized (point-like) laser–material interaction. This drawback is circumvented by rotating and translating the substrates.

Parallel to the experimental efforts related to MAPLE, theoretical considerations aiming at the explanation of the underlying processes were made. In this context, a model based on *complete evaporation* process was proposed for the interaction of the laser beam with the frozen target [88]. Namely, it is supposed that the laser energy is absorbed by the matrix, converted into thermal energy necessary for the solvent molecules to be vaporized. Next, the solute molecules are transported by collisions with solvent molecules and deposited on facing collector.

Later on, molecular dynamic simulations demonstrated that a consequence of the interaction between a laser beam and the cryogenic target is the formation of solvent-solute clusters due to overheating [89, 90]. In support of the proposed *explosive-boiling* model, it was shown that the ejection of material can be the result of explosive evaporation or spallation also described as "cold laser ablation", a phenomenon which becomes more evident when increasing the fluence. This results in the deposition of thin droplet coatings with high roughness morphology. Another mechanism based on *local overheating* of absorbing outmost surface layer of biomolecules was proposed [91]. More concretely, the solute molecules are also absorbing laser energy and their temperature is increasing. The heating is transferred to the solvent which, under vacuum, starts boiling just above melting point. The material ejection is consequently produced at lower temperature than the degradation threshold.

A *nonhomogeneous absorption* mechanism was proposed as well. This mechanism could account for the two cases: for low laser fluence the mass ejection is produced by surface evaporation while at higher fluencies hydrodynamic ablation mechanisms is responsible for the expulsion [92]. A frozen target composed of solute dissolved in a solvent also includes different phases such as ice cracks, air bubbles, or other defects. These phases were suggested to be involved in light absorption or scattering processes during laser irradiation of the heterogeneous frozen target [92]. Accordingly, the absorption was found to be higher in ice as compared to water. The laser absorption can be increased by the addition of other compounds in the solution, which introduce local modifications of material properties. An example is visible from Fig. 5.4 where the absorbance level at e.g. 248 nm is indicated for distilled water (d.w.) and solutions containing organic (TRIS–tris(hydroxymethyl)aminomethane), inorganic salts (NaCl) and proteins (bovine serum albumin–BSA) before and after freezing. These salts act in two ways: as protein stabilizers and absorption centers during laser-target interaction.

Based upon the above mentioned theoretical considerations, which confirm the important role of the chosen solvent regarding the properties of the ejected particles, films with quite small roughness and improved surface morphology were recently obtained [93–95]. In this view, properly selected solvent and a reduced solute concentration in the target allowed for an optimum absorption regime. Perfectly matching experimental conditions such as a proper selection of the solute-solvent mixture, suitable laser source (laser wavelength with respect to spectral absorption of the solute-solvent mixture), dynamic pressure, substrate temperature and target–substrate separation distance are important prerequisites for an optimized MAPLE transfer and deposition of the solute on the solid substrate.

Hence, a major issue within a MAPLE deposition process is related to the laser-target interaction, which essentially depends on the used solvent or solute. Moreover, the process becomes more complicated when introducing supplementary compounds in the solution. A MAPLE protocol, which defines appropriate experimental condi-

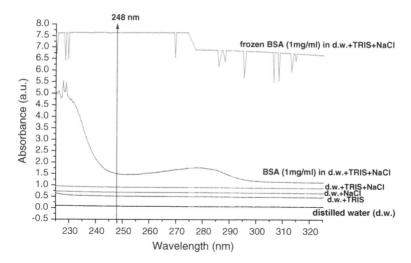

**Fig. 5.4** Optical absorption spectra of pure distilled water, distilled water + TRIS (50 mM), distilled water + NaCl (150 mM), distilled water + TRIS (50 mM) + NaCl (150 mM), distilled water + TRIS (50 mM) + NaCl (150 mM) + BSA (1 mg/ml) and frozen solution of distilled water + TRIS (50 mM) + NaCl (150 mM) + BSA (1 mg/ml)

tions, should therefore be elaborated for each investigated solvent-solute mixture and laser wavelength combination.

Thin coating processes are "wet" when the method involves solvent flow coating and air drying and "dry" when the liquid phase is avoided. A still open question is if the MAPLE process could be considered wet or dry. The concern is related to the possible transfer of solvent molecules along with the solute on the substrate. In this case, MAPLE could be considered wet. For example, in case of not very volatile solvents such as DMSO, some molecules are reaching the substrate and remain immobilized along with the solute. It is not dramatic in this case since DMSO is considered biocompatible. However, some further interaction with the biopolymer molecules or other film layer could affect morphology or even composition. MAPLE can be considered dry when using very volatile solvents such as toluene. This solvent is definitely undesirable, in particular for further biological investigations, but it is entirely pumped out by vacuum system and is not reaching the substrate. In case of proteins, water is completely eliminated during transfer but part of the buffer (inorganic and organic salts) is expected to reach the substrate and further act as stabilizers. We mention that nowadays, wet processes are employed in pharmaceutical production to coat drug tablets with a thin film which allows for the controlled release of the active substance. Dry processes are advantageous for multi-component coatings and are environmentally safe by avoiding the waste of organic gas or aqueous stream. The choice of a good solvent is therefore one important key to the successful transfer by MAPLE.

## 5.4.2 Reliability

Since polymers and proteins easily denaturate, precipitate or aggregate, special attention should be paid to the preparation of a homogenous starting solution. It is emphasized that this is a prerequisite for reproducible sample fabrication, especially in the field of biomimetics involving cell technology. Each material demands individual treatment and exposure conditions and is therefore discussed separately:

*Deposited Polymer Films*

Reproducible samples require well controlled film composition and morphology. The cluster formation after target irradiation by using a low solute concentration and low laser fluence should be avoided. This regime results on one hand in a reduced deposition rate and hence a slow film growth, but on the other hand yields a smoother film surface and composition conservation. Some morphological features specific to MAPLE deposited films were assumed to origin from evaporation of solvent remains in the coating, which accompanied the solute molecules to the substrate in case of explosive evaporation or spallation of target. This happens e.g. in case of not very volatile solvents like DMSO, for which the solvent molecules are reaching the substrate along with the solute. There, the solvent is slowly evaporated since the temperature of the substrate is moderately heated [96]. Heating carefully the substrate just below the degradation temperature of the solute one could improve the film assembling and morphology by better controlling the adhesion of the film and avoiding the formation of non-homogenous zones. Besides composition and morphology, the film thickness should be carefully monitored since MAPLE is not always an additive process. Thus, in some cases, when increasing the amount of solute deposited on the substrate it tends to become more compact then growing in thickness. This is not generally applicable and it happens for low quantities of transferred material. A combined effect of evaporation induced assembly after transfer of solute-solvent clusters with the specific linkage between the linear structures of polymers was found to influence film growth [93].

*Deposited Protein Films*

Proteins are sensitive biological macromolecules, consisting of one or more amino acids, which are held together by peptide bonds. While they are stable under normal ambient biological conditions (especially temperature), they are easily disrupted and denaturated at elevated temperatures [97]. Consequently, intense laser irradiation and the associated heating can cause irreversible structural changes of proteins, which affect drastically the protein or cell activity. Thus, a compromise must be sought between the protein stability in the solution, its freezing and the safe laser transfer and immobilization on the substrate. For this purpose, colloidal solutions containing organic and inorganic salts, which are beneficial for protein stability and increase the laser absorption of the matrix while protecting the protein, are used (Fig. 5.4). The aqueous solution is softly homogenized while its freezing is induced rapidly drop by drop by pouring in an already cooled copper container at liquid nitrogen temperature

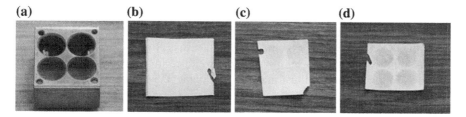

**Fig. 5.5** Ponceau Staining on nitrocellulose paper after MAPLE protein transfer. Staining of bovine serum albumin (BSA) on nitrocellulose through a mask with four circles of 12 mm diameters: the used mask (**a**), staining of BSA after the MAPLE transfer from solutions of 0.1 mg/ml (**b**), 0.25 mg/ml (**c**) and 1 mg/ml (**d**)

(to ensure target homogeneity in volume). During freezing the protein remains folded to preserve its structure. After laser transfer, proteins are usually adsorbed onto solid surfaces via electrostatic and hydrophobic forces. The protein recovering from frozen state on surface is preferred in order to get the desired bio-effect. To this aim, various functionalized substrates should be tested. To control the material spreading from target and the deposition distribution on substrate in order to uniformly collect the solute, experiments on nitrocellulose paper (known to exhibit a good protein affinity) can be carried out at different target-substrate separation distances which stand for an important issue in MAPLE transfer. The proteins are collected through masks on paper surfaces of identical areas with actual deposition substrates (Fig. 5.5).

The deposition area is visualized by staining with Ponceau S Solution [62, 98]. By monitoring the gradient of color intensity on the nitrocellulose paper, the optimum separation distance between the target and the collector is selected as the best compromise between deposition efficiency and distribution uniformity [21]. To quantify the amount of protein deposited by MAPLE, a bicinchoninic acid (BCA) assay can be performed using bovine serum albumin (BSA) dilutions as standard [99]. The spectrophotometry data indicate the amount of deposited protein on area of interest [63].

## 5.5 MAPLE: From the Origin to Biomimetics

### 5.5.1 Application to Organics

MAPLE provides a more gentle mechanism than PLD for transferring different compounds, including large molecular weight molecules. It is generally acknowledged that PLD is limited in case of organics due to the high laser intensity which can cause irreversible damage to polymer or protein chains. Nevertheless, there were a few attempts to apply PLD to polymers at very low laser fluence [100–103]. The technique failed, in particular, in case of more sensitive biopolymers [78, 104], enzymes or proteins [91, 92].

MAPLE was applied from the very beginning to organic biomaterials in order to develop structures, which mimic parts of organism [105]. The huge number of materials with potential interest for biomedical applications allowed for a fast expansion of the method which rapidly improved and developed in search for an appropriate niche. The following survey illustrates the development and application scenarios of MAPLE:

*Polymers for Drug Delivery Systems*

Biodegradable polymer coatings are applied to implants for controlled and local drug delivery. A poly(DL-lactide-co-glycolide) (PLGA) and polycaprolactone (PCL) composite in a multilayer configuration was deposited by spin-coating technique and dipyridamole was loaded as a drug into the surface nanopores [55]. This configuration was found effective as drug-delivery platform over 70 days for drug-eluting implants, in particular for cardiovascular stent applications. A PLGA coating with sirolimus (rapamycin—an immunosuppressant drug used to prevent rejection of an implant) was deposited by a dry-powder electrostatic process. During the 90 days of coating absorption, the drug was released and prevented inflammation [51].

Thin coatings of biocompatible and biodegradable polymers with potential as carrier vectors for drug delivery such as polyethylene glycol (PEG) [40], (PLGA) [96], mixtures PEG-PLGA [106, 107], poly(D,L-lactide) [108] or triacetate-pullulan [78] among others were obtained by MAPLE without noticeable chemical degradation. Among biomimetic "smart" solutions, some fascinating materials (e.g. polysaccharides such as chitin which is found in fungal and yeast cell walls or mollusk shells [109] or other biopolymers from microbial sources [110]) with exceptional properties are developed by biological organisms. In particular, Levan is an exopolysaccharide which can be used as food or feed additive and is distinguished from other polysaccharides by its low viscosity, high solubility in oil, compatibility with salts and surfactants, stability to heat, acid and alkali media, high holding capacity for water and chemicals, and good biocompatibility [111–113]. MAPLE application to obtain nanostructured thin films of Levan was reported for the first time in [93]. Thin coatings of desired thickness could be attractive to control the rate of dissolution for drug release and delivery applications. The nanostructure feature has the potential to enhance the biopolymer specific surface area for applications as carriers in drug delivery systems. An unusual ordered array was observed by AFM (Fig. 5.6), the most probably forming by the solvent (DMSO) evaporation induced nano-assembling combined with the specific linkages between the linear structures of polysaccharides. MAPLE samples exhibited a compact structure, with good adhesion to substrate and a homogenous nanostructured surface, fully compatible with potential use in biology or medicine [93].

*Proteins and Enzymes for Biosensors*

Biomimetic materials with sophisticated three-dimensional design, well-defined pattern and tunable properties used for drug carriers and tissue engineering can be also used to monitor biological microelectromechanical systems and diagnostics. They can respond to in vivo environmental changes and secure controlled parameters for

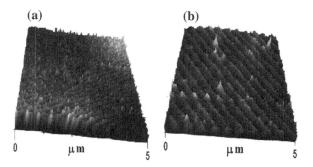

**Fig. 5.6** Typical AFM images of polymer surfaces for (**a**) Levan and (**b**) Oxidized Levan coatings by MAPLE on Si. Reprinted with permission from Biomacromolecules [93]

drug release, cell interaction, mechanical properties, or permeability [114]. The need to elucidate fundamental mechanisms of growth and the structure of biological systems in response to new biomaterials is one challenge for developing miniaturized protein or enzyme based biosensors [109]. MAPLE could provide in this respect an alternative to "wet" methods in view of obtaining patterns [105] or multi-structures since the solvent issue is avoided [94]. The accurate control of the expulsed material and layer thickness or the uniform and homogenous distributions of the material on the substrate are the real advantages of the method.

Insulin and horseradish peroxidase (HRP) were the first proteins deposited by MAPLE [115] as active biomolecules that could be used in biosensors with the goal of fabricating a functional microfluidic device. In the same study efforts have been paid to develop a polymer (poly(ethylene glycol -PEG)–protein(HRP) composite film with increased adhesion to surface. The structure and activity of the proteins were found unaffected and the method was proposed for depositing active biomolecules for sensor or microarray applications.

MAPLE was applied to immobilize urease (an enzyme which catalyses the hydrolysis of urea in biological systems, monitoring the nitrogen concentration of the human serum in the form of urea–a measure of the kidney function) on solid collectors with the aim to develop a sensor based on a biomimetic principle, a strategy that mimic natural processes, with interest in clinical applications [116]. The immunoassay, one of the most used analytical method based on the selective affinity of the biological antibody for its antigen, was applied to show that MAPLE-immobilized IgG films can be used as immunosensors for the detection of specific antigens in research or clinical investigations [83]. It is noted that IgG molecules are able to struggle with bacteria and viruses while a quantitative antibody test is clinically indispensable for autoimmune diseases, allergies and recurring infections [117]. To this prospective, the observed morphology change by the content of salts and lipids in MAPLE solution could open the door to reach the best compromise between the IgG content and surface condition over sensing capabilities, an essential step in developing personalized and miniaturized biosensors.

*Proteins for Tissue Engineering*

Sophisticated synthetic (tissue-engineered) and multi-(bio)functional surfaces or bone scaffolds were proposed to build up a microarchitecture that integrates different biological entities such as proteins, cells and cell processes [118]. An ideal synthetic engineered biomaterial should be biocompatible, biodegradable and to mimic the hierarchical structure of native tissue with the view of promoting actively desirable physiological responses (bioactive). This will avoid additional surgical procedures and reduce risks of infection [5]. In order to accelerate extracellular matrix (ECM) production, enhance cellular activity in the early stage of a material implantation and push its tissue integration, a common approach relies upon the presentation of the arginine-glycine-aspartic acid (RGD) adhesive sequence derived from fibronectin (FN) (glycoprotein present in ECM that interacts with cells to control cell adhesion, cytoskeletal organization and cellular signaling) [119, 120]. More specifically, in the field of biomaterials for bone reconstruction, FN has been proposed for enhancing osseointegration [121].

Deposition of FN by MAPLE from saline buffer-based cryogenic targets was reported. The aim was to transfer and immobilize a high molecular mass protein such as FN (~450 kDa) on a flat Si surface by a controllable approach in order to obtain a biologically active protein structure. Under these conditions, FN is exposing binding sites that promote intermolecular interactions and cell adhesion and consequently cell proliferation speeding up tissue formation around implant and a faster stabilization. A rather rough surface with a particulate-like morphology was observed. The particulates were uniformly distributed floating on a base film on substrates. The presence of particulates could be beneficial for e.g. orthopedic applications because they increase the specific surface area and thus the binding capacity of MAPLE coated implants to tissues. No noticeable changes were detected of FN structure after the MAPLE transfer. Different organization of intact protein, from small aggregates to fibril-like forms were observed while human osteoprogenitor cells grown on FN thin films exhibited a superior attachment as compared to controls [62] (Fig. 5.7). A similar cytoskeleton morphology was found in osteoblast-like cells grown on intact FN as compared to cells grown on FN fragments [122] suggesting that MAPLE-transferred FN forms patterns with non-denatured and functional cell binding domains. It was also demonstrated that FN adsorption on apatite/nanodiamond films or HA-coated solid substrates improved cellular attachment, adhesion and spreading [123, 124].

A layer-by-layer coating with heparin, growth factor and FN of titanium surfaces were found to improve cell proliferation while multilayer films significantly promoted cell attachment and growth [125].

Three ECM proteins (FN, vitronectin (VN), and collagen I) were in parallel tested and proved that they can play a role in wound and tissue repair [54]. VN is another glycoprotein present in serum and the extracellular matrix which promotes cell adhesion and spreading [126]. It was recently shown that a VN coated Ti implant improved primary fixation in vivo resulting in an increased osteointegration [127]. Human osteoprogenitor cells grown on MAPLE transferred VN on HA coated Ti samples exhibited an improved adherence, spreading and growth compared to cells grown on

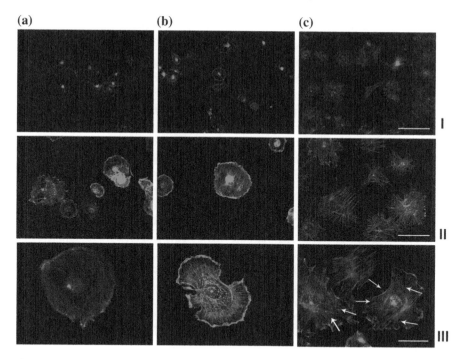

**Fig. 5.7** HOP cell actin filament staining on **a** standard cover slips, **b** silicon and **c** FN covered silicon by MAPLE after 3 h in cell culture. Cells were fixed, permeabilized, and stained for actin using Alexa Fluor 594-conjugated phalloidin (*red*). The mounting media contained DAPI (*blue*), which stained cell nuclei. On Fig. 7c FITC conjugated FN are marked in *green* (*arrows*). Scale bars are of 200 (I), 100 (II) and 50 (III) μm respectively. Reprinted with permission from Acta Biomaterialia [62]

Ti/HA samples, supporting a faster cell colonization and proving the physiological VN functionality after laser transfer [63]. Thin films of collagen obtained by MAPLE, with roughness controlled by experimental parameters were also reported [128].

A key advantage of ECM protein coatings obtained by MAPLE versus other simple adsorption methods is the accurate control of the expulsed material and coating thickness. Homogeneous distribution on the surface, in particular for small amounts (μg) of proteins, is beneficial, while the buffer salts (NaCl, TRIS) contained in the starting solution and deposited next to the protein are expected to act as a stabilizer. To enhance protein immobilization, one can easily use calcium phosphate or polymer films already deposited in an unique all-laser procedure [63] which demonstrated beneficial effects as shown in refs [129, 130]. One can thus develop ECM-mimetic biomaterial surfaces that could trigger protein organization into biologically active molecules. Protein-coated calcium phosphate layers, and in particular nanostructured thin films, are expected to provide a synergetic interface for biomimetic implant applications.

## 5.5.2 Application to Organic–Inorganic Composites

To mimic the multidimensional, hierarchical and complex structure of native tissues (e.g. bone) from the chemical and physical (morphology, structure, composition and functionality) points of view organic-inorganic composite coatings are the best choice to reconstruct the molecular architecture of the local environment and to trigger dynamic biomechanisms. Inorganic materials (e.g. ceramics, metals) are used to render the strength and to provide or compensate for the mineral phase of the tissue. Organic bioactive molecules are able to accelerate tissue integration (as e.g. ECM proteins) or to prevent undesired biological response (as e.g. drugs) in a well-controlled manner [131]. Biodegradable polymers or bioresorbable ceramics are used as scaffold materials as well as matrix carriers for drug release. As a soft laser deposition technique that minimizes the photochemical damage of an organic exposed to the laser light, MAPLE was naturally extended to organic-inorganic composites aiming to create three-dimensional structures for faster cell colonization and tissue regeneration. This could represent a benefit over other techniques including PLD (see Table 5.1) since most of the methods are efficiently applied to either inorganic or organic materials but not to both of them or to composites.

HA-sodium maleate (MP) copolymer thin coatings deposited by MAPLE on Ti substrates were tested in vitro. Osteoblast-like cells showed a higher proliferation when cultivated on these nanocomposite coatings in comparison with the cells grown on Ti coated with HA only (Ti-HA) (Fig. 5.8). This demonstrates that the polymer presence improved surface bio-adhesive characteristics, cell attachment, spreading and proliferation, which recommend the potential of Ti coated with HA-polymer nanocomposites as scaffolds in dental or orthopedic implantology [132].

HA-sodium maleate-vinyl acetate copolymer coatings were synthesized on Ti surfaces for specific biological investigations. Human primary osteoblasts spread and proliferated onto modified surface and formed groups of cells which during biosynthetic activity expressed osteoblast markers [133]. PMMA-bioglass composites were obtained by MAPLE as uniform thin layers onto chemically etched Ti from targets of mixtures containing PMMA reinforced with either 6P57 (lower silica content) or 6P61 (higher silica content) bioglass powders [134]. Osteoblast-like cells were found in both cases to entirely cover the MAPLE coatings with which they strongly interact, as proved by the pseudopodia deeply infiltrating into the composite material. The difference in density proves that cells find a more friendly living medium on glasses with lower silica content. In addition, the corrosion characteristics of these glass-polymer composite coatings on titanium were investigated [135]. An unexpected self-arrangement of a double layer nanostructure after immersion in SBF consisting of an inner barrier (polymer) and an outer porous layer (bioapatite) was revealed with potential effects for osseointegration capacity of the Ti implants. The authors suppose that the immersion in SBF triggered an intense ion exchange process between coating and solution leading to the formation of a rapidly increasing bioapatite layer, which proved very efficient in protection against corrosion. The process evolved faster in case of nanostructured 6P57+PMMA coatings but a better

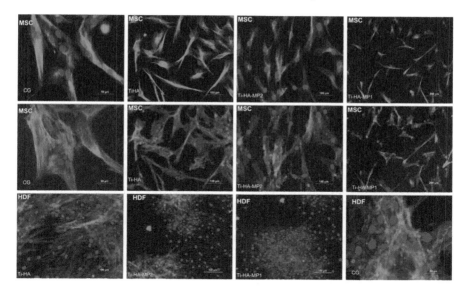

**Fig. 5.8** Cytoskeleton organization of mesenchymal stem cells and human dermal fibroblast cultured on different surfaces modified by MAPLE. Cells were grown for 24 h in direct contact with: Ti-HA, Ti-HA-MP1, or Ti-HA-MP2 (Ti-HA-MP1 and Ti-HA-MP2 corresond to solutions with 0.2 % and 1 % HA-MP powder); standard borosilicate cover glass (CG). Fixed cells were stained for actin (*red*), microtubules (*green*) and nuclei (*blue*) and analyzed by fluorescence microscopy; Reprinted with permission from J Mater Sci: Mater Med [132]

protection was reached for 6P61+PMMA coating when the corrosion was almost completely stopped.

MAPLE was also applied to obtain thin coatings of alendronate-hydroxyapatite composites [136] after nanocrystals' synthesis in aqueous medium with increasing bisphosphonate content (3.9, 7.1 % wt) [137]. For control, MAPLE was conducted with pure HA (0 % wt biphosphonate content) as well. The presence of alendronate in the MAPLE synthesized HA thin films had a positive effect on osteoblast viability and differentiation while inhibited osteoclast proliferation and differentiation, causing their apoptosis [136]. Similarly, a comparison was carried out between MAPLE deposited films of pure HA and silk fibroin mixed with HA thin structures for biomimetic implants [138]. The best results from physico-chemical and biological points of view were found for the composite HA-silk fibroin in comparison with MAPLE deposition of pure HA or fibroin films [138]. These were the first attempts to deposit HA by MAPLE technique to difference of PLD which is usually applied in this case [139]. We mention that in all cases the composite films presented superior mechanical and biological characteristics as compared to the films obtained by MAPLE from the respective pure materials.

### 5.5.3 Application to Inorganics

In view of reconstructing gradient inorganic layers or inorganic-organic multilayers by a single-experiment process, a significant attention was paid during the last 5 years to MAPLE application to inorganic compounds since the method was shown to produce nanoparticulate inorganic films [140, 141]. The single-experiment process of inorganic-organic multilayers could represent an actual advantage of MAPLE technique in respect with PLD which cannot be applied to sensitive organic polymers or proteins. Moreover, by using different solvents and deposition temperatures one can control the film assembling (growth and surface morphology) on substrate. When increasing the specific surface area a nanostructured or nanoparticulate film aspect (very often the MAPLE deposited coatings exhibit a high density of nano- and micron-sized droplets) could boost surface properties for specific applications such as sensors, drug delivery systems or biomimetic implants where a larger contact area is advantageous.

Following the application of MAPLE to organics, then to organic-inorganic composites, the method turned naturally to inorganics. Although the biological evaluation demonstrated that the composite films are presenting improved performances than pure inorganic coatings, as presented in Sect. 5.5.2, the opportunity to obtain inorganic-organic multilayers in a single-experiment process pushed the research toward experiments of MAPLE application to inorganics. Thus, MAPLE was employed to the deposition of calcium phosphates, in particular octocalcium phosphate (OCP) thin films, on titanium substrates [64] which is a challenge to OCP coatings fabricated by PLD [142]. It was demonstrated that the milder conditions of a MAPLE process ensured a higher degree of OCP crystallization with respect to PLD. This was in accordance with the presence of crystal fragments to the difference of OCP coatings deposited by PLD, which consisted of cauliflower-like aggregates and droplets only [64]. Next, Mg and Sr-doped OCP were deposited by MAPLE [64]. A remarkably uniform dopant distribution in films was evidenced. An enhanced cellular proliferation and differentiation on SrOCP and MgOCP in comparison with OCP films demonstrated that ion-doping improved the effect of OCP on bone cells, suggesting that such biomimetic coatings could be usefully applied for orthopedic use.

Nanoparticulate films obtained by MAPLE exhibited a good sensitivity in sensing, in particular, a good electrical response to acetone and ethanol vapors was evidenced for MAPLE deposited $TiO_2$ coatings starting from nanoparticle powders suspended in deionised water [140]. The good sensitivity was attributed to the nanoscale dimensions of the $TiO_2$ particles in the deposited films [141]. Recently, MAPLE was used to deposit single-wall carbon nanotubes functionalized with oxygen-containing groups without any alteration of the starting material [143].

## 5.6 Conclusion and Perspectives

During the last 15 years, MAPLE developed and improved fast and became a significant competitor to other deposition methods in the field of nanotechnology, in particular, for biomimetic applications such as drug delivery systems, biosensors or advanced implants. The method can ensure the control of film uniformity and thickness on either rough or flat substrates and permits the synthesis of coatings with increased adhesion.

MAPLE was initially proposed as an alternative to PLD for transferring and depositing thin organic materials. Extracellular matrix proteins, enzymes and polymers were transferred by MAPLE onto facing collectors and found to significantly improve their biological characteristics. It was possible to fabricate inorganic–organic composite coatings with improved bioactivity and controlled action in view of local release of some drugs to promote bone formation and prevent bone resorption. MAPLE turned recently unexpectedly to application of inorganic materials and transformed into a cryogenic PLD, a real competitor for PLD. Thus, MAPLE was for the first time applied to deposit inorganic coatings of ion-substituted OCP thin films. The obtained structures were found to support the growth and differentiation of osteoblast-like cells. In particular, an enhanced activity was demonstrated when cells were grown on ion-doped OCP coatings in comparison with pure OCP.

Because MAPLE was efficiently applied to either organic or inorganic material deposition it possesses actual advantages over other deposition techniques since most of them do not apply to both organic and inorganic or to composite materials. One single-experiment process of inorganic-organic multilayer deposition could be approachable as well by MAPLE technique. One can therefore foresee the potential use of appropriate masks in MAPLE experiments to manufacture well-controlled microsized, single or multilayer, organic-inorganic samples for advanced biomimetic applications.

**Acknowledgments** The authors acknowledge with thanks the financial support of UEFISCDI under the contracts PD 101/2012 and TE 82/2011 and of European Social Fund POSDRU 2007–2013 through the contract POSDRU/89/1.5/S/60746.

## References

1. Eason RW (2007) Pulsed laser deposition of thin films: applications-led growth of functional materials. Wiley, Hoboken
2. Piqué A, McGill RA, Chrisey DB, Leonhardt D, Mslna TE, Spargo BJ, Callahan JH, Vachet RW, Chung R, Bucaro MA (1999) Growth of organic thin films by the matrix assisted pulsed laser evaporation (MAPLE) technique. Thin Solid Films 355–356:536–541. doi:10.1016/s0257-8972(99)00376-x
3. Caricato AP, Luches A (2011) Applications of the matrix-assisted pulsed laser evaporation method for the deposition of organic, biological and nanoparticle thin films: a review. Appl Phys A Mater Sci Process 105(3):565–582. doi:10.1007/s00339-011-6600-0

4. Zhang L, Webster TJ (2009) Nanotechnology and nanomaterials: promises for improved tissue regeneration. Nano Today 4(1):66–80. doi:10.1016/j.nantod.2008.10.014
5. Porter JR, Ruckh TT, Popat KC (2009) Bone tissue engineering: a review in bone biomimetics and drug delivery strategies. Biotechnol Prog 25(6):1539–1560. doi:10.1002/btpr.246
6. Armentano I, Dottori M, Fortunati E, Mattioli S, Kenny JM (2010) Biodegradable polymer matrix nanocomposites for tissue engineering: a review. Polym Degrad Stab 95(11):2126–2146. doi:10.1016/j.polymdegradstab.2010.06.007
7. Censi R, Di Martino P, Vermonden T, Hennink WE (2012) Hydrogels for protein delivery in tissue engineering. J Controlled Release: Official J Controlled Release Soc 161(2):680–692. doi:10.1016/j.jconrel.2012.03.002
8. Gauvin R, Khademhosseini A (2011) Microscale technologies and modular approaches for tissue engineering: moving toward the fabrication of complex functional structures. ACS Nano 5(6):4258–4264. doi:10.1021/nn201826d
9. Geetha M, Singh AK, Asokamani R, Gogia AK (2009) Ti based biomaterials, the ultimate choice for orthopaedic implants–a review. Prog Mater Sci 54(3):397–425. doi:10.1016/j.pmatsci.2008.06.004
10. Shin H, Jo S, Mikos AG (2003) Biomimetic materials for tissue engineering. Biomaterials 24(24):4353–4364. doi:10.1016/s0142-9612(03)00339-9
11. Ma PX (2008) Biomimetic materials for tissue engineering. Adv Drug Deliv Rev 60(2):184–198. doi:10.1016/j.addr.2007.08.041
12. Bose S, Tarafder S (2012) Calcium phosphate ceramic systems in growth factor and drug delivery for bone tissue engineering: a review. Acta Biomater 8(4):1401–1421. doi:10.1016/j.actbio.2011.11.017
13. León B, Jansen JA (2009) Thin calcium phosphate coatings for medical implants. Springer, New York. Available via http://worldcat.org. http://public.eblib.com/EBLPublic/PublicView.do?ptiID=450789
14. Yan Y, Xiong Z, Hu Y, Wang S, Zhang R, Zhang C (2003) Layered manufacturing of tissue engineering scaffolds via multi-nozzle deposition. Mater Lett 57(18):2623–2628. doi:10.1016/s0167-577x(02)01339-3
15. Kokubo T, Kim H-M, Kawashita M (2003) Novel bioactive materials with different mechanical properties. Biomaterials 24(13):2161–2175. doi:10.1016/s0142-9612(03)00044-9
16. Davidenko N, Gibb T, Schuster C, Best SM, Campbell JJ, Watson CJ, Cameron RE (2012) Biomimetic collagen scaffolds with anisotropic pore architecture. Acta Biomater 8(2):667–676. doi:10.1016/j.actbio.2011.09.033
17. Ji C, Annabi N, Khademhosseini A, Dehghani F (2011) Fabrication of porous chitosan scaffolds for soft tissue engineering using dense gas CO2. Acta Biomater 7(4):1653–1664. doi:10.1016/j.actbio.2010.11.043
18. Yan L-P, Oliveira JM, Oliveira AL, Caridade SG, Mano JF, Reis RL (2012) Macro/microporous silk fibroin scaffolds with potential for articular cartilage and meniscus tissue engineering applications. Acta Biomater 8(1):289–301. doi:10.1016/j.actbio.2011.09.037
19. Seidlits SK, Drinnan CT, Petersen RR, Shear JB, Suggs LJ, Schmidt CE (2011) Fibronectin-hyaluronic acid composite hydrogels for three-dimensional endothelial cell culture. Acta Biomater 7(6):2401–2409. doi:10.1016/j.actbio.2011.03.024
20. Bigi A, Panzavolta S, Roveri N (1998) Hydroxyapatite-gelatin films: a structural and mechanical characterization. Biomaterials 19(7–9):739–744. doi:10.1016/s0142-9612(97)00194-4
21. Kim HW, Knowles JC, Kim HE (2005) Porous scaffolds of gelatin-hydroxyapatite nanocomposites obtained by biomimetic approach: characterization and antibiotic drug release. J Biomed Mater Res B Appl Biomater 74(2):686–698. doi:10.1002/jbm.b.30236
22. Strange DGT, Oyen ML (2011) Biomimetic bone-like composites fabricated through an automated alternate soaking process. Acta Biomater 7(10):3586–3594. doi:10.1016/j.actbio.2011.06.025
23. Zhou H, Lawrence JG, Bhaduri SB (2012) Fabrication aspects of PLA-CaP/PLGA-CaP composites for orthopedic applications: a review. Acta Biomater 8(6):1999–2016. doi:10.1016/j.actbio.2012.01.031

24. Jegal S-H, Park J-H, Kim J-H, Kim T-H, Shin US, Kim T-I, Kim H-W (2011) Functional composite nanofibers of poly(lactide-co-caprolactone) containing gelatin-apatite bone mimetic precipitate for bone regeneration. Acta Biomater 7(4):1609–1617. doi:10.1016/j.actbio.2010.12.003

25. Seidi A, Ramalingam M, Elloumi-Hannachi I, Ostrovidov S, Khademhosseini A (2011) Gradient biomaterials for soft-to-hard interface tissue engineering. Acta Biomater 7(4):1441–1451. doi:10.1016/j.actbio.2011.01.011

26. Stigter M, de Groot K, Layrolle P (2002) Incorporation of tobramycin into biomimetic hydroxyapatite coating on titanium. Biomaterials 23(20):4143–4153. doi:10.1016/s0142-9612(02)00157-6

27. Liu P, Smits J, Ayers DC, Song J (2011) Surface mineralization of Ti6Al4V substrates with calcium apatites for the retention and local delivery of recombinant human bone morphogenetic protein-2. Acta Biomater 7(9):3488–3495. doi:10.1016/j.actbio.2011.05.025

28. Garripelli VK, Kim J-K, Son S, Kim WJ, Repka MA, Jo S (2011) Matrix metalloproteinase-sensitive thermogelling polymer for bioresponsive local drug delivery. Acta Biomater 7(5):1984–1992. doi:10.1016/j.actbio.2011.02.005

29. Ariga K, Nakanishi T, Michinobu T (2006) Immobilization of biomaterials to nano-assembled films (self-assembled monolayers, Langmuir-Blodgett films, and layer-by-layer assemblies) and their related functions. J Nanosci Nanotechnol 6(8):2278–2301

30. Andre R, Tahir MN, Natalio F, Tremel W (2012) Bioinspired synthesis of multifunctional inorganic and bio-organic hybrid materials. Febs J 279(10):1737–1749

31. Park BW, Yoon DY, Kim DS (2010) Recent progress in bio-sensing techniques with encapsulated enzymes. Biosens Bioelectron 26(1):1–10

32. Frantz C, Stewart KM, Weaver VM (2010) The extracellular matrix at a glance. J Cell Sci 123(Pt 24):4195–4200

33. Alberts B (2010) Cell biology: the endless frontier. Mol Biol Cell 21(22):04–0334

34. Kaplan FS, Hayes WC, Keaveny TM, Boskey A, Einhorn TA , Iannotti JP(1994) In: Simon SR (ed) Orthopedic basic science. American Academy of Orthopaedic Surgeons, Rosemont, pp 127–185

35. Anselme K (2000) Osteoblast adhesion on biomaterials. Biomaterials 21(7):667–681

36. Anselme K, Davidson P, Popa AM, Giazzon M, Liley M, Ploux L (2010) The interaction of cells and bacteria with surfaces structured at the nanometre scale. Acta Biomater 6(10):3824–3846. doi:10.1016/j.actbio.2010.04.001

37. Aizenberg J, Fratzl P (2009) Biological and biomimetic materials. Adv Mater 21(4):387–388. doi:10.1002/adma.200803699

38. Vallet-Regí M, Ruiz-Hernández E (2011) Bioceramics: from bone regeneration to cancer nanomedicine. Adv Mater 23(44):5177–5218. doi:10.1002/adma.201101586

39. Lee JS, Suarez-Gonzalez D, Murphy WL (2011) Mineral coatings for temporally controlled delivery of multiple proteins. Adv Mater 23(37):4279–4284. doi:10.1002/adma.201100060

40. Wu PK, Ringeisen BR, Callahan J, Brooks M, Bubb DM, Wu HD, Piqué A, Spargo B, McGill RA, Chrisey DB (2001) The deposition, structure, pattern deposition, and activity of biomaterial thin-films by matrix-assisted pulsed-laser evaporation (MAPLE) and MAPLE direct write. Thin Solid Films 398–399:607–614. doi:10.1016/s0040-6090(01)01347-5

41. Li L, Mao C, Wang J, Xu X, Pan H, Deng Y, Gu X, Tang R (2011) Bio-inspired enamel repair via glu-directed assembly of apatite nanoparticles: an approach to biomaterials with optimal characteristics. Adv Mater 23(40):4695–4701. doi:10.1002/adma.201102773

42. Tampieri A, Sandri M, Landi E, Pressato D, Francioli S, Quarto R, Martin I (2008) Design of graded biomimetic osteochondral composite scaffolds. Biomaterials 29(26):3539–3546. doi:10.1016/j.biomaterials.2008.05.008

43. Martin PM (2009) Handbook of Deposition Technologies for Films and Coatings: Science, Applications and Technology. Elsevier, Oxford

44. Falconnet D, Csucs G, Michelle Grandin H, Textor M (2006) Surface engineering approaches to micropattern surfaces for cell-based assays. Biomaterials 27(16):3044–3063. http://dx.doi.org/10.1016/j.biomaterials.2005.12.024

45. Palacios M, Garcia O, Rodriguez-Hernandez J (2013) Constructing robust and functional micropatterns on polystyrene surfaces by using deep UV irradiation. Langmuir 11:11
46. Hench LLAO (1993) Bioactive glass coatings. In: Hench LL WJ (ed) An introduction to bioceramics, World Scientific, Singapore pp 239–259
47. Stan GE, Pasuk I, Husanu MA, Enculescu I, Pina S, Lemos AF, Tulyaganov DU, El Mabrouk K, Ferreira JM (2011) Highly adherent bioactive glass thin films synthetized by magnetron sputtering at low temperature. J Mater Sci Mater Med 22(12):2693–2710
48. Sima LE, Stan GE, Morosanu CO, Melinescu A, Ianculescu A, Melinte R, Neamtu J, Petrescu SM (2010) Differentiation of mesenchymal stem cells onto highly adherent radio frequency-sputtered carbonated hydroxylapatite thin films. J Biomed Mater Res Part A 95(4):1203–1214
49. LA Hong Z, Chen L, Chen X, Jing X (2009) Preparation of bioactive glass ceramic nanoparticles by combination of sol-gel and coprecipitation method. J Non-Cryst Solids 355(6):368–372
50. Boccaccini AR, Keim S, Ma R, Li Y, Zhitomirsky I (2010) Electrophoretic deposition of biomaterials. J Roy Soc Interface/Roy Soci 7(Suppl 5):S581–613. doi:10.1098/rsif.2010.0156. focus
51. Carlyle WC, McClain JB, Tzafriri AR, Bailey L, Zani BG, Markham PM, Stanley JRL, Edelman ER (2012) Enhanced drug delivery capabilities from stents coated with absorbable polymer and crystalline drug. J Controlled Release 162(3):561–567. doi:10.1016/j.jconrel.2012.07.004
52. Peter MBN, Soumya S, Nair SV, Furuike T, Tamura H, Jayakumar R (2010) Nanocomposite scaffolds of bioactive glass ceramic nanoparticles disseminated chitosan matrix for tissue engineering applications. Carbohydr Polym 79(2):284–289
53. Couto DS, Alves N, Mano JF (2009) Nanostructured multilayer coatings combining chitosan with bioactive glass nanoparticles. J Nanosci Nanotechnol 9:1741–1748
54. Thibault MM, Hoemann CD, Buschmann MD (2007) Fibronectin, vitronectin, and collagen I induce chemotaxis and haptotaxis of human and rabbit mesenchymal stem cells in a standardized transmembrane assay. Stem Cells Dev 16(3):489–502
55. Karagkiozaki V, Vavoulidis E, Karagiannidis PG, Gioti M, Fatouros DG, Vizirianakis IS, Logothetidis S (2012) Development of a nanoporous and multilayer drug-delivery platform for medical implants. Int J Nanomed 7:5327–5338
56. Fujiwara I, Ohnishi M, Seto J (1992) Atomic force microscopy study of protein-incorporating Langmuir-Blodgett films. Langmuir 8(9):2219–2222. doi:10.1021/la00045a025
57. Chen XN, Gu YX, Lee JH, Lee WY, Wang HJ (2012) Multifunctional surfaces with biomimetic nanofibres and drug-eluting micro-patterns for infection control and bone tissue formation. Eur Cells Mater 24:237–248
58. Doraiswamy A, Narayan RJ, Harris ML, Qadri SB, Modi R, Chrisey DB (2007) Laser microfabrication of hydroxyapatite-osteoblast-like cell composites. J Biomed Mater Res Part A 80(3):635–643
59. Guillemot F, Souquet A, Catros S, Guillotin B, Lopez J, Faucon M, Pippenger B, Bareille R, Remy M, Bellance S, Chabassier P, Fricain JC, Amedee J (2010) High-throughput laser printing of cells and biomaterials for tissue engineering. Acta Biomater 6(7):2494–2500
60. Colina M, Serra P, Fernandez-Pradas JM, Sevilla L, Morenza JL (2005) DNA deposition through laser induced forward transfer. Biosens Bioelectron 20(8):1638–1642
61. Dinca V, Ranella A, Farsari M, Kafetzopoulos D, Dinescu M, Popescu A, Fotakis C (2008) Quantification of the activity of biomolecules in microarrays obtained by direct laser transfer. Biomed Microdevices 10(5):719–725
62. Sima F, Davidson P, Pauthe E, Sima LE, Gallet O, Mihailescu IN, Anselme K (2011) Fibronectin layers by matrix-assisted pulsed laser evaporation from saline buffer-based cryogenic targets. Acta Biomater 7(10):3780–3788. doi:10.1016/j.actbio.2011.06.016
63. Sima F, Davidson P, Pauthe E, Gallet O, Anselme K, Mihailescu I (2011) Thin films of vitronectin transferred by MAPLE. Appl Phys A Mater Sci Process 105(3):611–617. doi:10.1007/s00339-011-6601-z

64. Boanini E, Torricelli P, Fini M, Sima F, Serban N, Mihailescu IN, Bigi A (2012) Magnesium and strontium doped octacalcium phosphate thin films by matrix assisted pulsed laser evaporation. J Inorg Biochem 107(1):65–72. doi:10.1016/j.jinorgbio.2011.11.003
65. Mironov V, Boland T, Trusk T, Forgacs G, Markwald RR (2003) Organ printing: computer-aided jet-based 3D tissue engineering. Trends Biotechnol 21(4):157–161. doi:10.1016/s0167-7799(03)00033-7
66. Yan Y, Wang X, Pan Y, Liu H, Cheng J, Xiong Z, Lin F, Wu R, Zhang R, Lu Q (2005) Fabrication of viable tissue-engineered constructs with 3D cell-assembly technique. Biomaterials 26(29):5864–5871. doi:10.1016/j.biomaterials.2005.02.027
67. Billiet T, Vandenhaute M, Schelfhout J, Van Vlierberghe S, Dubruel P (2012) A review of trends and limitations in hydrogel-rapid prototyping for tissue engineering. Biomaterials 33(26):6020–6041. doi:10.1016/j.biomaterials.2012.04.050
68. Tan W, Desai TA (2004) Layer-by-layer microfluidics for biomimetic three-dimensional structures. Biomaterials 25(7–8):1355–1364. doi:10.1016/j.biomaterials.2003.08.021
69. Sill TJ, von Recum HA (2008) Electrospinning: applications in drug delivery and tissue engineering. Biomaterials 29(13):1989–2006. doi:10.1016/j.biomaterials.2008.01.011
70. Townsend-Nicholson A, Jayasinghe SN (2006) Cell electrospinning: a unique biotechnique for encapsulating living organisms for generating active biological microthreads/scaffolds. Biomacromolecules 7(12):3364–3369. doi:10.1021/bm060649h
71. Gaebel R, Ma N, Liu J, Guan J, Koch L, Klopsch C, Gruene M, Toelk A, Wang W, Mark P, Wang F, Chichkov B, Li W, Steinhoff G (2011) Patterning human stem cells and endothelial cells with laser printing for cardiac regeneration. Biomaterials 32(35):9218–9230
72. Sima LE, Buruiana EC, Buruiana T, Matei A, Epurescu G, Zamfirescu M, Moldovan A, Petrescu SM, Dinescu M (2013) Dermal cells distribution on laser-structured ormosils. J Tissue Eng Regenerative Med. 7(2):129–138. doi:10.1002/term.507
73. Matei A, Zamfirescu M, Radu C, Dinescu M, Buruiana E, Buruiana T, Sima L, Petrescu S (2011) Laser processing of ormosils for tissue engineering applications. Appl Phys A Mater Sci Proces 104(3):821–827. doi:10.1007/s00339-011-6421-1
74. Califano V, Bloisi F, Vicari LRM, Colombi P, Bontempi E, Depero LE (2008) MAPLE deposition of biomaterial multilayers. Appl Surf Sci 254(22):7143–7148. doi:10.1016/j.apsusc.2008.05.295
75. Caricato AP, Cesaria M, Gigli G, Loiudice A, Luches A, Martino M, Resta V, Rizzo A, Taurino A (2012) Poly-(3-hexylthiophene)/6,6 -phenyl-C-61-butyric-acid-methyl-ester bilayer deposition by matrix-assisted pulsed laser evaporation for organic photovoltaic applications. Appl Phys Lett 100(7):073306-1-073306-4. doi:10.1063/1.3685702
76. Pique A (2011) The matrix-assisted pulsed laser evaporation (MAPLE) process: origins and future directions. Appl Phys A Mater Sci Process 105(3):517–528. doi:10.1007/s00339-011-6594-7
77. Jelinek M, Kocourek T, Remsa J, Cristescu R, Mihailescu IN, Chrisey DB (2007) MAPLE applications in studying organic thin films. Laser Phys 17(2):66–70. doi:10.1134/s1054660x0702003x
78. Cristescu R, Stamatin I, Mihaiescu DE, Ghica C, Albulescu M, Mihailescu IN, Chrisey DB (2004) Pulsed laser deposition of biocompatible polymers: a comparative study in case of pullulan. Thin Solid Films 453–454:262–268. doi:10.1016/j.tsf.2003.11.145
79. Cristescu R, Popescu C, Popescu AC, Socol G, Mihailescu I, Caraene G, Albulescu R, Buruiana T, Chrisey D (2012) Pulsed laser processing of functionalized polysaccharides for controlled release drug delivery systems. In: Vaseashta A, Braman E, Susmann P (eds) Technological innovations in sensing and detection of chemical, biological, radiological, nuclear threats and ecological terrorism. NATO science for peace and security series A: chemistry and biology. Springer, The Netherland, pp 231–236. doi:10.1007/978-94-007-2488-4_25
80. Guha S, Adil D, Ukah NB, Gupta RK, Ghosh K (2011) MAPLE-deposited polymer films for improved organic device performance. Appl Phys A Mater Sci Process 105(3):547–554. doi:10.1007/s00339-011-6596-5

81. Palla-Papavlu A, Dinca V, Dinescu M, Di Pietrantonio F, Cannata D, Benetti M, Verona E (2011) Matrix-assisted pulsed laser evaporation of chemoselective polymers. Appl Phys A Mater Sci Process 105(3):651–659. doi:10.1007/s00339-011-6624-5

82. Motoc MM, Axente E, Popescu C, Sima LE, Petrescu SM, Mihailescu IN, Gyorgy E (2013) Active protein and calcium hydroxyapatite bilayers grown by laser techniques for therapeutic applications. J Biomed Mater Res Part A 101(9):2706-2711. doi:10.1002/jbm.a.34572

83. Sima F, Axente E, Ristoscu C, Mihailescu IN, Kononenko TV, Nagovitsin IA, Chudinova G, Konov VI, Socol M, Enculescu I, Sima LE, Petrescu SM (2011) Tailoring immobilization of immunoglobulin by excimer laser for biosensor applications. J Biomed Mater Res Part A 96(2):384–394. doi:10.1002/jbm.a.32991

84. Purice A, Schou J, Kingshott P, Pryds N, Dinescu M (2007) Characterization of lysozyme films produced by matrix assisted pulsed laser evaporation (MAPLE). Appl Surf Sci 253(15):6451–6455. doi:10.1016/j.apsusc.2007.01.066

85. Shepard KB, Priestley RD (2013) MAPLE deposition of macromolecules. Macromol Chem Phys 214(8):862–872. doi:10.1002/macp.201200621

86. Guo Y, Morozov A, Schneider D, Chung JW, Zhang C, Waldmann M, Yao N, Fytas G, Arnold CB, Priestley RD (2012) Ultrastable nanostructured polymer glasses. Nat Mater 11(4):337–343. http://www.nature.com/nmat/journal/v11/n4/abs/nmat3234.html

87. Jelinek M, Kocourek T, Remsa J, Cristescu R, Mihailescu I, Chrisey D (2007) MAPLE applications in studying organic thin films. Laser Phys 17(2):66–70. doi:10.1134/s1054660x0702003x

88. Chrisey DB, Pique A, McGill RA, Horwitz JS, Ringeisen BR, Bubb DM, Wu PK (2003) Laser deposition of polymer and biomaterial films. Chem Rev 103(2):553–576. doi:10.1021/cr010428w

89. Leveugle E, Zhigilei LV (2007) Molecular dynamics simulation study of the ejection and transport of polymer molecules in matrix-assisted pulsed laser evaporation. J Appl Phys 102(7). doi:10.1063/1.2783898

90. Kokkinaki O, Georgiou S (2007) Laser ablation of cryogenic films: implications to matrix-assisted pulsed laser deposition of biopolymers and dedicated applications in nanotechnology. Digest J Nanomater Biostructures 2(2):221–241

91. Smausz T, Megyeri G, Kékesi R, Vass C, György E, Sima F, Mihailescu IN, Hopp B (2009) Comparative study on pulsed laser deposition and matrix assisted pulsed laser evaporation of urease thin films. Thin Solid Films 517(15):4299–4302. doi:10.1016/j.tsf.2008.11.141

92. György E, del Pino PA, Sauthier G, Figueras A (2009) Biomolecular papain thin films grown by matrix assisted and conventional pulsed laser deposition: a comparative study. J Appl Phys 106(11):114702. doi:10.1063/1.3266670

93. Sima F, Mutlu EC, Eroglu MS, Sima LE, Serban N, Ristoscu C, Petrescu SM, Oner ET, Mihailescu IN (2011) Levan nanostructured thin films by MAPLE assembling. Biomacromolecules 1(6):2251–2256. doi:10.1021/bm200340b

94. Canulescu S, Schou J, Fæster S, Hanse, KV, Conseil H (2013) Deposition of matrix-free fullerene films with improved morphology by matrix-assisted pulsed laser evaporation (MAPLE). Chem Phys Lett (in press). doi:10.1016/j.cplett.2013.09.047

95. Pervolaraki M, Sima F, Socol G, Teodorescu CM, Gheorghe NG, Socol M, Mihailescu IN, Moushi EE, Tasiopoulos AJ, Athanasopoulos GI, Viskadourakis Z, Giapintzakis J (2012) Matrix assisted pulsed laser evaporation of Mn12(Propionate) thin films. Appl Surf Sci 258(23):9471–9474. doi:10.1016/j.apsusc.2011.10.136

96. Mercado AL, Allmond CE, Hoekstra JG, Fitz-Gerald JM (2005) Pulsed laser deposition vs. matrix assisted pulsed laser evaporation for growth of biodegradable polymer thin films. Appl Phys A Mater Sci Process 81(3):591–599. doi:10.1007/s00339-004-2994-2

97. Pauthe E, Pelta J, Patel S, Lairez D, Goubard F (2002) Temperature-induced $\beta$-aggregation of fibronectin in aqueous solution. Biochim Biophys Acta (BBA) Protein Struct Mol Enzymol 1597(1):12–21. doi:10.1016/s0167-4838(02)00271-6

98. Salinovich O, Montelaro RC (1986) Reversible staining and peptide mapping of proteins transferred to nitrocellulose after separation by sodium dodecylsulfate-polyacrylamide gel electrophoresis. Anal Biochem 156(2):341–347. doi:10.1016/0003-2697(86)90263-0

99. Smith PK, Krohn RI, Hermanson GT, Mallia AK, Gartner FH, Provenzano MD, Fujimoto EK, Goeke NM, Olson BJ, Klenk DC (1985) Measurement of protein using bicinchoninic acid. Anal Biochem 150(1):76–85. doi:10.1016/0003-2697(85)90442-7

100. Kecskemeti G, Smausz T, Kresz N, Tóth Z, Hopp B, Chrisey D, Berkesi O (2006) Pulsed laser deposition of polyhydroxybutyrate biodegradable polymer thin films using ArF excimer laser. Appl Surf Sci 253(3):1185–1189. doi:10.1016/j.apsusc.2006.01.084

101. Bubb DM, Toftmann B, Haglund RF, Horwitz JS, Papantonakis MR, McGill RA, Wu PW, Chrisey DB (2002) Resonant infrared pulsed laser deposition of thin biodegradable polymer films. Appl Phys A Mater Sci Process 74(1):123–125. doi:10.1007/s003390101010

102. Suske E, Scharf T, Schaaf P, Panchenko E, Nelke D, Buback M, Kijewski H, Krebs HU (2004) Variation of the mechanical properties of pulsed laser deposited PMMA films during annealing. Appl Phys A Mater Sci Process 79(4–6):1295–1297. doi:10.1007/s00339-004-2754-3

103. Cristescu R, Socol G, Mihailescu IN, Popescu M, Sava F, Ion E, Morosanu CO, Stamatin I (2003) New results in pulsed laser deposition of poly-methyl-methacrylate thin films. Appl Surf Sci 208–209:645–650. doi:10.1016/s0169-4332(02)01415-0

104. Jelinek M, Cristescu R, Kocourek T, Vorlicek V, Remsa J, Stamatin L, Mihaiescu D, Stamatin I, Mihailescu IN, Chrisey DB (2007) Thin films growth parameters in MAPLE-application to fibrinogen. In: Hess WP, Herman PR, Bauerle D, Koinuma H (eds) Cola'05: 8th international conference on laser ablation. J Phys Conf Ser, vol 59. Iop Publishing Ltd, Bristol, pp 22–27. doi:10.1088/1742-6596/59/1/005

105. Wu PK, Ringeisen BR, Callahan J, Brooks M, Bubb DM, Wu HD, Pique A, Spargo B, McGill RA, Chrisey DB (2001) The deposition, structure, pattern deposition, and activity of biomaterial thin-films by matrix-assisted pulsed-laser evaporation (MAPLE) and MAPLE direct write. Thin Solid Films 398:607–614. doi:10.1016/s0040-6090(01)01347-5

106. Paun IA, Moldovan A, Luculescu CR, Dinescu M (2011) Biocompatible polymeric implants for controlled drug delivery produced by MAPLE. Appl Surf Sci 257(24):10780–10788. doi:10.1016/j.apsusc.2011.07.097

107. Paun IA, Ion V, Moldovan A, Dinescu M (2012) MAPLE deposition of PEG:PLGA thin films. Appl Phys A Mater Sci Process 106(1):197–205. doi:10.1007/s00339-011-6548-0

108. Califano V, Bloisi F, Vicari LR, Bretcanu O, Boccaccini AR (2008) Matrix-assisted pulsed laser evaporation of poly(D, L-lactide) for biomedical applications: effect of near infrared radiation. J Biomed Opt 13(1):014028

109. Meyers MA, Chen P-Y, Lin AY-M, Seki Y (2008) Biological materials: structure and mechanical properties. Prog Mater Sci 53(1):1–206. doi:10.1016/j.pmatsci.2007.05.002

110. Donot F, Fontana A, Baccou JC, Schorr-Galindo S (2012) Microbial exopolysaccharides: main examples of synthesis, excretion, genetics and extraction. Carbohydr Polym 87:951–962

111. Kang SA, Jang K-H, Seo J-W, Kim KH, Kim YH, Rairakhwada D, Seo MY, Lee JO, Ha SD, Kim C-H, Rhee S-K (2009) Levan: applications and perspectives. In: BHA R (ed) Microbial production of biopolymers and polymer precursors. Caister Academic Press

112. Liu JLJ, Ye H, Sun Y, Lu Z, Zeng X (2010) In vitro and in vivo antioxidant activity of exopolysaccharides from endophytic bacterium Paenibacillus polymyxa EJS-3. Carbohydr Polym 82:1278–1283

113. Esawy MA, Ahmed EF, Helmy WA, Mansour NM, El-Senousy WM, El-Safty MM (2011) Production of levansucrase from novel honey Bacillus subtilis isolates capable of producing antiviral levans. Carbohydr Polym 86(2):823–830. doi:10.1016/j.carbpol.2011.05.035

114. Ratner BD, Bryant SJ (2004) BIOMATERIALS: where we have been and where we are going. Annu Rev Biomed Eng 6(1):41–75. doi:10.1146/annurev.bioeng.6.040803.140027

115. Ringeisen BR, Callahan J, Wu PK, Pique A, Spargo B, McGill RA, Bucaro M, Kim H, Bubb DM, Chrisey DB (2001) Novel laser-based deposition of active protein thin films. Langmuir 17(11):3472–3479. doi:10.1021/la0016874

116. Gyorgy E, Sima F, Mihailescu IN, Smausz T, Megyeri G, Kekesi R, Hopp B, Zdrentu L, Petrescu SM (2009) Immobilization of urease by laser techniques: synthesis and application to urea biosensors. J Biomed Mater Res Part A 89(1):186–191. doi:10.1002/jbm.a.31963

117. Pier GB, Lyczak JB, Wetzler LM (2004) Immunology, Infection, and Immunity. ASM Press, Washington, D.C
118. Kasemo B, Gold J (1999) Implant surfaces and interface processes. Adv Dent Res 13:8–20
119. Roy DC, Hocking DC (2012) Recombinant fibronectin matrix mimetics specify integrin adhesion and extracellular matrix assembly. Tissue Eng Part A 1:1
120. Akiyama SK (1996) Integrins in cell adhesion and signaling. Hum Cell 9(3):181–186
121. Jimbo R, Sawase T, Shibata Y, Hirata K, Hishikawa Y, Tanaka Y, Bessho K, Ikeda T, Atsuta M (2007) Enhanced osseointegration by the chemotactic activity of plasma fibronectin for cellular fibronectin positive cells. Biomaterials 28(24):3469–3477
122. Dalton BA, McFarland CD, Underwood PA, Steele JG (1995) Role of the heparin binding domain of fibronectin in attachment and spreading of human bone-derived cells. J Cell Sci 108(Pt 5):2083–2092
123. Hristova K, Pecheva E, Pramatarova L, Altankov G (2011) Improved interaction of osteoblast-like cells with apatite-nanodiamond coatings depends on fibronectin. J Mater Sci Mater Med 22(8):1891–1900
124. Pendegrass CJ, El-Husseiny M, Blunn GW (2012) The development of fibronectin-functionalised hydroxyapatite coatings to improve dermal fibroblast attachment in vitro. J Bone Joint Surg Br 94(4):564–569
125. Wang HG, Yin TY, Ge SP, Zhang Q, Dong QL, Lei DX, Sun DM, Wang GX (2012) Biofunctionalization of titanium surface with multilayer films modified by heparin-VEGF-fibronectin complex to improve endothelial cell proliferation and blood compatibility. J Biomed Mater Res Part A 3(10):34339
126. Felding-Habermann B, Cheresh DA (1993) Vitronectin and its receptors. Curr Opin Cell Biol 5(5):864–868. doi:10.1016/0955-0674(93)90036-p
127. Cacchioli A, Ravanetti F, Bagno A, Dettin M, Gabbi C (2009) Human vitronectin-derived peptide covalently grafted onto titanium surface improves osteogenic activity: a pilot in vivo study on rabbits. Tissue Eng Part A 15(10):2917–2926
128. Cristescu R, Mihaiescu D, Socol G, Stamatin I, Mihailescu IN, Chrisey DB (2004) Deposition of biopolymer thin films by matrix assisted pulsed laser evaporation. Appl Phys A Mater Sci Process 79(4–6):1023–1026. doi:10.1007/s00339-004-2619-9
129. Pellenc D, Berry H, Gallet O (2006) Adsorption-induced fibronectin aggregation and fibrillogenesis. J Colloid Interface Sci 298(1):132–144
130. Salmeron-Sanchez M, Rico P, Moratal D, Lee TT, Schwarzbauer JE, Garcia AJ (2011) Role of material-driven fibronectin fibrillogenesis in cell differentiation. Biomaterials 32(8):2099–2105
131. Stevens MM, George JH (2005) Exploring and engineering the cell surface interface. Science 310(5751):1135–1138
132. Negroiu G, Piticescu RM, Chitanu GC, Mihailescu IN, Zdrentu L, Miroiu M (2008) Biocompatibility evaluation of a novel hydroxyapatite-polymer coating for medical implants (in vitro tests). J Mater Sci Mater Med 19(4):1537–1544. doi:10.1007/s10856-007-3300-6
133. Sima LE, Filimon A, Piticescu RM, Chitanu GC, Suflet DM, Miroiu M, Socol G, Mihailescu IN, Neamtu J, Negroiu G (2009) Specific biofunctional performances of the hydroxyapatite-sodium maleate copolymer hybrid coating nanostructures evaluated by in vitro studies. J Mater Sci Mater Med 20:20
134. Sima F, Ristoscu C, Popescu A, Mihailescu IN, Kononenko T, Simon S, Radu T, Ponta O, Mustata R, Sima LE, Petrescu SM (2009) Bioglass -polymer thin coatings obtained by MAPLE for a new generation of implants. J Optoelectron Adv Mater 11(9):1170–1174
135. Floroian L, Sima F, Florescu M, Badea M, Popescu AC, Serban N, Mihailescu IN (2010) Double layered nanostructured composite coatings with bioactive silicate glass and poly-methylmetacrylate for biomimetic implant applications. J Electroanal Chem 648(2):111–118. doi:10.1016/j.jelechem.2010.08.005
136. Bigi A, Boanini E, Capuccini C, Fini M, Mihailescu IN, Ristoscu C, Sima F, Torricelli P (2009) Biofunctional alendronate-hydroxyapatite thin films deposited by matrix assisted pulsed laser evaporation. Biomaterials 30(31):6168–6177. doi:10.1016/j.biomaterials.2009.07.066

137. Boanini E, Torricelli P, Gazzano M, Giardino R, Bigi A (2008) Alendronate-hydroxyapatite nanocomposites and their interaction with osteoclasts and osteoblast-like cells. Biomaterials 29(7):790–796. doi:10.1016/j.biomaterials.2007.10.040

138. Miroiu FM, Socol G, Visan A, Stefan N, Craciun D, Craciun V, Dorcioman G, Mihailescu IN, Sima LE, Petrescu SM, Andronie A, Stamatin I, Moga S, Ducu C (2010) Composite biocompatible hydroxyapatite-silk fibroin coatings for medical implants obtained by matrix assisted pulsed laser evaporation. Mater Sci Eng B 169(1–3):151–158. doi:10.1016/j.mseb.2009.10.004

139. Sima F, Ristoscu C, Caiteanu D, Mihailescu CN, Stefan N, Mihailescu IN, Prodan G, Ciupina V, Palcevskis E, Krastins J, Sima LE, Petrescu SM (2011) Biocompatibility and bioactivity enhancement of Ce stabilized ZrO(2) doped HA coatings by controlled porosity change of Al(2) O(3) substrates. J Biomed Mater Res Part B Appl Biomater 96(2):218–224. doi:10.1002/jbm.b.31755

140. Caricato AP, Manera MG, Martino M, Rella R, Romano F, Spadavecchia J, Tunno T, Valerini D (2007) Uniform thin films of TiO2 nanoparticles deposited by matrix-assisted pulsed laser evaporation. Appl Surf Sci 253(15):6471–6475. doi:10.1016/j.apsusc.2007.01.113

141. Caricato AP, Luches A, Rella R (2009) Nanoparticle thin films for gas sensors prepared by matrix mssisted pulsed laser evaporation. Sensors 9(4):2682–2696. doi:10.3390/s90402682

142. Socol G, Torricelli P, Bracci B, Iliescu M, Miroiu F, Bigi A, Werckmann J, Mihailescu IN (2004) Biocompatible nanocrystalline octacalcium phosphate thin films obtained by pulsed laser deposition. Biomaterials 25(13):2539–2545. doi:10.1016/j.biomaterials.2003.09.044

143. del Pino PA, György E, Cabana L, Ballesteros B, Tobias G (2012) Deposition of functionalized single wall carbon nanotubes through matrix assisted pulsed laser evaporation. Carbon 50(12):4450–4458. doi:10.1016/j.carbon.2012.05.023

# Chapter 6
# Laser Additive Manufacturing of Metals

**Claus Emmelmann, Jannis Kranz, Dirk Herzog and Eric Wycisk**

**Abstract**  Laser Additive Manufacturing (LAM) is based on a repeating layer wise manufacturing process which uses a laser beam to melt and solidify material in a powder-bed according to slices of a corresponding three dimensional computer aided design (3D-CAD) model. The stepwise production causes a reduction of complex three-dimensional (3D) geometries into simpler two-dimensional (2D) manufacturing steps [1, 2]. Thus LAM offers great potential in design and production due to its high freedom of geometry and flexibility. The basic manufacturing process is tool less. Its preferred field of application is the one-step manufacturing of complex geometries in low lot sizes, ideally with lot size 1 where conventional machining would require a longer overall production time due to a high number of processing steps [3]. This chapter gives an overview over the process basics, its parameters and the major influences on the quality of parts manufactured by LAM. It ends with design guidelines, current applications of LAM and future developments.

## 6.1 Process Basics

Laser Additive Manufacturing (LAM) embraces all laser based additive manufacturing processes. Processable materials include polymers, ceramics and metals. During the recent years especially LAM of metal alloy has gained attention for the

J. Kranz (✉) · C. Emmelmann
TUHH Technische Universität Hamburg-Harburg, 17 (L) Denickestr,
D-21073 Hamburg, Germany
e-mail: jannis.kranz@tuhh.de

J. Kranz · C. Emmelmann
iLAS Institut für Laser-und Anlagensystemtechnik, Hamburg, Germany

D. Herzog · E. Wycisk · C. Emmelmann
LZN Laser Zentrum Nord GmbH, Am Schleusengraben 14, D-21029 Hamburg, Germany
e-mail: dirk.herzog@lzn-hamburg.de
e-mail: eric.wycisk@lzn-hamburg.de

V. Schmidt and M. R. Belegratis (eds.), *Laser Technology in Biomimetics*,
Biological and Medical Physics, Biomedical Engineering,
DOI: 10.1007/978-3-642-41341-4_6, © Springer-Verlag Berlin Heidelberg 2013

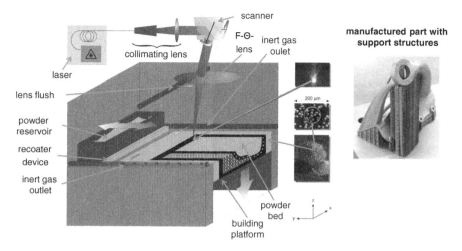

**Fig. 6.1** *Left* laser additive manufacturing: machine setup; *Right* part with support structures

manufacturing of geometrically highly complex functional parts. LAM of metals is an iterative, layer wise and powder bed based manufacturing process (Fig. 6.1). Depending on the machine other designations are Laser-Cusing, Laser-Sintering or Selective Laser Melting for example. The process basis is a 3D-CAD-Model of the part to be manufactured. The first steps of data preparation include the 3D-CAD file conversion according to the machine requirements and the part orientation in the building chamber. Manufacturing multiple parts with different shapes in the building chamber is possible [1, 2].

Even though the geometrical freedom of LAM is very high compared to conventional processes, there is one mayor manufacturing restriction that must be obeyed when designing and preparing parts for manufacturing. If a part would be built directly in the powder bed with no connection to the building platform, two major problems would occur:

Firstly the powder beds poor carrying capabilities can lead to part imperfections at overhangs because of a melt flow in the underlying unexposed powder layer. Secondly temperature gradients between the melting pool and the surrounding, already melted and solidified part sections can lead to different shrinkages and thus internal stresses during the cooling process. Possible outcomes are a reduction of form stability during the cooling process, a distortion of the part and / or the delamination of single layers. The latter can lead to a collision between the recoater device, see Fig. 6.1, and the parts topmost layer leading to a misalignment of the part.

Adequately placed support structures consisting of fine lattice structures connected with the building platform can help minimizing these effects, see Fig. 6.1. Based on the 3D-CAD data the automatic addition of support structures to the part geometry at sections with overhanging structures can be initiated [1, 2, 4, 5].

That followed the part and support structures will be separated in the process specific layers which is designated as slicing-process. Hence a complex three-dimensional problem is reduced into simpler two-dimensional manufacturing steps. After defining the process parameters the actual manufacturing process starts. Manufacturing the part is performed on a building platform which is connected to a lifting table guaranteeing the parts fixation. The first step of the manufacturing process is lowering the lifting table by one layer thickness and the application of a powder layer by the recoater device, see Fig. 6.1. Typical layer thicknesses vary between 30 and 60 $\mu$m, depending on material and machine setup. The applied powder is then locally exposed and completely melted by the laser beam according to the slicing data. Usually Ytterbium-fibre-lasers with output-powers between 200 and 400 W (state of the art) and beam spots of 100–500 $\mu$m are in use. Typical building rates depend on the material as well as the machine setup and differ between 2 and 20 cm$^3$h. Usually, single component metal are being used [6–9]. The solidification occurs by heat transfer. By applying an inert gas to the process chamber the molten metal's oxidation will be minimized. Moreover, the chamber heating reduces the temperature gradient during solidification thus leading to a reduction of internal stresses [1, 5, 10].

Repeating these steps leads to a layer wise melting, solidification and connection of the basic powder material by a defined energy insertion. The iterative process ends with the final layer's exposure. Thereupon the not molten powder and the building platform with the part will be removed from the machine. Subsequently the not molten powder needs to be sieved in order to remove process contaminants like spilling from the exposure process. After the parts separation from the building platform and removing the support structures, an additional end machining can be necessary in order to obtain sufficient surface qualities or to realize exact holes and threads. Especially at filigree parts special attention to the support structure dimensioning must be given in order to guarantee a damage free separation from the part itself [1, 2, 4, 5, 11].

Beside the ability of manufacturing highly complex parts, the high recycling rate of additive manufacturing leads to a reduction of material waste during machining, compared to conventional subtractive manufacturing processes like shape cutting. Furthermore, the raw material cost savings, handling smaller billets of raw feedstock, by-products or scrap produced by machining due to netshape manufacturing is easier, Additionally, this toolless manufacturing process benefits from significantly reduced tool costs especially when processing highly abrasive materials such as titanium.

On first sight, LAM is predestined for the production of highly complex parts in small or single lot sizes, due to the described tool-less and layer wise production process [2]. However, considering restrictions and other additional benefits of the LAM process the assessment for an economical production with LAM is more complex. Possible limitations for economic LAM production and geometrical possibilities are among others high machine hours, high powder costs compared to wrought material, the need for support structures, the limitations of the building chamber, limited surface qualities due to powder adhesion and potential need for finishing (see Sect. 6.3). However, major benefits from new engineering possibilities arise: LAM allows converting structurally optimized geometries and bionic structures without adaptation into the part design which results in new lightweight constructions. Additionally, functions can be integrated, which could only be realized by multiple parts in conventional designs. Due to the freedom of design LAM facilitates the combination of parts conventionally designed in assemblies into one single part or complex new product features like internal cooling ducts. Commercially used LAM machines show a huge diversity, even though they are all based on the same physical concept. Varying specifications include laser type and power, building chamber and powder handling concepts as well as machine specific digital process chains. An overview of the available machines for LAM, their denotations and specifications can be seen in Gebhardt et al. [1, 2, 12, 13].

## 6.2 Process Parameters

Laser Additive Manufacturing depends on manifold process parameters that must be matched with each other depending on the material and machine type to be used in order to obtain parts with a density comparable to conventional materials. In his studies Rehme [3] was able to identify 157 single parameters influencing the LAM process. The essential ones shall be shortly described in the following. The parameters and procedures may slightly vary between different machine types. Thus they are described in a generic way in order to show the basic coherences according to Rehme et al. [1–5, 14]. An overview over future developments incorporating the aspects described in the following is given in Sect. 6.5.

### 6.2.1 Building Chamber Dimensions

Currently available machines for powder bed LAM based metal parts are limited to a maximum available part size of approximate $250 \times 250 \times 300\,mm^3$. Dimensions can slightly differ depending on the machine concept. Current machine generations limitation is primarily the optical concept. Future machine setups including new optical concepts show a trend towards larger building chambers [1, 2, 7, 9, 15].

### 6.2.2 Layer Thickness

The layer thickness ($T_L$) significantly determines the geometry's resolution of the part and has an impact on the manufacturing time as well as on the realizable surface qualities. The layer wise production process leads to a staircase effect at angled surfaces due to a limited geometry resolution (Fig. 6.2). Thus only an approximation of the ideal part contour is possible. Compared to high layer thicknesses low thicknesses facilitate a higher geometry resolution, but lead to an increase of manufacturing time

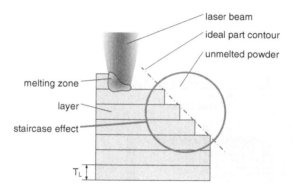

**Fig. 6.2** Layer composition by laser additive manufacturing, simplified

by an increase of the overall layers to be build and to be exposed. Typical layer thicknesses vary between 20 and 60 μm and depend on the material specific powder grain size distribution. Moreover the machines recoater concept also influences the suitable layer thickness. [1, 5, 6].

### 6.2.3 Scanspeed

The powder melting is correlated to the material-specific amount of energy deposited in the melting zone, which is described by the energy per length $E_L$ or energy density

$$E_L = \frac{P_L}{v_s} \tag{6.1}$$

The adjustable parameters are the laser power $P_L$ and the scan speed $v_s$. Since $E_L$ depends on the melting temperature, the absorptivity and the materials thermal conductivity, laser power and scan speed can be tuned for a best possible melting of the material. A high energy density at preferably high scan speeds is desired in order to minimize the manufacturing time. Therefore, the applied laser power must be maximized without vaporizing the material in order to avoid spillings and degradation of the part properties due to erroneous layers [5].

### 6.2.4 Laser Beam Power

It is required to apply a high laser beam power to obtain high part density and thus optimum part properties. However, the material specific maximal applicable power at a constant spot size and scan speed is limited by material evaporation. This leads to spattering powder particles, reduced density and increased surface roughness due to

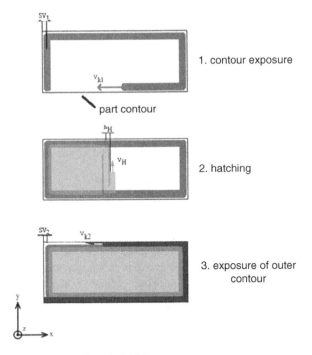

1. contour exposure

part contour

2. hatching

3. exposure of outer
   contour

**Fig. 6.3** Typical exposure procedures in LAM

powder particles adhering to part's surfaces. On one hand this constricts the smallest possible size of features that can be fabricated. On the other hand neglecting the changed melt pool size will result in inaccurate part dimensions since the laser beam focus diameter typically increases with rising power yielding to a melt pool growth as well as an increase of thermal induced powder adhesion [3, 5].

## 6.2.5 Exposure and Scan Strategy

As described in Sects. 6.1 and 6.2 the process basis is the exposure of single layers that assemble the part to be manufactured. A single layer's exposure can be separated into three single steps, see Fig. 6.3. The first step comprises the exposure of the parts contour. In this regard the spot shift plays an important role for an exact layer build up.

At the first contour exposure the spot shift $S_V$ conducts slightly more than half of a melted scan line width. Its purpose is the contour's enlargement compensation due to the laser spot shape by slightly moving the scanlines inwards of the area to be exposed [5, 16]. During the second step the inner surface will be exposed and filled by multiple scan lines, also called hatching. The third step is a final contour exposure with an exact spot shift of the half spot size in order to assure exact part geometries [16].

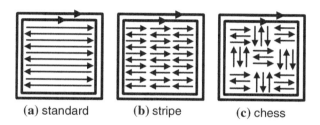

(a) standard          (b) stripe          (c) chess

**Fig. 6.4**  Typical hatching variants in laser additive manufacturing

## 6.2.6 Hatching

There are three common scan strategies for the inner area's exposure, also called hatching, see Figs. 6.3 and 6.4. The standard strategy fills the whole inner area of a slice by line scanning the laser focus in x- or y-direction, beginning in one corner.

The stripe strategy divides the area to be exposed in stripes, beginning in one corner as well. The chess strategy divides the areas to be exposed in singles squares. The squares exposure and the exposures direction can be of a defined or stochastic nature. The stripe and chess strategies are recommended for large areas due to reduced thermal induced stresses compared to the classical strategy. The hatch distance ($h_H$) is the separation of adjacent lines, which is crucial for an optimal connection of every single line and a fully exposed part layer. A correct setting of the hatch distance is essential: Both, a too large and a too small hatch distance with respect to the laser spot size lead to a degradation of the final part.

A hatch distance equal or larger than the laser's spot size results in an insufficient overlap of adjacent lines. This leaves unexposed powder particles between the scan tracks and hence a degradation of the overall density respective part strength. On the other hand, the same results can be seen with hatching distances that are too small. A too narrow hatch distance draws powder from adjacent unmelted areas into the melt pool. Consequently, the next scan track suffers from too little unmelted powder, which in turn will reduce the part density [3]. Further details are given in Meiners et al. [3, 5].

## 6.2.7 Available Metals and Alloys for LAM

Currently available materials for LAM of metals include aluminum, titanium and steel alloys. They are being processed as a single component powder with varying grain size compositions depending on the material and supplier. Generally speaking, weldable materials can be processed by LAM. The list of available materials is steadily increasing and usually provided by the machine manufacturers [1–3]. Material properties achieved by LAM, especially its strength, are comparable to conventional manufacturing processes [17–19]. Past applications of LAM were focused on the fields of tooling and the medical implants thus leading to a broad field of available steel alloys as well as biocompatible titanium alloys. The current trends towards lightweight applications lead to the exploitation of further lightweight alloys like titanium alloys or aluminum scandium alloys.

## 6.3 Part Quality

In order to be suitable for functional parts in serial production LAM-parts must fulfill certain quality criteria when being compared with conventionally manufactured counterparts. According to Rehme [3] the major factors are the parts "density in terms of pore-free microstructure, strength (ultimate stress, yield stress, Young's modulus, break strain), hardness as a value for resistance to wear, residual stresses (leading to warping or cracking of parts during the process or after), the accuracy to shape and size as well as the surface quality". This section introduces the main factors influencing the quality of parts manufactured by LAM.

### 6.3.1 Density

The bulk density of solid LAM parts should be the most relevant issue of all initial examinations for new materials since other part quality properties such as high strength and hardness require low porosity. Ideally, the density should always be near the theoretical maximum, i.e. it should be above 99% of the value obtained for rolled material [3]. As previously described the density can be influenced by parameters such as the scan speed, the laser power as well as the exposure type and its settings. Moreover, the density is influenced by the powders grain-size fraction. The smaller the grain-size, the more consistently is the melt-pool, which leads to a higher density [5].

### 6.3.2 Strength

Parts manufactured by LAM exhibit a structural orthotropy (=different material properties or strengths in different orthogonal directions) due to the layer wise manufacturing [1–3, 17, 20]. Concerning the yield strength and tensile strength a maximum at a part orientation of 0° and a minimum at 90° with respect to the building plane can be observed, compare Fig. 6.1. Their maximum values are comparable to the ones of wrought material. The extreme values of the breaking elongation act analogous to the strengths in regard to the orientation. Moreover, the Young's modulus shows no dependency in orientation and is comparable to conventional counterparts [1, 3]. Up to now, analyses on the fatigue strength are barely available due to the fact that the process was mainly used in applications with static loadings [1, 20].

## 6.3.3 Hardness

The hardness of a material highly depends on the density as it is defined as the resistance against indentation, which is in fact another strength property. Moreover, hardness is often found to follow similar behavior as strength properties. Typically, hardness values above those of rolled materials are obtained due to the extremely short solidification times of local material areas in the LAM process [3]. Studies by Santos investigated the Vickers hardness of LAM titanium samples [21]. The variation of the laser beam power and scan speeds showed a hardness increase with a rising energy density in the melting zone (cf. 6.1).

## 6.3.4 Residual Stresses

Residual stresses are defined as mechanical stresses inside a closed system despite the absence of outer forces. Two major reasons can be identified for the existence of such stresses in parts manufactured by LAM in particular [3–5]. The first reason inducing stress is a volume change due to a variation in the materials microstructure during solidification after melting the powder. The second reason is the influence of shrinkage at the just exposed and melted layer adverse the already solidified layer, which leads to residual stresses acting perpendicularly to the building direction that can lead to the parts deformation. In general the building platform, respectively the already solidified layers, countervails a free formation of the parts deformation. At overhangs the latter is not the case, which leads to the necessity for additional support structures preventing the part's deformation [3]. The deformations and residual stresses' intensity strongly depends on the material and process properties.

## 6.3.5 Accuracy Grade

Both, the dimensional accuracy and the accuracy of shape have to be regarded concerning the accuracy grade of LAM parts [5]. Both are primarily influenced by thermal induced residual stresses as well as the machine setup. Residual stresses can lead to inaccuracies due to part warping either during the process or after the process when the occurring thermal stresses are relieved [3]. Depending on the machine setup zero point deviations of the scanner as well as a tolerance in the lifting table can lead to form deviations. Furthermore the positioning accuracy of the lifting table as well as the abrasive wear of the recoater device influence the accuracy in building direction [5]. Due to the residual stresses and processes complex nature the realizable tolerances strongly depend on the parts geometry and the chosen material.

### 6.3.6 Subsurface Quality

The surface quality usually is being defined by the factor $R_a$ [5], which constitutes the arithmetic average of the absolute values regarding the surface texture. It shows a clear dependency on the part orientation, the powder composition and the chosen layer thickness [5, 22]. The best surface qualities can be achieved perpendicular to the working plane and in the working plane [3, 5]. Due to the layer wise manufacturing process part slopes exhibit a stair case effect, see Fig. 6.2, resulting in poor surface qualities. On the other hand the roughness of surfaces perpendicular and parallel to the working plane depends on the contour exposure. Moreover powder particles adhered on the part during the melting process can lower the surface quality additionally [3, 5]. Overall, a rising surface roughness from the horizontal plane to an angle of 90° to is the result. Detailed studies on the influencing parameters and dependencies of the surface quality were done by Meiners [5]. Depending on the machine type and material typical surface roughnesses $R_z$ vary between 40 and 90 $\mu$m [6–8]. Requiring high surface qualities usually causes the necessity of a conventional final machining [1–3, 23].

## 6.4 Designs for LAM

Due to the processes reduction of complex three-dimensional topologies into simplified two-dimensional manufacturing steps, compare Sect. 6.1, LAM facilitates new impulses for the design of functional parts by having the potential of manufacturing topologies that are impossible to manufacture by conventional processes, compare Sect. 6.6. This geometrical freedom enables the designer to realize complex hollow structures, overhangs and lattice structures. Albeit the potential for manufacturing complex geometries by LAM is currently essentially not exploited in practice as the knowledge of its capabilities, especially its design-freedom, are still limited to experts. Available design guidelines are currently fragmentary at most. Available design rules give recommendations regarding the basis process inherent restrictions influences, part design as well as basics for post processing. Process inherent restrictions include available building chamber dimensions influencing the parts allowable dimensions, the surface qualities dependence on the parts orientation versus the building platform as well as available materials and their basic characteristics. Process specific guidelines for part design are mostly of a twofold nature. Firstly, conceptual recommendations include basics for the design of complex geometries as well as the integration of functions in order to benefit from the processes ability to produce highly complex geometries. Secondly, restrictions are given that enable the designer to consider the process inherent thermal induces stresses in form of a tolerant part design. Finally, post processing restrictions are outlines including recommendations for support design as well as powder removal after the building process. Table 6.1 summarizes the available guidelines and is based on the studies of Aumund-Kopp et al. [3, 5, 6, 22].

**Table 6.1**  Essential design guidelines for LAM

| Design - guidelines | |
|---|---|
| | **building chamber dimensions**:<br>- currently max. 250 x 250 x 300 mm$^3$<br>- larger parts can be split |
| ca. 60° | **part overhangs**:<br>- from ca. 60° to the vertical support structures necessary<br>- depending on material larger angles possible |
| bad        good | **notch effect**:<br>- thermal induced notch effect can deform and damage the part during<br>  manufacturing process<br>- avoid strongly bent part changeovers |
| | **prefere hollow structures**:<br>- reduction of manufacturing time and part weight due to reduction of the parts<br>  volume<br>- reduction of thermal induced stresses by avoiding an accumulation of material |
| building<br>platform | **integration of the building platform where possible**:<br>- reduction of end machining<br>- reduction of manufacturing time |
| | **integration of functions**:<br>- enables new approaches for part design not possible up to now<br>- reduction of assembly time |
| AlSi12<br>Ti<br>CoCrMo<br>TiAl6V4<br>... | **material**:<br>- currently limited lineup of materials available<br>- consider process-specific material properties! |
| bad<br>optimum | **strength parameters**:<br>- highest strength in building plane<br>- consider flux of force during part operation<br>- consider process-specific material properties! |
| | **layer thickness**:<br>- defines smallest geometry resolution<br>- possible layer thicknesses depend on material and machine<br>- typical thicknesses: 20 - 60 µm |
| $R_a$<br>$R_{max}$<br>$R_z$ | **surface quality**:<br>- consider material and machine dependency!<br>- typical values: $R_z$ 40 - 90 µm<br>- surface roughnesses worst at down facing areas<br>- surfaces with high demands on surface quality to be positioned upward |
| bad        good | **supports structures**:<br>necessary at:<br>- overhangs<br>- fine strutures that could be damaged by the recoater device<br>- mostly manually removed, thus guarantee accessability |
| | **removing of powder**:<br>- holes for powder removal must be considered |

## 6.5 Future Development

Considering the possibilities of LAM, in terms of high geometrical freedom or low process waste and thus high ecological efficiency, the process shows great potential for manifold applications. Yet, due to the currently only marginally spread knowledge about LAM, holistic research approaches concerning the process chain are necessary. Thus, the whole value chain including the part design, material development, manufacturing process inclusive pre and post processing, machine development and finally the process chain qualification must be in the focus of future research. Beside other research institutions, the iLAS of the Hamburg University of Technology is investigating these fields within versatile research activities. Furthermore the Laser Zentrum Nord GmbH concentrates on an industrial approach in order to analyze new markets, feasible business cases and to provide production and qualification services for new factory concepts and personal training.

Especially the machines and machine hours are still expensive compared to conventional processes [1, 2, 24, 25]. Thus, the optimization of the process productivity is of uttermost importance. For example the latest commercially available machine generations show an increase in laser power compared to previous generations, raising the laser power from 200 to 400 W leading to higher buildup rates. Moreover, new exposure and layer strategies are in the focus of research: currently, a system capable of varying the spot size and layer thicknesses in process with laser powers up to 1 kW in order to optimize the buildup rate and thus the overall process productivity is under development. The material being evaluated is aluminum AlSi10Mg. Parts are being separated in hull and core segments with layer thicknesses of $50\,\mu m$ in the hull and $200\,\mu m$ in the core area. Compared to state of the art LAM machines, an increase of the buildup rate of the factor 4 to 5 was realized [24, 25].

Moreover the optimization of the process repeatability and quality assurance by a process control gains more and more attention. Kruth et al. are currently developing a camera and photodiode based system that is capable of analyzing the melt pool. First trials with the focus of optimizing the downfacing surface of overhangs were very promising [24–26].

The rising field of application with differing demands leads to a steady increase of the available materials. Due to its high geometrical freedom LAM's potential for lightweight design is obvious. Commonly available lightweight materials are limited to titanium and low strength aluminum alloys. Therefore the development of new lightweight alloys offering a high specific strength is very promising. One important material development in progress is the aluminum alloy ScalmalloyRP by EADS with more than 500 MPa engineering strength after a precipitation hardening treatment thus offering a viable alternative to classical, well-established 7xxx plate material [18]. Regarding rapid tooling, copper has been evaluated for manufacturing by LAM due to its good thermal conductivity [24]. For medical applications dental bridges made of gold and ceramics are already scientifically investigated [24, 27]. The latter required a new machine concept which incorporated an "in-process" heating system in order to avoid micro cracks in the ceramic material due to thermal stresses [24, 28].

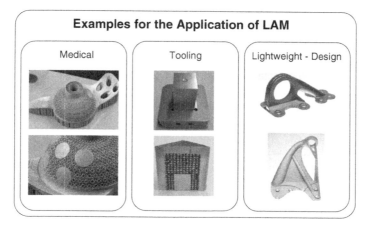

**Fig. 6.5** Application of LAM: hip implant, tooling insert, lightweight brackets

Other studies focus on virtual material development processes based on analytical and numerical calculation of the temperature distribution and the related residual stress analysis versus effective process factors in order to identify and optimize acceptable operating conditions for the LAM process. This strategy shall lead to a rather small number of required experiments for developing new materials and process parameters thus reducing the overall expense [29–31].

## 6.6 Biomimetic Application Areas of LAM

In general, the application of LAM is feasible if the amortization of fix tool costs is impeded by small lot sizes. Furthermore, if the parts complexity leads to high conventional manufacturing costs as they are primarily depending on the parts volume and not its topology, compare Sect. 6.2 [2, 20, 32, 33]. Additionally, LAM eases consistent product revisions as the process is tool-less and the need for time as well as cost intensive adaptations of the manufacturing equipment no longer exists. Thus LAM facilitates the realization of mass customized products of a high complexity and a high diversity of variants in the product line-up [1–3].

Beside the field of rapid prototyping several end-use applications manufactured by LAM are already in service (cf. Fig. 6.5). Currently a major field of application is the medical sector. Due to the demand for individual solutions for patients this market is predestined to take advantage of the process specific advantages. Especially the field of lattice based osseointegrative interfaces conventionally not manufacturable show an increase in application [34]. Already realized in an industrial scale is the manufacturing of patient individual dental bridges or implants by LAM [24].

A second major field of application is rapid tooling that takes advantage of the geometric freedom of LAM by facilitating tool designs with complex internal cooling

Choice of Design

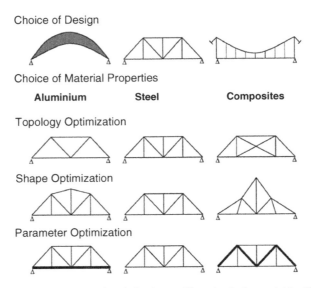

Choice of Material Properties

**Aluminium**          **Steel**          **Composites**

Topology Optimization

Shape Optimization

Parameter Optimization

**Fig. 6.6** Differentiation of structural optimization problems by design variables [38]

ducts conventionally not manufacturable. This results in shorter cycle times and a reduction of part warpage in the injection moulding process [1, 2, 24].

Especially in the automotive sector LAM is used to manufacture prototype parts for pre-series cars with mechanical properties identical to their conventionally manufactured counterparts [35]. A part of the liquid oxygen systems tubing for the Eurofighter has successfully been applied in the military aviation sector. It facilitated a reduction of the system leakage and an optimized flow with an only marginal raise of the parts costs [36, 37].

Moreover LAM facilitates nearly unrestricted geometrical processing and hence is very promising for lightweight design based on structural optimization and bionics, which in general require intricate part architectures with a high geometric complexity. Usually, the resulting geometries are a combination of complex freeform surfaces, hollow structures and undercuts. Conventional manufacturing processes with their inherent restrictions do not support the direct conversion of theoretical optimized structural designs. Therefore deviations from the optimal structures to less optimal structures imposed by manufacturing issues are required and the fully exploitation of the lightweight potential is missed.

Methods for structural optimization are manifold, see Fig. 6.6. Therefore just a brief summary will be given exemplarily with the help of a simple example: a bridge. The "choice of design" includes the consideration of truss construction or compound structures. Furthermore it has to be decided whether the parts design shall be integral or differential. During the "choice of materials" the materials themselves are the design variables which have to be chosen according to the case specific requirements. The preceding methods of structural optimization were of an analytical nature. With the advent of affordable and highly capable personal computers, numerical methods

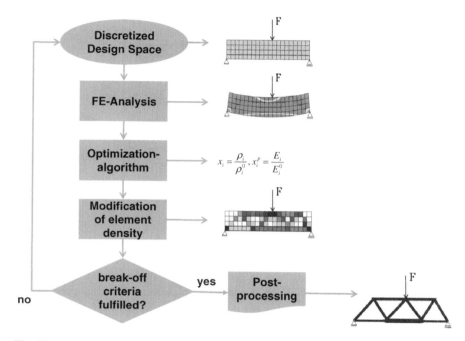

**Fig. 6.7** Process of topology optimization according to the SIMP-method [39]

for structural optimization are being increasingly used. The major methods will be described briefly in the following.

The "topology optimizations" goal is the determination of the fundamental optimum part structure in form of the distribution of the structural relevant material. The "shape optimizations" basis is the adaptation of local geometry according to the part load in order to reduce local stress-peaks. In contrast, the "parameter optimization" is the simplest numerical method. Its goal is the adaptation of wall thicknesses or profiles according to the parts loads. Because the results of the topology optimization are the most radical ones, it is the decisive method in combination with LAM and will be described in more detail.

The applied method of topology optimization is the "solid-isotropic-material-with-penalization" method, see Fig. 6.7, which will be described in the following with the help of a simple bar. It is based on an iterative process that begins with the discretization of design space which includes the modeling of design space and the definition of the FEM-problem (e.g. forces, mesh, constraints, etc.) [39]. The definition of the optimization problem is then carried out by the declaration of design variables, the definition of objective function e.g. weighted compliance or a volume reduction of 85 %, and the definition of manufacturing constraints. Thereon an FE-analysis is being performed in order to determine the parts stress distribution. Basis of the optimization algorithm is the adaption of the element density by a fictional material law during the iterative optimization process. Elements of high

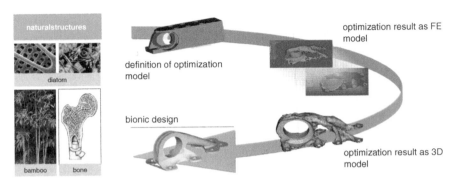

**Fig. 6.8** Application of topology optimization and potential bionic lightweight structures

structural relevance will be assigned a density of $<1$ and elements of negligible structural relevance $>0$. During the break-off criteria examination the model is checked weather the objective functions change lies beneath a specific value, otherwise a further iteration will start. The consequence of this process is a constant evolving of the structural relevant volume during the iterations.

Figure 6.8 shows an application of the topology optimization and potential biological lightweight structures. With the help of the topology optimizations implementation in the development process of "design-for-LAM" the following part characteristics can be achieved [40]:

- minimization of weight
- reduction of peak stresses
- homogeneous stress distribution
- reduction of production waste.

Using the explained methods iLAS (Institut für Laser- und Anlagensystemtechnik) optimized the part design and production of an aluminium bracket in an A380 Airbus, see Fig. 6.9.

This bracket is a part of the fixation of the crew rest compartment with the primary aircraft hull. Overall, more than 25.000 brackets fulfil similar functions in a typical aircraft. In contrast to the shown counterparts manufactured by LAM the majority is milled with intensive machining load resp. chipping removal. A high material waste and energy consumption during the part production is the result.

The incorporation of bionic inspired bamboo-beams facilitated a very stiff and light bracket design that is not manufacturable by any other process, see Fig. 6.9. An overall weight reduction of 50 % could be realized due to the design freedom of LAM and a material change to titanium [40].

Even though a simple manufacturing process substitution combined with a redesign can achieve a tremendous weight reduction, the LAM potential for lightweight design goes far beyond this and needs far more radical approaches in order to be fully exploited. In addition to bionic and optimized design further lightweight potential could be successfully shown by an integration of functions with the help of

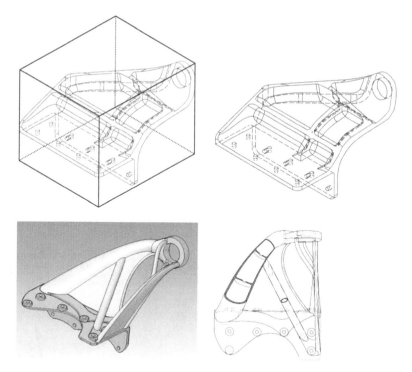

**Fig. 6.9** *Top* raw part and conventionally manufactured bracket. *Bottom* topology optimized bracket with bamboo structure [41]

another A350 bracket design which was initially directly attached onto the honeycomb, see Fig. 6.10. The force off-set from the neutral axis generated a momentum and additional loads that had to be regarded during the design phase. Aiming for an improved lightweight design and an optimized load input the initial bracket and the honeycomb insert plate were combined, following the bionic analogy of the tree and its load transfer from branches to the trunk, see Fig. 6.10. By directly gluing the part into the sandwich compound a direct load induction into the neutral axis is achieved.

Figure 6.11 summarizes the results of weight saving for the different designs of the analyzed A380 bracket. The initial design was manufactured by milling and weighs 330 g, the insert plate including fasteners weighs ca. 1.400 g. In comparison the bionic bracket designed for a manufacturing process substitution shows a weight saving of 41 % whereas the radical approach of integrated design saves even more than 80 %.

For a comprehensive overview over the state of the industry of additive manufacturing especially see the regular reports of Wohlers [12, 13].

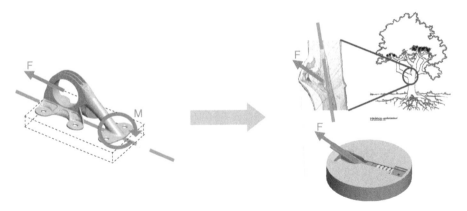

**Fig. 6.10** Adapted bracket design incorporating integration of functions and loading description

| | original design | bionic design | integrated design ("bracket" directly glued into honeycomb) |
|---|---|---|---|
| bracket weight | 330 g | 195 g | 0 g |
| assembly weight | 1.400 g incl. fiber mount and HiLocks | 1.265 g incl. fiber mount and HiLocks | 300 g |
| dimensioning load case | 35 kN | 35 kN | 35 kN |
| weight saving | | **135 g** per bracket **-41%** | **1.100 g** per assembly **> -80%** |

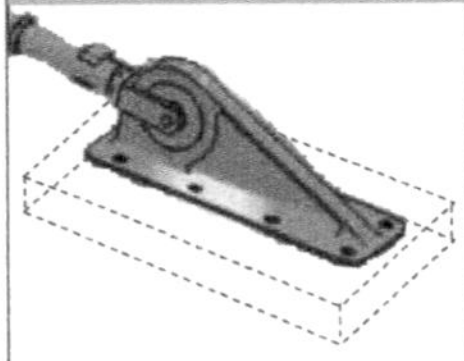

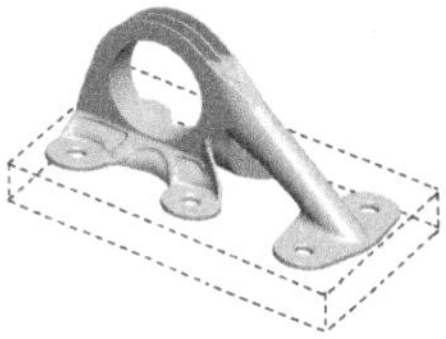

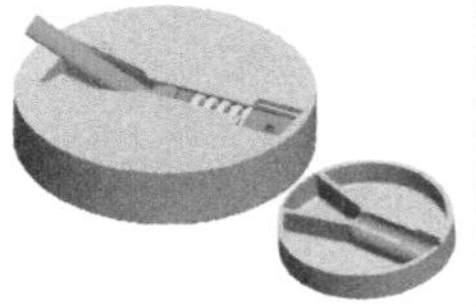

**Fig. 6.11** Comparison of different design alternatives of aircraft brackets

## 6.7 Summary and Conclusion

This chapter gave an overview over LAM of metals which is an emerging additive manufacturing technology. Outlined were the processes basics, significant process parameters, influences on the part quality for LAM, a brief summary of basic design guidelines and current biomimetic application areas of LAM. Primarily being used as a manufacturing process for rapid prototyping in the past, recent developments show a trend towards functional parts. Even though it will not make conventional manufacturing superseded, the shown applications underline the potential of this manufacturing technology for new approaches in product design. Presented applications show that especially highly complex parts can benefit from the processes geometrical freedom. Yet, in order to fully exploit the processes potential designers must leave conventional paths of product development in terms of design for manufacturing. As shown in the example parts biomimetic can be a solution for designers to unlock new approaches for product design and development. In the end costs and the designer's creativity will judge weather this manufacturing process will lead to a new industrial revolution as it is often called [20].

# References

1. Gebhardt A (2007) Generative Fertigungsverfahren. Carl Hanser Verlag, München
2. Zäh MF (2006) Wirtschaftliche Fertigung mit Rapid-Technologien. Carl Hanser Verlag, München
3. Rehme O (2008) Cellular design for laser freeform fabrication. Dissertation, Technische Universität Hamburg-Harburg
4. Over C (2003) Generative Fertigung von Bauteilen aus Werkzeugstahl X38CrMoV5-1 und Titan TiAl6V4 mit Selective Laser Melting. Dissertation, RWTH Aachen
5. Meiners W (1999) Direktes Selektives Laser Sintern einkomponentiger metallischer Werkstoffe. Dissertation, RWTH Aachen
6. Aumund-Kopp C Petzold F (2008) Laser sintering of parts with complex internal structures. Paper presented at the 2008 world congress on powder metallurgy and particulate materials, Washington, 8–12 June 2008
7. Concept Laser GmbH (2008) LaserCUSING Materialdatenblatt für Leichtbau- und Hochtemperaturwerkstoffe. Concept Laser GmbH, Lichtenfels
8. GmbH EOS (2008) EOS Titanium Ti64 für EOSINT M 270-Systeme-Werkstoffdatenblatt. EOS GmbH-Electro Optical Systems, Krailling / München
9. MTT Technologies GmbH (2008) SLM Materialien. MTT Technologies GmbH, Lübeck
10. Kruth JP et al (2005) Binding mechanisms in selective laser sintering and selective laser melting. Rapid Prototyp J 11(1):26–36
11. Leistner M (2004) Herstellung von Funktionsprototypen und Werkzeugen mit serienidentischen Eigenschaften durch Selective Laser Melting. Forschungszentrum Karlsruhe GmbH, Dresden
12. Wohlers T (2008) Wohlers report 2008-state of the industry. Wohlers Associates, Fort Collins Colo
13. Wohlers T (2010) Wohlers Report 2010 state of the industry. Wohlers Associates, Fort Collins Colo
14. Wagner C (2003) Untersuchungen zum Selektiven Lasersintern von Metallen. Dissertation, RWTH Aachen
15. GmbH EOS (2009) Technisches Datenblatt-Laser-Sinter-System EOSINT M 270. Electro Optical Systems, Krailling / München
16. Eisen MA (2010) Optimierte Parameterfindung und prozessorientiertes Qualitätsmanagement für das Selective Laser Melting Verfahren. Dissertation, RWTH Aachen
17. Wirtz TP (2005) Herstellung von Knochenimplantaten aus Titanwerkstoffen durch Laserformen. Dissertation, RWTH Aachen
18. Schmidtke K Palm F Hawkins A Emmelmann C (2011) Process and mechanical properties: application of a scandium modified Al-alloy for laser additive manufacturing. In: Proceedings of the international WLT-Conference on lasers in manufacturing, München, Germany, May 23–26, 2011
19. Wycisk E, Kranz J, Emmelmann C (2012) Fatigue strength of light weight structures produced by laser additive manufacturing in TiAl6V4. In: Proceedings of 1st international conference of the international journal of structural integrity, Porto, June 15–28, 2012

20. Hopkinson N, Hague RJM, Dickens PM (2006) Rapid manufacturing-an industrial revolution for the digital age. Wiley, Chichester
21. Santos EC, Abe F, Kitamura Y et al (2002) Mechanical properties of pure titanium models processed by selective laser melting. In: Bourell DL, Crawford RH, Beaman JJ, Wood KL, Marcus HL (eds). In: Proceedings of the 13th solid freeform fabrication 2002. University of Texas, Austin, pp 180–186
22. Castillo L (2005) Study about the rapid manufacturing of complex parts of stainless steel and titanium. TNO Industrial Technology, Delft
23. VDI (2007) 3404-Entwurf: Generative Fertigungsverfahren-Rapid-Technologien (Rapid Prototyping). Grundlagen, Begriffe, Qualitätskenngrößen, Liefervereinbarungen. Beuth, Berlin
24. Meiners W (2011) Selective laser meltuing-additive manufacturing for series production of the future?. Paper presented at the Intermat 2011, Luxemburg
25. Bremen S, Buchbinder D, Meiners W, Wissensbach K (2011) Mit Selective Laser Melting auf dem Weg zur Serienproduktion? Laser-Technik-J 8(6):24–28
26. Kruth JP Mercelis P Van Vaerenbergh J Craeghs T (2007) Feedback Control of Selective Laser Melting. In: Proceedings of the 3rd international conference on advanced research in virtual and rapid prototyping, Leiria, Portugal, Sept 24–29, 2007
27. Khan M, Dickens P (2011) Selective laser melting of gold. Rapid Prototyp J 18(1):81–94
28. Hagedorn Y-C, Dierkes S (2011) Generative manufacturing of all-ceramic frameworks. Digital Dental News 5. Jahrgang, Marz
29. Petersen M (2006) Lasergenerieren von Metall-Keramik-Verbundwerkstoffen. Dissertation, TU Hamburg-Harburg
30. Petersen M, Emmelmann C (2005) Theoretical and Experimental Studies of Direct Laser Generating of Ceramic Metal Composites. In: Proceedings of the International WLT-Conference on "Lasers in Manufacturing", München, Germany, June 13–16, 2005
31. Emmelmann C, Petersen M, Kranz J, Wycisk E (2011) Bionic lightweight design by laser additive manufacturing (LAM) for aircraft industry. In: Proceedings of the international WLT-conference on lasers in manufacturing, München, Germany, May 23–26, 2011
32. DeGrange J (2006) Boeing's Vision for Rapid Progress between Dream and Reality. Paper presented at the Euro-uRapid 2006, Frankfurt, Nov 27–28, 2006
33. DeGrange J (2006) Steps to Improve Direct Manufacturing Readiness Levels. Paper presented at the euromold 2006, Frankfurt, 29 Nov–2 Dec 2006
34. Emmelmann C Scheinemann P Munsch M Seyda V (2011) Laser additive manufacturing of modified implant surfaces with osseointegrative characteristics. In: Proceedings of the international WLT-conference on lasers in manufacturing, München, Germany, May 23–26, 2011
35. Skrynecki N (2010) Kundenorientierte Optimierung des generativen Strahlschmelzprozesses. Dissertation, Universität Duisburg-Essen
36. Grund M, Emmelmann C (2011) Methodische Unterstützung zur Implementierung von Rapid Manufacturing Technologien. Paper presented at the RapidTech 2011, Erfurt, May 24–25, 2011
37. SAE International (2010) Aerospace engineering and manufacturing, vol 2 Number 29.
38. Schumacher A (2005) Optimierung mechanischer Strukturen-Grundlagen und industrielle Anwendungen. Springer, Berlin
39. Bendsøe P, Sigmund O (2003) Topology optimization: theory. In: Methods and applications, Springer, Berlin
40. Emmelmann C, Sander P, Kranz J, Wycisk E (2011) Laser additive manufacturing and bionics: redefining lightweight design. In: Proceedings of the international WLT-conference on lasers in manufacturing, München, Germany, May 23–26, 2011
41. Wycisk E, Kranz J, Emmelmann C (2010) Leichtbaupotenzial durch das Lasergenerieren bionischer Strukturen. Paper presented at the RapidTech 2010, Erfurt, May 18–19, 2010

# Chapter 7
# Biomimetic Coatings by Pulsed Laser Deposition

**Carmen Ristoscu and Ion N. Mihailescu**

**Abstract** The study of high-intensity laser radiation interacting with solid materials started at the beginning of laser era, i.e. more than 50 years ago. This interaction was initially described as vaporization, sputtering, desorption, etching or laser ablation. Ablation was used for the first time in connection with lasers for introducing material expulsion by visible-infrared (VIS-IR) sources. The advent of the short pulsed sources in visible and especially ultra-violet has made possible laser ablation deposition, an extremely versatile processing technique. When a high intensity laser pulse hits a solid material, the photons absorption can initiate the melting and local vaporization of the outer layer. A cloud of substance described as plasma plume, consisting of photons, electrons, ions, atoms, molecules, clusters, liquid and/or solid particles, is generated. Next, the plume expands from target surface with high velocity and can either be used to grow a film on a nearby substrate or to analyze its composition by using various spectroscopic techniques. In materials science, pulsed laser action opened a door towards various applications, such as localized melting, laser annealing, surface cleaning by desorption and ablation, and surface hardening by rapid quench. After 1988 pulsed laser deposition (PLD) technologies were applied for synthesizing high quality nanostructured thin films. This chapter reviews important applications of PLD and recent work in the field of biomimetic coatings. Furthermore, technical limitations and possible solutions are outlined. The general characteristics of PLD relevant to solid-state physics, e.g. the initial ablation processes, plume formation and plume characteristics are discussed as well.

C. Ristoscu · I. N. Mihailescu (✉)
Lasers Department, National Institute for Lasers, Plasma and Radiations Physics,
409 Atomistilor street, P.O. Box MG-54, RO-077125, Magurele-Ilfov, Romania
e-mail: carmen.ristoscu@inflpr.ro

I. N. Mihailescu
e-mail: ion.mihailescu@inflpr.ro

V. Schmidt and M. R. Belegratis (eds.), *Laser Technology in Biomimetics*,
Biological and Medical Physics, Biomedical Engineering,
DOI: 10.1007/978-3-642-41341-4_7, © Springer-Verlag Berlin Heidelberg 2013

## 7.1 Methods and Biomaterials

### 7.1.1 PLD Definition and Concept

As a materials processing technique, laser ablation was introduced for the first time in the 1960s, after the invention of the ruby laser [1]. Nevertheless, as a thin film growth method it did not attract much research interest until the late 1980s, when it was applied for growing high temperature superconductor films [2, 3]. Since then, the development of the pulsed laser deposition (PLD) technique has highly been accelerated and the research devoted to this topic has increased dramatically. Laser radiation provides many unique properties such as narrow frequency bandwidth, spatial and temporal coherence and high power density. Intense laser light is able to vaporize the hardest and most heat resistant materials. Besides, due to high precision, reliability and spatial resolution, it is widely used in the industry for controlled machining of thin films, modification of materials, material surface heat treatment, welding, micropatterning and nanostructuring [4–10]. Moreover, multicomponent materials can be ablated and deposited onto substrates to form stoichiometric thin films. Accordingly, this application of the laser is called pulsed laser deposition (PLD) [11]. This denomination was finally chosen among many other terms, such as laser evaporation, laser assisted deposition and annealing, laser flash evaporation, laser molecular beam epitaxy (MBE), hydrodynamic sputtering, laser ablation, laser ablation deposition, or photonic sputtering [12].

The concept of PLD is simple. A pulsed laser beam is focused onto the surface of a solid target. The absorption of the intense electromagnetic radiation by the solid leads to rapid evaporation of the target material. The evaporated material consists of neutrals and excited ionized particles, which appear as a glowing plasma plume just in front of the target surface.

**PLD advantages**
Two major advantages of PLD are the simplicity and versatility of the experiment. By using high-intensity pulsed UV-lasers and a vacuum chamber, a stoichiometric film can be grown in a reactive background gas without need for further processing. PLD is not limited to special classes of compounds, because nearly any type of material can be ablated by choosing appropriate irradiation conditions [13, 14].

One key feature of PLD is the retention of the stoichiometry of the target in the deposited films due to the extremely high heating rate of the target surface ($10^8$ K/s) under short (ns) pulsed laser irradiation. It leads to the congruent evaporation of the target irrespective of the evaporating points of each constituent element or compound in target composition. Because of the high heating rate induced by the laser, PLD of crystalline film requires a much lower substrate temperature than other film growing techniques [15, 16]. Several important parameters must be controlled during the PLD-process: incident laser fluence, $F = E_0/S_s$ (where $E_0$ is the laser pulse energy and $S_s$ spot area), ambient gas nature and pressure, deposition geometry, cleaning and heating procedure of the substrate, possible application of external electric and/or magnetic fields. The targets used in PLD are rather small compared to the large size required for other conventional sputtering techniques. Multilayered films of different materials can easily be fabricated by subsequent ablation of different targets. Using a carousel system, where targets of different compositions are loaded, multilayer films can be obtained without opening the deposition chamber. The material combinations are nearly unlimited and novel composite materials with improved and challenging properties can be synthesized [17].

The number of pulses and the sequential nature of the PLD process enable a very precise control of the processed film thickness ($\sim 10^{-2}$–$10^{-1}$ Å/pulse) and determine whether the ablated substance is deposited on the substrate as thin film or as isolated nanoparticles. By accurately monitoring the number of pulses during the PLD process, even an atomic monolayer can be achieved [18].

In contrast to sputtering, the material processing by light does not contaminate the target and the receiving substrate during the deposition procedure. By varying the deposition parameters, macroscopically and microscopically differing structures can be obtained regarding the crystalline status and physical-chemical properties. This feature is treated in detail later in the chapter.

**PLD disadvantages**

One major concern in case of PLD is the presence of the particulates or droplets on the surface or embedded into the film [11, 14]. The main physical mechanisms leading to the formation of particulates are: (1) explosive dislocation of substance caused by local subsurface overheating of the target; (2) gas phase condensation of the evaporated material (clustering); (3) liquid phase expulsion under the action of the recoil pressure of the ablated substance; (4) blast-wave explosion at the liquid (melt)-solid interface; and (5) hydrodynamic instabilities on target surface [19]. The size of particulates may reach a few micrometers for ns laser pulses. Such particulates affect the growth and spoil the quality of the subsequent layers, as well as the electrical and optical properties of the films. However, for some applications in biomedicine or chemical catalysis, the presence of particulates is not disadvantageous, since they can improve the film quality due to extended surface area. This aspect will be discussed in detail for biomimetic coatings in Sect. 7.3.

The presence of droplets in films could be drastically reduced by properly choosing the laser wavelength, and/or minimizing the presence of liquid phase inside the crater. Additionally, electric and/or magnetic fields perpendicular to the expansion direction are applied in order to deflect the particulates from their trajectory towards the film. A second laser beam directed parallel to the target surface may intersect and eliminate particulates [20–22].

With respect to film uniformity on larger substrates, PLD has a fairly narrow angular distribution of the ablated species, which is connected to the adiabatic expansion of laser generated plasma plume and the pitting of the target surface. By rotating and translating the target and the substrate larger uniform films are obtained [23].

PLD cannot be extended to the deposition of organic complex molecules, since intense laser pulses break the long organic chains and the deposited material is irreversibly damaged and/or altered as compared to the original target material.

Therefore, matrix-assisted pulsed laser evaporation (MAPLE) was invented as a complementary method to PLD, adequate for delicate (organic and/or biologic) material transfer. This technique is presented and discussed with representative examples for biomimetic coatings in Chap. 5 of this book.

## 7.1.2 Inorganic Biomaterials

PLD based processes aim for the fabrication of complex oxide thin films for superconductors [24, 25], thin films for optical components [26], laser active media [27], wide band-gap compounds for electronics [28, 29], oxide sensor devices [30–35], protective coatings and barriers [36–39], and inorganic biomaterial thin films, which is discussed in detail in the following section.

Inorganic biomaterials are often used to repair and reconstruct diseased or damaged part of the musculo-skeletal system of vertebrates. It is noted that the materials are considered biocompatible if they exhibit a minimal biological response, i.e. not being toxic or injurious and not causing immunological rejection [40].

The reason of PLD method applied to inorganic biomaterials is motivated by the ability to grow high quality, pure, crystalline and stoichiometric films [41–46], combined with a technical implementation that facilitates easy control of morphology, phase, crystallinity and chemical composition. Generally metal substrates (mainly Ti or Ti alloys) are used as coated implants. The substrate temperature during biocompatible thin films deposition by PLD is typically kept within the range 350–600 °C, thus ensuring the growth of a highly crystalline and single phase coating on implants [14]. Depending on the application, a lower or higher substrate temperature is chosen to grow coatings with a different fine texture and roughness. A thermal treatment (typically in water vapor) after the deposition process is crucial for restoring the stoichiometry of the synthesized compound and to improve the overall crystallinity of the coating.

Biomaterial thin films are nowadays used for medical prostheses to modify the implant surface. A hydroxyapatite (HA) coating on medical implants deposited by conventional thermal plasma spraying, a popular and commercially used method, was intended to function as an intermediate layer between human tissue and the metallic implant [47]. However, this method produced too thick films, which crack, peel or dissolve in biological fluids. Because of the very high temperatures reached during the process, the crystallinity and solubility were also altered [48]. To surpass these shortcomings it was suggested the use of alternative coating techniques for the deposition of films, such as liquid plasma spraying [49], ultrasonic spray pyrolysis [50], radio-frequency magnetron sputtering [49, 51], direct current magnetron sputtering [49], ion implantation [49, 52], ion beam sputtering [49], ion beam-assisted deposition [49], PLD [49, 52], MAPLE [cf. [53] and Chap. 5 of this book], or combination of different techniques [49].

A comprehensive review of plasma-assisted methods for calcium phosphate-based coatings fabrication is available in [49]. Results obtained in the last decade in synthesizing biomimetic thin films of inorganic biomaterials by PLD comprise e.g. hydroxyapatite (HA–$Ca_{10}(PO_4)_6(OH)_2$), carbonated HA doped with $Mn^{2+}$ ions (Mn–CHA; HA with (0.4–2)% $Mn^{2+}$ and (2–6)% $CO_3^{2-}$), Ce-stabilized $ZrO_2$-doped HA (Ce–$ZrO_2$:HA), Ag-doped HA, Sr-doped HA, octacalciumphosphate (OCP–$Ca_8(HPO_4)_2(PO_4)_4 \cdot 5H_2O$), and bioactive glasses (BG) for applications as coatings on metallic implants.

### Hydroxyapatite (HA)

HA is a hydrated calcium phosphate with the chemical composition and crystallographic structure similar to the mineral part of bone and teeth [54]. The natural mineral component of bone ($\sim$50% weight and $\sim$70% volume) basically consists of HA.

The presence of $Mn^{2+}$ ions in the HA should increase the ligand binding affinity of integrins [55]. Manganese is a common structural component of bones and cartilages. $(CO_3)^{2-}$ ions, also present in biological apatite, enter the $(PO_4)^{3-}$ sites in HA.

The addition of a metal oxide dopant to HA, of the type $ZrO_2$, has been proposed to reinforce the biomimetic layer [56] and improve its mechanical performances. The new compound was stabilized with Ce against stress and wear.

Embedding silver (Ag) into calcium phosphates (CaPs) one could expect the enhancement of the antimicrobial performances of coatings for load bearing implants. The drug strontium ranelate has been shown to reduce the incidence of fractures in osteoporotic patients [57, 58]. In-vitro, it increases the number of osteoblasts and reduces the number and activity of osteoclasts [59, 60], while in-vivo, Sr inhibits bone resorption and improves bone formation [61–63].

### Octacalciumphosphate (OCP)

OCP is a basic calcium phosphate which hydrolyzes in aqueous solution to the more stable HA phase [64]. OCP structural resemblance to HA and its higher solubility [65, 66] recommend it as a promising alternative to HA for metallic implants coating.

### Bioactive glasses (BG)

BG are a class of biomaterials that can be used as surgical bone replacement material in otorhinolaryngology, oral surgery, orthopedics, and dentistry [67]. Silica based glasses in the system $SiO_2$–$Na_2O$–$K_2O$–$CaO$–$MgO$–$P_2O_5$ have active surfaces and can form mechanically strong bonds with bone. An optimal composition of the silica-based glass results in a suitable compromise between bioactivity and solubility in which the percentage of silica content plays a key role.

**Table 7.1** Commonly used laser systems in PLD [68]

| Laser type | Pulse duration | Repetition frequency | Wavelength range (nm) | Fluence | Intensity |
|---|---|---|---|---|---|
| Excimer | 10–30 ns | 200 Hz | 308, 248, 193 | High | Modest |
| Nd:YAG | 5–20 ns | 30 Hz | 1064, 532, 355, 266 | High | Modest |
| Nd:YVO$_4$ | 5–10 ns | 20 kHz | 1064, 532, 355, 266 | Modest | Low |
| Yb:YAG | 10–100 ps | 1–5 kHz | 1064, 532, 355, 266 | Modest | High |
| Ti:sapphire oscillator | 10–100 fs | 100 MHz | 790–820 | Low | High |
| Ti:sapphire amplifier | 100–150 fs | 5 kHz | 790–820 | High | High |
| Yb:YAG + OPA | 10–100 ps | 1–5 kHz | 200–1500 | Modest | High |

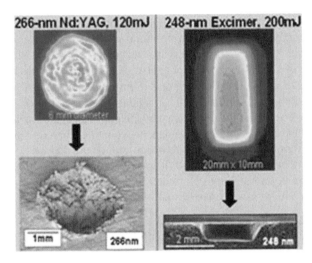

**Fig. 7.1** Comparison of ablation craters with Nd:YAG and excimer laser pulses [69]

## 7.2 PLD Background

### 7.2.1 Laser Sources

The pulsed laser sources for material processing can be classified according to their pulse energy and duration, repetition frequency, and wavelength. Table 7.1 summarizes this information for the laser systems commonly used in PLD.

Excimer lasers have a much higher power output and typically a more uniform power distribution across the beam profile, which is usually described as "top hat" (Fig. 7.1). Solid-state Nd:YAG systems are mainly employed because of relatively low investment costs, little maintenance requirements, and easy incorporation into small commercial PLD systems.

In general, the preferred laser wavelength for the growth of thin films is in the range 200–400 nm, because most materials for deposition exhibit strong absorption

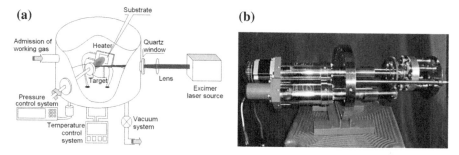

**Fig. 7.2** Typical PLD set-up (**a**), detail with carousel (**b**)

bands within this spectral region. The absorption coefficient generally increases at shorter wavelengths and the penetration depth decreases [70, 71]. Consequently, the ablation threshold is correspondingly lower.

Therefore, most PLD research involved excimer lasers as well as 3$^{rd}$ (355 nm) and 4$^{th}$ (266 nm) harmonics of Nd:YAG lasers, which usually generate ns pulses. Until now, such lasers were commonly used in PLD research for the deposition of biomimetic coatings [72, 73] and hence most of the considerations and examples in the following refer to these types of lasers.

For the sake of completeness, it is noted that recently Ti:Sapphire sources were also introduced in materials processing, in particular for PLD of biocompatible coatings [74]. For example, HA was deposited by means of a Ti:Sapphire (wavelength around 800 nm) laser, which generated pulses in the range of 50 fs–1 ps at a pulse energy in the sub-mJ range and a repetition rate of 1 kHz [75].

## 7.2.2 Targets and Set-Ups

A typical PLD set-up is depicted schematically in Fig. 7.2a.

Presently, most of the PLD targets are commercially available as pressed powders, sintered pellets, cast materials, single crystals or metal foils, having the shape, size and purity required by users. The alternative is to manufacture the appropriate targets from raw powders using moulds and presses. In order to obtain a denser target, it should be sintered for several hours at temperatures ranging from 300 to 1400 °C, depending on component materials [76]. If necessary, in order to minimize the presence of the redeposited particulates, the targets are to be grinded or polished before each deposition. A uniform deposited layer without target piercing is achieved by rotating it or scanning the laser over the irradiated surface. Usually, the target is mounted in a holder which can be part of a carousel system. A carousel typically houses several different targets (see Fig. 7.2b). Before each deposition, the target is generally cleaned with a series of laser pulses. A shutter, placed between target and substrate, prevents deposition on the collecting surface during cleaning.

**Fig. 7.3** Combinatorial PLD
setup. Reproduced with
permission from [78]

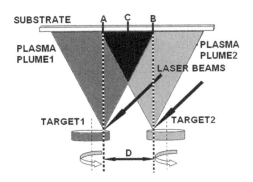

One recent development in the field is the Combinatorial-PLD (C-PLD) [77, 78]. In C-PLD, the targets are located in two different positions and ablated (Fig. 7.3).

The combinatorial depositions have been performed at room temperature by alternative ablation of the two targets of HA and Ag using a KrF* laser source with a repetition rate of 10 Hz in a continuous flow of $O_2$ at 10 Pa dynamic pressure. The laser fluence was set to 1 J/cm². After splitting, the two laser beams are incident on the targets and the separation distance between the two laser spots is D = 50 mm (see Fig. 7.3). As depicted in Fig. 7.3, positions A and B on the substrate correspond to mirror positions of the laser spots on the target 1 and target 2, respectively. All coatings were grown at a target-substrate separation distance of 5 cm by applying 5000 subsequent laser pulses on each target. The films were deposited on typical microscope glass slides (26 × 76 mm²). In order to obtain comparable results, the positions of the targets and substrates in respect to the laser beam focus points were identical in all experiments. Positions A and B are correspond to the substrate coordinates 0 and 50 mm, respectively. With this C-PLD geometry, films with a well-defined composition gradient across the length of the substrate were deposited.

## 7.2.3 Plasma Plume

### Laser ablation, plasma ignition and expansion

Absorbed photons of a high intensity laser beam can initiate the fusion and local vaporization of the outer material layer. A cloud (plasma plume) is then generated in front of the surface, consisting of photons, electrons, ions, atoms, molecules, clusters, liquid or solid particles. The material ejection under the action of a laser beam is known in literature as laser ablation. With respect to the predominant physical phenomenon, the laser ablation can be classified as [79]:

- *thermal ablation*: the heating induced by the laser pulse and the thermal vaporization are prevalent;
- *photo-physical ablation*: the non-thermal excitations directly influence the ablation rate;

- *photo-chemical ablation*: the laser beam breaks chemical bonds by photo-dissocia-
  tion or by indirect energy transfer via defects and impurities.

Several theoretical models have been proposed to describe material removal from
a solid target by laser irradiation [80–84]. A current description involves several
successive stages:

- coupling of the laser energy to the target material;
- melting of the surface;
- vaporization, ionization and formation of a plume consisting of material from the
  thin upper layer of molten surface;
- photon absorption from the laser beam by vaporized species, which reduces the
  laser fluence reaching the target surface;
- propagation of the plume in the direction perpendicular to the target surface;
- return to the initial state at the end of the pulse, to a resolidified surface.

The laser-matter energy transfer is described by the heat equation. For a one dimensional case and if heat losses are neglected, the target enthalpy as a function of the
local temperature is:

$$\frac{dH(T)}{dT} = \frac{\partial\left(K \cdot \frac{\partial T}{\partial z}\right)}{\partial z} + f(z,t) \tag{7.1}$$

where $T$ denotes the temperature, $H(T)$ is the enthalpy which accounts for the phase
changes (melting and boiling), $z$ is the coordinate normal to the sample surface, $K$
is the thermal conductivity of the target material and $f(z,t)$ represents the volumetric
laser energy absorption. $f(z,t)$ acts as a volumetric heat source and is determined by
the incident laser intensity and optical properties of the target:

$$f(z,t) = (1 - R)I_0(t)\alpha e^{-\alpha z} \tag{7.2}$$

where $\alpha = 1/\delta_{opt} = 4\pi k/\lambda$ is the reciprocal of the optical penetration depth, $\delta_{opt}$, $\lambda$
is the laser wavelength, $k$ is the imaginary component of the complex refractive index
(also known as extinction coefficient), $I_0(t)$ is the laser pulse intensity at $z = 0$ and
$R$ is the optical reflectivity.

The laser pulse duration and the corresponding intensity profile, the attenuation
of the incident beam by the vaporized species and the target characteristics (surface
roughness, porosity and density) have a drastic influence on the coupling of the
laser energy to the target surface. The fusion and vaporization processes appear only
when the laser beam intensity reaches a threshold value. The ablation threshold is
quantified by the threshold fluence, $F_{th}$, defined as the minimal energy of the laser
pulse per surface unit that initiates plasma ignition.

As a rule, high intensity laser pulses congruently ablate small volumes of material
[85]. Congruent is referred to as the same chemical composition of the deposited film
and the irradiated target. By contrast, incongruent transfer of material form target to
collector can appear because of insufficient laser intensity, film sputtering induced
by incoming ablated species or different sticking coefficients [86].

**Fig. 7.4** Schematic of the
gas cloud expansion after
target irradiation by a laser
pulse with energy superior
to the ablation threshold.
Reproduced with permission
from [87]

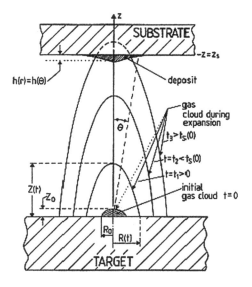

The angular distribution of the expulsed material is centered around the normal
direction to the target surface (Fig. 7.4). The ablated material can be collected on a
nearby substrate with the same chemical composition as the irradiated target.

This distribution is reasonably described by a law $h(\theta) \sim (1 + k^2 tg^2\theta)^{-3/2}$,
where $\theta$ (also called deposition angle) is the angle between the normal to surface and
the ablated species propagation direction. For $\theta$ around the normal, $h(\theta) \sim cos^n\theta$,
with $n$ being an integer. The larger the laser fluence, the closer is the distribution
of the expulsed material flux to the orthogonal direction. $Z_0$ it is defined as initial
length of the plasma, $R_0$ is initial width of the plasma (which practically coincides
with the laser spot dimension) and $h(z_s, \theta)$ is the profile of the deposited film [87].
Correspondingly, at deposition, $z = z_s$ and $R = R_s$ (the dimension of the deposition
spot).

If the laser heating of the target is at equilibrium or quasi-equilibrium, the target
surface is melted and vaporized similar to conventional thermal processes. The evaporated matter is strongly excited and ionized and forms a very dense plasma. The
plasma is fed with energy from the laser pulse. It starts to heat via the absorption of
photons by electrons and collisions between atoms. After becoming more and more
ionized, the main mechanism of plasma heating is the collisions between electrons
and ions. The ablation plasma reaches typical temperatures of $10^4$ K. The absorption
coefficient of plasma, $\alpha_p$, was approximated by Singh and Narayan [88] as:

$$\alpha_p \approx 3.7 \cdot 10^8 \left( Z^3 \frac{n_i^2}{T_p^{0.5} v^3} \right) \left( 1 - e^{-\frac{h\nu}{kT_p}} \right) \tag{7.3}$$

where $Z$ is the mean ion charge, $n_i$ the ionic density and $T_p$ the plasma temperature. The term $(1 - e^{-\frac{h\nu}{kT_p}})$ corresponds to losses by stimulated emission. For UV laser pulses, this term is approximately equal to unity when $T_p < 4 \times 10^4$ K and it can be approximated by $h\nu/kT_p$ for $T_p > 4 \times 10^4$ K. Accordingly, the temperature dependency of $\alpha_p$ is $T_p^{-0.5}$ for the generated plasma and $T_p^{-1.5}$ for the two plasma temperatures regime [89–91]. For excimer lasers (see Table 7.1), the plasma temperatures reach values of $7 \times 10^3 - 2 \times 10^4$ K, meaning a dependence of the plasma absorption coefficient according to $T_p^{-0.5}$.

The charge density rapidly decreases with the distance from the target surface because the plasma expansion takes place with velocities of $10^5$–$10^6$ cm/s. Conversely, the plasma is very dense ($10^{19}$–$10^{20}$ cm$^{-3}$) and extremely absorbent in a thin area close to target surface, known as Knudsen layer [92, 93], which is a continuous fluid. Thus, an exponential attenuation of particle density and plasma pressure along the $z$ direction (i.e. perpendicular to target surface) is expected. Because the initial plasma dimensions are much larger on target surface (up to a few mm) as compared to the $z$ direction (sub-micron), the anisotropic plasma expansion evolves mainly in the $z$ direction according to the respective pressure gradients.

At the end of the ns laser pulse, the particle evaporation and plasma feeding stop [85]. Conversely, ultrashort laser pulses (fs) are usually terminated before the first particles are ejected from the target. In this case, photons are mainly absorbed by free electrons, which are thermalized within femtoseconds through electron-electron scattering. The energy transfer to the lattice through electron-phonon coupling occurs on a larger timescale, typically from a few ps to some hundred ps. Thus, the time required for the electron-lattice thermalisation is much larger than the laser pulse duration [89, 90, 94].

The plasma expands rapidly in an adiabatic regime. The heat is gradually converted into kinetic energy. Species acceleration depends on temperature, plasma dimension, and atomic mass. The maximum velocity is reached along the smallest dimension of plasma. Consequently, the expansion continues mainly in the $z$ direction.

For intensities exceeding $10^8$ W/cm$^2$, plasma becomes strongly ionized and its characteristic frequency, $\omega_p$ is larger than the frequency of the incident laser beam. Then, plasma absorbs the laser beam in a very thin layer, the temperature reaches values of $10^5$ K and the propagation becomes explosive. In this regime, the absorption wave is called detonation wave, the propagation velocity, $v_d$ being approximately $v_d \sim \sqrt[3]{I}$. The front of the detonation wave compresses the surrounding gas, determining the generation of a shock wave. If the laser intensity $I$, is augmented ($10^9$ W/cm$^2$) over the threshold value of plasma initiation $I_{th}$, a decoupling of the plasma from the target can occur. This phenomenon is due to an absorption wave determined by the difference between the energy absorbed in plasma and the energy losses by thermal conduction and radiative emission of particles. Absorption waves propagating with subsonic velocities are behaving like combustion waves. The propagation velocity of the combustion wave, $v_c$ varies with the laser intensity according to the law $v_c \sim \sqrt[3]{I}$ [85].

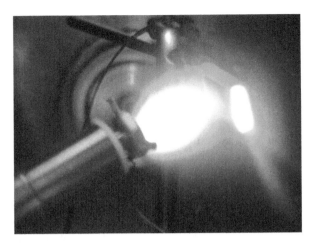

**Fig. 7.5** Photograph of plasma generated when ablating a HA target with an KrF* excimer laser source (25 ns, 248 nm) in 50 Pa $H_2O$

If plasma expands in a gaseous medium, the propagation is attenuated. Plasma is influenced by the gas pressure and the interaction mechanism (elastic or inelastic collisions) between the ejected and gas atoms. In a reactive ambient atmosphere new compounds can be formed in plasma. Their concentration depends on the reactivity of the target atoms and the ambient pressure.

By placing a collector parallel and at a distance of a few centimeters in front of the target (see Fig. 7.2a), adherent inorganic thin films can be obtained, with possible applications in many fields. The thin films synthesis method commonly associated to laser ablation is PLD. In PLD plasma exerts the role of a piston, pushing the expulsed substance from target to substrate and is therefore essential for the entire process [12].

**Plasma diagnostic**

The plasma generated from a HA ($Ca_{10}(PO_4)_6(OH)_2$) target under the action of ablative laser pulses (Fig. 7.5) allows for a better insight into the fundamental mechanisms of the laser-matter interaction [72, 95–98]. The characteristics of the deposited films depend crucially on the properties of the ablation plume, e.g. the degree of ionization, and the velocity distribution of the various species present in the plume. The plume characteristics can be analyzed by wavelength, time- and spatially-resolved optical emission spectroscopy (OES). This technique allows quantitative determination of the velocity distribution of the various emitting species in the plume, and permits estimations of the relative degree of ionization of the ablated material.

In case of laser ablation in vacuum of a HA target at 3 J/cm$^2$ using a Ti–Sapphire laser source (100 fs, 800 nm), the imaging data show a quasi-symmetric expansion of the plasma [99]. Four typical images of the entire plasma plume for different delays between the ablating laser pulse and the observation gate of 10 ns duration, are given in Fig. 7.6. The target position is indicated by a black bar. Beside the

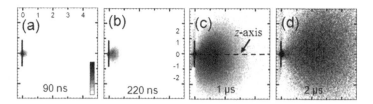

**Fig. 7.6** Plasma images recorded during HA ablation in vacuum with 3 J/cm² using a Ti–Sapphire laser source (100 fs, 800 nm). The target position is indicated by a *black* bar. Reproduced with permission from [99]

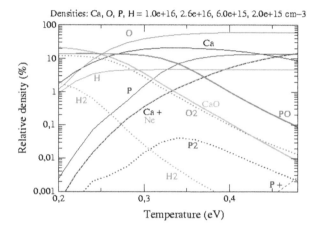

**Fig. 7.7** Computed density of the plasma species for the obtained electron temperatures within the range 0.2–0.5 eV

expanding plume, a strongly luminous zone around the laser impact spot and the photoluminescence of the entire target was visible for times t ≥ 1μs.

Expansion velocities in the range $(1.1–4.7) \times 10^5$ cm/s are deduced from Fig. 7.6, in good agreement with the theoretical considerations in the previous section. Assuming the local thermal equilibrium (LTE) and using Boltzmann equation for excitation equilibrium and Saha equation for ionization equilibrium [85], the inferred electron temperatures at a distance of about 0.1 mm apart from the target are of about 3,500 K (0.30 eV) for 20 ns, and of 2,500 K (0.21 eV) for 50 ns delay time after the laser pulse.

Time- and space-resolved emission spectroscopic analyses showed that lines of neutral Ca dominate the spectra. $Ca^+$ ionic lines and very weak emissions from O and H neutral atoms were also detected. N or P (neutral or ionised) neither ionised O lines were identified. Figure 7.7 is a simulation of the density of the species present in the plasma, within the range of the obtained electron temperatures, (0.2–0.5) eV. These values were computed taking into account the ionization and dissociation potentials and the chemical reactions [100]. It has to be mentioned that PO group has no spectral lines within the range 350–800 nm.

The plasma plume generated by ArF* excimer laser (20 ns, 193 nm) ablation of a HA target was studied in ambient water vapor at a pressure of 20 Pa [101]. A velocity of the plasma front of $1.75 \times 10^6$ cm/s close to the target was measured, which decreases to $\sim 2 \times 10^5$ cm/s at a distance of about 25 mm from the target. The electron density was of $\sim 1.2 \times 10^{18}$ cm$^{-3}$ near the target and of $\sim 4.5 \times 10^{15}$ cm$^{-3}$ at about 18 mm apart from the target. The electron temperature was of 11500 K close to the target and decreased to 4500 K at a distance of 29 mm from the target. The authors inferred from plasma front velocities the kinetic energies of Ca, P and O atoms of 64, 49 and 25 eV, respectively. These energies dropped to 0.47, 0.3 and 0.19 eV only, at a distance of 25 mm from the target. It was concluded that the required energy to deposit crystalline structures must be sustained by heating of the substrate. As known, HA and other CaPs phases should be at least partly crystalline in order to boost bone cells growth and proliferation [52, 73].

## 7.2.4 Thin Films Growth and Characterization

The laser wavelength, pulse duration and fluence determine whether the target material will be thermally or non-thermally evaporated. Whichever process rules the ratio between neutral and ionized species in a plume is different. The repetition rate of the laser fixes the degree of supersaturation of the evaporated material on the substrate surface. As known, one can control crystallization process and product quality by carefully monitoring the level of supersaturation. Supersaturation is therefore a critical parameter because it is the driving force for crystal nucleation and growth. Nucleation represents the formation of new crystal nuclei, which can occur spontaneously (primary nucleation) or in the presence of existing crystals (secondary nucleation). These two mechanisms are in competition during growth process and eventually determine the final crystal size distribution [102]. High levels of supersaturation can often lead to a nucleation-dominated crystallization with small crystal growth. Accordingly, the supersaturation has a direct influence on the growth mode of the deposited film.

There are four different growth modes for films on a substrate: 2D layer-by-layer growth (Frank-van der Merwe); 3D island growth (Volmer-Weber); layer-by-layer followed by an island growth (Stranski-Krastanov); [103] and step-flow growth [15]. The substrate temperature is important to ensure a sufficient surface mobility of the arriving species to support crystalline growth and influence on the deposition rate during pulsed laser deposition in ambient gases [104]. To obtain the desired film composition and structure, the background gas, nature and pressure should be controlled [12].

After deposition, the biomimetic inorganic biomaterial films are submitted to various complementary physical-chemical and biological characterizations, amongst which are: optical microscopy (OM), atomic force microscopy (AFM), scanning and transmission electron microscopy (SEM and TEM); profilometry; Fourier Transform Infrared Spectroscopy (FTIR); X-ray diffraction (XRD); X-ray photoelectron spectroscopy (XPS); energy-dispersive X-ray spectroscopy (EDS); tribological, electrical and magnetical measurements; in-vitro biological assays (toxicity, viability, proliferation, cells growth); and eventually in-vivo tests.

## 7.3 Biomimetic Inorganic Biomaterial Thin Films

Biomaterials are currently used for the repair and reconstruction of diseased or damaged parts of the musculo-skeletal system of vertebrates. Metals, ceramics, carbon, glasses, and composites have so far emerged among biocompatible/biomimetic materials [105], and they can be used in the form of powder, single crystals, polycrystalline, glass, glass-ceramics, and/or composites, thin films or multistructures.

### 7.3.1 Ca Phosphates (CaPs)

Nanostructured biomimetic CaP coatings demonstrated a significant potential for improving the performance of titanium implants in bone. The composition and structure of the CaP layers have a considerable influence on their biological effects.

In a recent work, thin HA films were deposited at a constant temperature of 400 °C in 13–50 Pa water vapors [106]. The deposition substrates were heated / cooled with a constant rate of 6 °C/min. The samples were allowed to cool down in the same atmosphere that had been used for deposition. In order to improve hydroxylation and crystallinity status, all coatings were subjected to 6 h post-deposition heat treatment in water vapor enriched atmosphere at the deposition temperature. The dark field images in Fig. 7.8 clearly show the presence of microcrystalline domains. It appeared that the deposition parameters which control the obtaining of an adherent HA-based layer on Ti are (i) the vapour pressure of water during ablation and (ii) the annealing temperature of ~400 °C to avoid the delamination and cracking.

The film surface is covered with droplets, some of them filled with nano-crystalline material as visible in the two dark field images (Fig. 7.8a, c). The presence of particulates seems to be beneficial for the growth and proliferation of cells because they favor the better anchorage of cells roots or cytoplasmatic extensions [73]. More details are visible in Fig. 7.8b, which shows the 211 ring, while in Fig. 7.8d 002 spots are visible, which reveal a microcrystalline spherical layer of HA wrapping nano-crystalline material.

Figure 7.9 presents a high resolution XTEM micrograph of the HA–TiN interface. The image shows the succession of the (100) atomic planes characteristic to crystalline network of HA. The interplanar distance measured from the image was 0.812 nm, a value very close to the theoretical one for the (100) HA plane (0.817 nm). The image also shows that a very thin film of amorphous HA is growing on the TiN buffer layer. This amorphous layer, of about 7 nm thickness, facilitates the transition from the cubic TiN structure to hexagonal HA. The appearance of the coatings

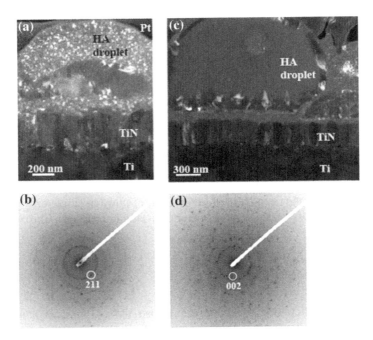

**Fig. 7.8** TEM dark field images of droplets (**a**, **c**) with the corresponding SAED patterns (**b**, **d**) for a HA/TiN/Ti sample deposited at 400 °C in 13 Pa $H_2O$ vapour

**Fig. 7.9** HRTEM micrograph showing the interface HA film-TiN buffer layer

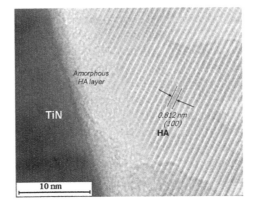

consisting of HA nanocrystals embedded in an amorphous matrix is quite similar to the bone structure [107].

EDS results were consistent with the data provided by the structural analyses. The Ca/P atomic ratio varied in the range $(1.66–1.85) \pm 5\%$ for HA films, depending on substrate temperature during deposition (Fig. 7.10), which is very close to the theoretical value 1.67 for the Ca/P ratio.

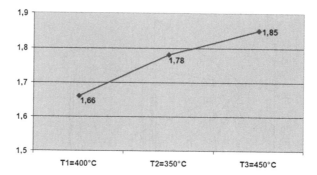

**Fig. 7.10** Evolution of the Ca/P atomic ratio of films versus deposition temperature. Values were obtained by EDS

The films exhibited a hardness H $\cong$ (1–1.5) GPa, and a Young's modulus, $E$, between (50–80) GPa. Films processed at high temperature had a larger Young's modulus: E $\cong$ 70–80 GPa, which can be the result of a better densification of the structure.

In-vitro tests indicated that primary human osteoblasts cultured on the surface of PLD CaPs thin films for up to 21 days, had a normal morphology, a very high proliferation rate, and a significantly improved viability as compared with those growing on uncoated Ti [45, 73, 104]. Consequently, the functionalized biomimetic titanium implants coated with nanostructured CaP layers were tested in-vivo [106]. Eight weeks after implantation in the tibia bones of 6-month-old New Zealand White female rabbits the attachment between bone and implants was investigated by a pull-out test. The pull-out force was found to increase about twice in the case of CaPs-coated Ti coins, in respect with the uncoated ones. Furthermore, compared with the commercial HA-coated implants, the pull-out force increased by 25 % for implants with $Mn^{2+}$-doped carbonated HA, and 10 % for octacalcium phosphate-coated implants.

## 7.3.2 Multistructures

In multistructures, a buffer bio-inert layer is introduced in order to enhance the adhesion onto substrate and improve the crystallinity of the HA outer coating.

Recently, functional multistructures of HA and bioactive glasses (BG) were obtained by PLD in water vapor and low pressure oxygen, respectively, onto Ti substrates [108]. BG in the system $SiO_2$–$Na_2O$–$K_2O$–$CaO$–$MgO$–$P_2O_5$ containing 61 wt. % (BG61) or 57 wt. % (BG57) silica and nanosized HA raw powders were used to prepare the targets. Two series of multistructures were prepared: HA/BG61/Ti denoted HA-BG61 and HA/BG57/Ti denoted HA-BG57.

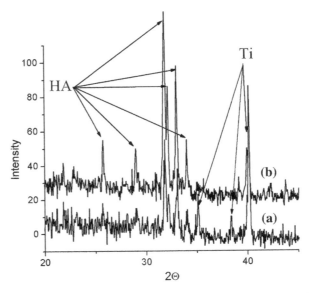

**Fig. 7.11** GIXRD pattern at 2.58 incidence angle for the coatings HA-BG61 (**a**) and HA-BG57 (**b**). Modified with permission from [108]

The crystallinity of samples was analyzed by grazing incidence XRD (GIXRD). The diffraction peaks at 25.88°, 28.98°, 31.88°, 32.28°, 32.98°, and 34.08° were assigned to HA (Fig. 7.11), while peaks belonging to the Ti substrate are visible at 35.18°, 38.48°, and 40.28°. At deposition temperatures of 400 °C, the diffraction lines intensity is higher for HA-BG57 than for HA-BG61 samples. One can therefore assume that the crystallization process is favored for the structures with a buffer layer of BG57.

Figure 7.12 shows typical FTIR spectra for samples HA-BG61 and HA-BG57. All spectra contain the bands characteristic to the $PO_4^{3-}$ group. More precisely, the band at 1200–1000 cm$^{-1}$ is assigned to the asymmetric stretching mode, the band around 960 cm$^{-1}$ corresponds to the symmetric stretching vibration and the band around 560 cm$^{-1}$ is attributed to the asymmetric bending vibration. Based on the resemblance between the FTIR spectra of the HA-BG61, HA-BG57 samples and the original HA powder, it was concluded that the coatings have the same structure and composition with the starting HA material.

Representative SEM micrographs of the sample surface of the two multistructures are given in Fig. 7.13. An important note is that both films mainly consist of spherical droplets of different size within the range (0.2–5 μm). Another observation is that the geometrical shape of the deposited layers generally follows the surface microrelief of the chemically etched Ti substrate. Thus, the pinholes that are sometimes visible on the coatings originate from the pitting of Ti surface. The deposited structures very strongly adhered to the Ti substrates, and the high density of particulates is expected to enhance cell growth and proliferation.

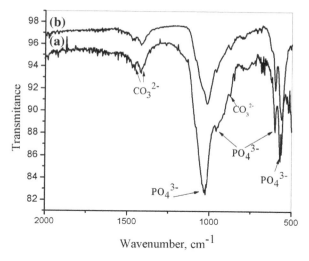

**Fig. 7.12** Typical FTIR spectra of the coatings HA-BG61 (**a**) and HA-BG57 (**b**). Reproduced with permission from [108]

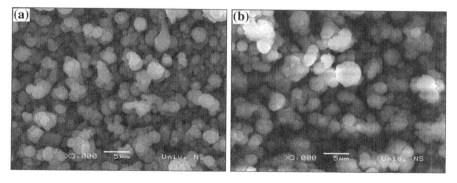

**Fig. 7.13** Representative scanning electron micrographs of the surface of coatings HA-BG61 (**a**) and HA-BG57 (**b**) (scale 5 μm, magnification 3000X). Reproduced with permission from [108]

## 7.3.3 Composite Nanostructures

Recent studies revealed the necessity to reproduce not only the composition in perfectly compatible and active layers but also the structure, morphology, and eventually the functionality of the hard tissues [73]. The use of composite multistructures appeared to be the most appropriate solution with respect to combining the biocompatibility of CaP thin coatings and the strength of substrate.

The influence of porous $Al_2O_3$ substrates on Ce-stabilized $ZrO_2$-doped HA (Ce–$ZrO_2$:HA) thin films morphology pulsed laser deposited on their surface was studied [109, 110]. The $Al_2O_3$ substrates were sintered at 1400 °C (substrate A), 1500 °C (substrate B) or 1600 °C (substrate C). The deposition was conducted at

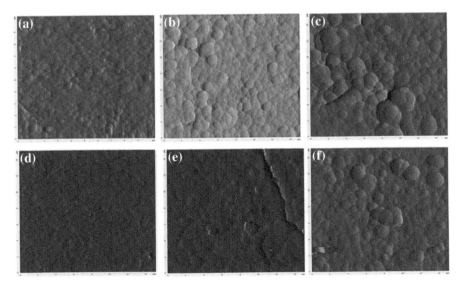

**Fig. 7.14** AFM images of substrate A (**a**) and C (**d**) before PLD and of Ce–ZrO$_2$:HA thin films deposited on A substrate with 5000 (**b**) and 10000 (**c**) laser pulses and on C substrate with 5000 (**e**) and 10000 (**f**) pulses

a water vapor pressure of 50 Pa. The surface morphology of synthesized composite nanostructures was investigated by AFM (Fig. 7.14). Ce–ZrO$_2$:HA thin films deposited with 5000 laser pulses generally copy the surface microrelief of the substrates, while the ones obtained after 10,000 pulses exhibited no differences between surfaces. A Ca/P atomic ratio within the range 1.67–1.70 was found by EDS.

The films were seeded with mesenchymal stem cells (MSCs) for in-vitro tests. Deposited coatings proved compatible with MSC growth: cells attached, spread and covered uniformly the whole surface (Fig. 7.15).

Differences have been observed in the efficiency of developing long filopodia in order to ensure an optimal intracellular organization. On substrates B and C large stem cells colonies were visible, while the substrate A exhibited uniform cell coverage. This can be indicative for enhanced cells division. The exact number of cells covering the Ce–ZrO$_2$:HA films was therefore quantified via a DNA content assay 90 min after seeding. It resulted that around 6000 cells were attached to the film deposited on substrate A, while 10000 were present on the coating on the substrate C [109]. The difference was attributed to the pore size effect rather than to porosity of the substrate [111]. Apparently, the surface structure and the open porosity play a key role in cell attachment and next proliferation and differentiation.

It was recently reported on the antibacterial properties of Ag-doped HA layers obtained by PLD [112]. HA layers doped with 0.06 at%, 1.2 at%, 4.4 at%, 8.3 at% and 13.7 at% silver were prepared. The antibacterial in-vitro measurements were carried out for *Escherichia coli* and Gram-positive *Bacillus subtilis* [113]. It was observed that the antibacterial efficacy increased with Ag doping from 71.0 to

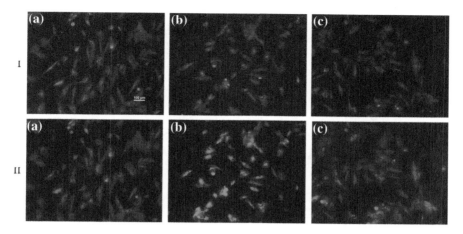

**Fig. 7.15** MSCc 48 h after seeding on Ce–ZrO$_2$:HA thin films deposited with 5,000 laser pulses. Cells were labeled with MitoTracker Red (I), and NBD C6 Ceramide (II): **a**-film on A substrate, **b**-film on B substrate and **c**-film on C substrate (10X). Scale bar = 100 μm. Reproduced with permission from [110]

99.9 %. The best results were obtained for doped HA layers with silver concentration higher than 1.2 at %. For *E. coli*, it was noticed that 2 h of treatment were sufficient irrespective of the Ag doping. Moreover, no difference in antibacterial efficacy was found for amorphous or crystalline layers.

C-PLD technique was used to evaluate whether the cellular morphology is influenced by the variation of Ag content, as well as by the topological features of Ag-doped HA coatings [114]. To this aim, compositional libraries of Ag-doped HA films were deposited using the set-up depicted schematically in Fig. 7.3. Sample properties were analyzed along the AB direction (see Fig. 7.3). In order to establish the Ag content profile, EDS measurements were carried out (Fig. 7.16). The Ag content was the highest in position "A" (0 mm) with value of 0.88 at % and the lowest in "B" (50 mm apart) with a value of 0.14 at %.

AFM and SEM analyses revealed differences in surface topography along the AB direction, with distinctive features. An increase of particulates density from A to B was observed. This specific topology consisting of grains of tens or hundreds of nanometers in size combined with the presence of irregular particulates with dimensions in the sub-micronic and micronic range is considered beneficial for adhesion and proliferation of MSC cells. An Ag content of no more than 0.6 at % into HA coatings was proved nontoxic for MSCs [114].

One should therefore chose for the best compromise between good antibacterial activity (1.2 at %) and toxicity (0.6 at %) against tissue cells when designing the most appropriate coatings for biomimetic metallic implants.

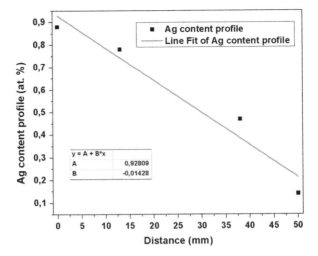

**Fig. 7.16** Ag content profile determined by EDS for film areas between A and B points on the surface of Ag-doped HA sample obtained by C-PLD. Reproduced with permission from [114]

Pulsed laser deposited HA thin films with different extent of Sr substitution for Ca (0, 1, 3, or 7 at%) [115] were synthesized on Ti substrates. The XRD spectra evidenced the presence of HA as the sole crystalline phase. The slight broadening of the diffraction peaks was consistent with a shortening of the perfect crystalline domains when strontium concentration was increased. The presence of Sr significantly improved osteoblast adhesion in early culture phases, while cell adhesion was increasing with Sr content in the coatings. Appreciably higher values of alkaline phosphatase (ALP), osteocalcin (OCN), collagen type I production (CICP), and osteoprotegerin (OPG)/Trance ratio were found in case of thin HA films with relatively high Sr content (of 3 and 7 at%). The increased values of the OGP/Trance ratio in case of coatings with 3 and 7 at% Sr indicated that strontium concentration larger than 3 at% inhibits osteoclast production and differentiation. Accordingly, the investigation of osteoclast proliferation at 21 days after seeding on deposited layers showed a significant reduction in cells number, in correlation with the Sr content increase in the HA coating (Figs. 10.17a–d).

These findings suggest that the presence of strontium in HA thin films can enhance the positive effect of hydroxyapatite biomimetic coatings on osteointegration and bone regeneration, and prevent undesirable bone resorption, and can prove efficient in the treatment of osteoporosis.

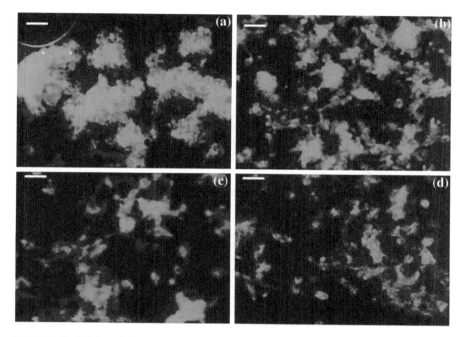

**Fig. 7.17** Phalloidin staining and proliferation (WST1 test) of osteoclast culture 21 days after seeding on **a** HA (3.285 ± 0.021); **b** 1 at % Sr in HA (3.252 ± 0.047); **c** 3 at % Sr in HA (3.211 ± 0.008*); and **d** 7 at % Sr in HA (3.193 ± 0.019*). * (p<0.05). Bars: 50 μm. Reproduced with permission from [115]

## 7.4 Conclusions and Outlook

Biomimetic thin films and multistructures were synthesized by pulsed laser deposition from simple or composite inorganic biomaterials. They proved identical in composition, structure and morphology with the starting material, very likely preserving their functionality and biological activity. The doping and substituting elements were neither eliminated nor segregated after PLD, and films with a uniform composition could be deposited. The recent developments of combinatorial PLD open the door towards synthesizing compositional libraries between different compounds, thus offering the chance of rapid and consistent tests for drugs dosage and delivery and for the development of advanced coatings for a new generation of implants. It may be concluded that PLD technique is now well renowned to be applied for biomedical scale-up.

**Acknowledgments** The authors acknowledge with thanks the financial support of UEFISCDI under the contract ID304/2011 and of the European Social Fund through the contract POS-DRU/89/1.5/S/60746.

# References

1. Maiman TH (1960) The first experimental LASER: stimulated optical emission in ruby. Nature 187:493–494
2. Bednorz JG, Muller KA (1986) Possible high $T_c$ superconductivity in the Ba-La-Cu-O system. Z. Physik B 64 (1): 189–193
3. Wu MK, Ashnuru JR, Torng CJ, Hor PH, Meng RL, Gao L, Huang ZJ, Wang YQ, Chu CW (1987) Superconductivity at 93 K in a new mixed-phase Y-Ba-Cu-O compound system at ambient pressure. Phys Rev Lett 58:908–910
4. Hecht J (2012) Ultrafast lasers make ultraprecise tools. Laser Focus World 48(3):39–42
5. Gaković B, Radak B, Radu C, Zamfirescu M, Trtica M, Petrović S, Stašić J, Panjan P, Mihailescu IN (2012) Selective single pulse femtosecond laser removal of alumina ($Al_2O_3$) from a bilayered $Al_2O_3$/TiAlN/steel coating. Surf Coat Technol 206(24):5080–5084
6. Gakovic B, Radu C, Zamfirescu M, Radak B, Trtica M, Petrovic S, Panjan P, Zupanic F, Ristoscu C, Mihailescu IN (2011) Femtosecond laser modification of multilayered TiAlN/TiN coating. Surf Coat Technol 206(2–3):411–416
7. Ristoscu C, Ghica C, Papadopoulou EL, Socol G, Gray D, Mironov B, Mihailescu IN, Fotakis C (2011) Modification of AlN thin films morphology and structure by temporally shaping of fs laser pulses used for deposition. Thin Solid Film 519:6381–6387
8. Kononenko TV, Nagovitsyn IA, Chudinova GK, Mihailescu IN (2011) Clean, cold, and liquid-free laser transfer of biomaterials. Laser Phys 21(4):823–829
9. Ulmeanu M, Jipa F, Radu C, Enculescu M, Zamfirescu M (2012) Large scale microstructuring on silicon surface in air and liquid by femtosecond laser pulses. Appl Surf Sci 258(23):9314–9317
10. Bogue R (2010) Fifty years of the laser: its role in material processing. Assem Automat 30(4):317–322
11. Chrisey DB, Hubler GK (eds) (1994) Pulsed laser deposition of thin film. Wiley, New York
12. Mihailescu IN, Gyorgy E (1999) Pulsed laser deposition: an overview. In: Asakura T (ed) International trends in optics and photonics ICO IV. Springer, Berlin
13. Chrisey DB, Hubler GK (eds) (1994) Pulsed laser deposition of thin film (Chaps. 14–25). Wiley, New York
14. Eason R (ed) (2007) Pulsed laser deposition of thin films-applications-led growth of functional materials. Wiley, USA
15. Rijnders G, Blank DHA (2007) Growth kinetics during Pulsed laser deposition. In: Eason R (ed) Pulsed laser deposition of thin films-applications-led growth of functional materials. Wiley, USA
16. Ristoscu C, Cultrera L, Dima A, Perrone A, Cutting R, Du HL, Busiakiewicz A, Klusek Z, Datta S, Rose S (2005) $SnO_2$ nanostructured films obtained by pulsed laser ablation deposition. Appl Surf Sci 247(1–4):95–100
17. Ortega N, Bhattacharya P, Katiyar RS (2006) Enhanced ferroelectric properties of multilayer $SrBi_2Ta_2O_9$/$SrBi_2Nb_2O_9$ thin films for NVRAM applications. Mat Sci Eng B 1–3:36–40
18. Dorcioman G, Ebrasu D, Enculescu I, Serban N, Axente E, Sima F, Ristoscu C, Mihailescu IN (2010) Metal oxide nanoparticles synthesized by pulsed laser ablation for proton exchange membrane fuel cells. J Power Source 195(23):7776–7780
19. Mihailescu IN, Gyorgy E, Teodorescu VS, Neamtu J, Perrone A, Luches A (1999) Characteristic features of the laser radiation-target interactions during reactive pulsed laser ablation of Si targets in ammonia. J Appl Phys 86(12):7123–7128
20. Yoshitake T, Nagayama K (2004) The velocity distribution of droplets ejected from Fe and Si targets by pulsed laser ablation in a vacuum and their elimination using a vane-type velocity filter. Vacuum 74(3–4):515–520
21. Yoshitake T, Shiraishi G, Nagayama K (2002) Elimination of droplets using a vane velocity filter for pulsed laser ablation of FeSi2. Appl Surf Sci 197–198:379–383

22. György E, Mihailescu IN, Kompitsas M, Giannoudakos A (2004) Deposition of particulate-free thin films by two synchronized laser sources: effects of ambient gas pressure and laser fluence. Thin Solid Film 446(2):178–183
23. Greer J (2007) Large-area commercial pulsed laser deposition. In: Eason R (ed) Pulsed laser deposition of thin films-applications-led growth of functional materials. Wiley, USA
24. Develos-Bagarinao K, Yamasaki H, Nakagawa Y, Endo K (2004) Relationship between composition and surface morphology in YBCO films deposited by large-area PLD. Physica C Superconductivity 412–414:1286–1290
25. Sudakar C, Subbanna GN, Kutty TRN (2003) Hexaferrite-FeCo nanocomposite particles and their electrical and magnetic properties at high frequencies. J Appl Phys 94:6030–6033
26. Zakery A, Ruan Y, Rode AV, Samoc M, Luther-Davies B (2003) Low-loss waveguides in ultrafast laser-deposited $As_2S_3$ chalcogenide films. J Opt Soc Am B 20:1844–1852
27. Stanoi D, Popescu A, Ghica C, Socol G, Axente E, Ristoscu C, Mihailescu IN, Stefan A, Georgescu S (2007) Nanocrystalline Er:YAG thin films prepared by pulsed laser deposition: an electron microscopy study. Appl Surf Sci 253:8268–8272
28. Ohtomo A, Kawasaki M, Sakurai Y, Ohkubo I, Shiroki R, Yoshida Y, Yasuda T, Segawa Y, Koinuma H (1998) Fabrication of alloys and superlattices based on ZnO towards ultraviolet laser. Mater Sci Eng B 56:263–266
29. Ohta H, Orita M, Hirano M, Nakahara K, Maruta H, Tanabe T, Kamiya M, Kamiya T, Hosono H (2003) Fabrication and photoresponse of a pn-heterojunction diode composed of transparent oxide semiconductors, p-NiO and n-ZnO. Appl Phys Lett 83:1029–1031
30. Starke TKH, Coles GSV, Ferkel H (2002) High sensitivity $NO_2$ sensors for environmental monitoring produced using laser ablated nanocrystalline metal oxides. Sens Actuat B 85: 239–245
31. Gyorgy E, Socol G, Axente E, Mihailescu IN, Ducu C, Ciuca S (2005) Anatase phase $TiO_2$ thin films obtained by pulsed laser deposition for gas sensing applications. Appl Surf Sci 247:429–433
32. György E, Socol G, Mihailescu IN, Ducu C, Ciuca S (2005) Structural and optical characterization of $WO_3$ thin films for gas sensor applications. J Appl Phys 97:093527
33. Mazingue Th, Escoubas L, Spalluto L, Flory F, Socol G, Ristoscu C, Axente E, Grigorescu S, Mihailescu IN, Vainos NA (2005) Nanostructured ZnO coatings grown by pulsed laser deposition for optical gas sensing of butane. J Appl Phys 98:074312
34. Mazingue Th, Escoubas L, Spalluto L, Flory F, Jacquouton P, Perrone A, Kaminska E, Piotrowska A, Mihailescu IN, Atanasov P (2006) Optical characterizations of ZnO, $SnO_2$, and $TiO_2$ thin films for butane detection. Appl Optic 45:1425–1435
35. Ristoscu C, Mihailescu IN, Caiteanu D, Mihailescu CN, Mazingue Th, Escoubas L, Perrone A, Du H (2008) Nanostructured thin optical sensors for detection of gas traces. In: Vaseashta A, Mihailescu IN (eds) Functionalized nanoscale materials, devices, and systems. In: Proceedings of NATO advanced study institute. Functionalized nanoscale materials, devices, and systems for chem.-bio sensors, photonics, and energy generation and storage, June 4–15, 2007, Sinaia, Romania. Springer Science + Business Media B.V, pp. 27–50
36. Mosaner P, Bonelli M, Miotello A (2003) Pulsed laser deposition of diamond-like carbon films: reducing internal stress by thermal annealing. Appl Surf Sci 208–209:561–565
37. Marotta V, Orlando S, Parisi GP, Santagata A (2003) Boron nitride thin films deposited by RF plasma reactive pulsed laser ablation. Appl Surf Sci 208–209:575–581
38. Pelletier H, Carradò A, Faerber J, Mihailescu IN (2011) Microstructure and mechanical characteristics of hydroxyapatite coatings on Ti/TiN/Si substrates synthesized by pulsed laser deposition. Appl Phys A 102(3):629–640
39. Craciun D, Socol G, Stefan N, Mihailescu IN, Bourne G, Craciun V (2009) High-repetition rate pulsed laser deposition of ZrC thin films. Surf Coat Tehnol 203(8):1055–1058
40. http://www.merriam-webster.com/medical/biocompatibility
41. Cotell CM (1994) Pulsed laser deposition of biocompatible thin films. In: Chrisey DB, Hubler GK (eds) Pulsed laser deposition of thin film. Wiley, New York

42. Fernández-Pradas JM, Sardin G, Clèries L, Serra P, Ferrater C, Morenza JL (1998) Deposition of hydroxyapatite thin films by excimer laser ablation. Thin Solid Film 317:393–396
43. Nelea V, Ristoscu C, Chiritescu C, Ghica C, Mihailescu IN, Pelletier H, Mille P, Cornet A (2000) Pulsed laser deposition of hydroxyapatite thin films on Ti-5Al-2.5Fe substrates with and without buffer layers. Appl Surf Sci 168(1–4):127–131
44. Nelea V, Pelletier H, Iliescu M, Verckmann J, Craciun V, Mihailescu IN, Ristoscu C, Ghica C (2002) Calcium phosphate thin film processing by pulsed laser deposition and in situ assisted ultraviolet pulsed laser deposition. J Mater Sci Mater Med 13:1167–1173
45. Bigi A, Bracci B, Cuisinier F, Elkaim R, Fini M, Mayer I, Mihailescu IN, Socol G, Sturba L, Torricelli P (2005) Human osteoblast response to pulsed laser deposited calcium phosphate coatings. Biomaterials 26:2381–2385
46. Nelea V, Mihailescu IN, Jelinek M (2007) Biomaterials: new issues and breakthroughs for biomedical applications. In: Eason R (ed) Pulsed laser deposition of thin films-applications-led growth of functional materials. Wiley, USA
47. Renghini C, Girardin E, Fomin AS, Manescu A, Sabbioni A, Barinov SM, Komlev VS, Albertini G, Fiori F (2008) Plasma sprayed hydroxyapatite coatings from nanostructured granules. Mater Sci Eng B 152:86–90
48. Massaro C, Baker MA, Consentino F, Ramires PA, Klose S, Milella E (2001) Surface and biological evaluation of hydroxyapatite-based coatings on titanium deposited by different techniques. J Biomed Mater Res 58:651–657
49. Sumarev RA (2012) A review of plasma-assisted methods for calcium phosphate-based coatings fabrication. Surf Coat Technol 206:2035–2056
50. Ye G, Troczynski T (2008) Hydroxyapatite coatings by pulsed ultrasonic spray pyrolysis. Ceram Int 34:511–516
51. Socol G, Macovei AM, Miroiu F, Stefan N, Duta L, Dorcioman G, Mihailescu IN, Petrescu SM, Stan GE, Marcov DA, Chiriac A, Poeata I (2010) Hydroxyapatite thin films synthesized by pulsed laser deposition and magnetron sputtering on PMMA substrates for medical applications. Mater Sci Eng B 169:159–168
52. León B, Jansen J (eds) (2009) Thin calcium phosphate coatings for medical implants. Springer Science + Business Media, New York
53. Miroiu FM, Socol G, Visan A, Stefan N, Craciun D, Craciun V, Dorcioman G, Mihailescu IN, Sima LE, Petrescu SM, Andronie A, Stamatin I, Moga S, Ducu C (2010) Composite biocompatible hydroxyapatite-silk fibroin coatings for medical implants obtained by matrix assisted pulsed laser evaporation. Mater Sci Eng B 169:151–158
54. Elliott JC (1994) Structure and chemistry of the apatites and other calcium orthophosphates. Elsevier, Amsterdam
55. Armulik A, Svinberg G, Wennerberg K, Fässler R, Johansson S (2000) Expression of integrin subunit $\beta$1B in Integrin $\beta$1-deficient GD25 cells does not interfere with $\alpha$V$\beta$3 Functions. Exp Cell Res 254:53–55
56. Fu L, Khor KA, Lim JP (2000) Yttria stabilized zirconia reinforced hydroxyapatite coatings. Surf Coat Technol 127:66–75
57. Ammann P (2005) Strontium ranelate: a novel mode of action leading to renewed bone quality. Osteoporos Int 16(1):S11–15
58. Marie PJ (2007) Strontium ranelate: new insights into its dual mode of action. Bone 40(5):S5–S8
59. Canalis E, Hott M, Deloffre P, Tsouderos Y, Marie PJ (1996) The divalent strontium salt S12911 enhances bone cell replication and bone formation in vitro. Bone 18(6):517–523
60. Chang W, Tu C, Chen T, Komuwes L, Oda Y, Pratt S, Miller S, Shoback D (1999) Expression and signal transduction of calcium-sensing receptors in cartilage and bone. Endocrinology 140(12):5883–5893
61. Ammann P, Shen V, Robin B, Mauras Y, Bonjour JP, Rizzoli R (2004) Strontium ranelate improves bone resistance by increasing bone mass and improving architecture in intact female rats. J Bone Miner Res 19(12):2012–2020

62. Grynpas MD, Hamilton E, Cheung R, Tsouderos Y, Deloffre P, Hott M, Marie PJ (1996) Strontium increases vertebral bone volume in rats at a low dose that does not induce detectable mineralization defect. Bone 18(3):253–259
63. Marie PJ, Hott M, Modrowski D, De Pollak C, Guillemain J, Deloffre P, Tsouderos Y (1993) An uncoupling agent containing strontium prevents bone loss by depressing bone resorption and maintaining bone formation in estrogen-deficient rats. J Bone Miner Res 8:607–615
64. Zhang J, Nancollas GH (1992) Kinetics and mechanisms of octacalcium phosphate dissolution at 37°C. J Phys Chem 96:5478–5483
65. Mathew M, Brown WE, Schroeder LW, Dickens B (1988) Crystal structure of octacalcium bis(hydrogenphosphate) tetrakis(phosphate)pentahydrate, $Ca_8(HPO_4)_2(PO_4)_4.5H_2O$. J Crystallograph Spectros Res 18(3):235–250
66. Bigi A, Boanini E, Bracci B, Falini G, Rubini K (2003) Interaction of acidic poly-amino acids with octacalcium phosphate. J Inorg Biochem 95:291–296
67. González P, Serra J, Liste S, Chiussi S, León B, Pérez-Amor M (2002) Ageing of pulsed-laser-deposited bioactive glass films. Vacuum 67:647–651
68. Haglund R (2012) Fundamentals I: types of lasers and laser optics. In: Invited lecture at 3rd international school on lasers in materials science, Isola di San Servolo, Venice, Italy, July 8–15, 2012
69. http://www.coherent.com
70. Green SM, Pique A, Harshavardhan KS, Bernstein J (1994) Equipment. In: Chrisey DB, Hubler GK (eds) Pulsed laser deposition of thin film. Wiley, New York
71. Prokhorov AM, Konov VI, Ursu I, Mihailescu IN (1990) Laser heating of metals. IOP Publishing Ltd, New York
72. León B (2009) Pulsed laser deposition of thin film calcium phosphate coatings. In: León B, Jansen J (eds) Thin calcium phosphate coatings for medical implants. Springer Science+Business Media, New York
73. Mihailescu IN, Ristoscu C, Bigi A, Mayer I (2010) Advanced biomimetic implants based on nanostructured coatings synthesized by pulsed laser technologies. In: Miotello, Antonio; Ossi, Paolo M (eds) Laser-surface interactions for new materials production tailoring structure and properties. In: Springer Series in Materials Science, vol 130, pp. 235–260
74. Perrière J, Millon E, Fogarassy E (eds) (2006) Recent advances in laser processing of materials. Elsevier, Amsterdam
75. Kamata M, Imahoko T, Ozono K, Obara M (2004) Materials processing by use of a Ti:Sapphire laser with automatically-adjustable pulse duration. Appl Phys A 79:1679–1685
76. CRC Handbook of chemistry and physics (2008) CRC Press, Taylor and Francis Group LLC
77. Craciun D, Socol G, Stefan N, Miroiu M, Mihailescu IN, Galca AC, Craciun V (2009) Structural investigations of ITO-ZnO films grown by the combinatorial pulsed laser deposition technique. Appl Surf Sci 255(10):5288–5291
78. Socol G, Galca AC, Luculescu CR, Stanculescu A, Socol M, Stefan N, Axente E, Duta L, Mihailescu CN, Craciun V, Craciun D, Sava V, Mihailescu IN (2011) Tailoring of optical, compositional and electrical properties of the $In_xZn_{1-x}O$ thin films obtained by combinatorial Pulsed Laser Deposition. Dig J Nanomater Biostruct 6(1):107–115
79. Bauerle D (2011) Laser processing and chemistry, 4th edn (Chap. 1). Spinger, Berlin
80. Chan CL, Mazunder J (1987) One-dimensional steady-state model for damage by vaporization and liquid expulsion due to laser-material interaction. J Appl Phys 62:4579–4586
81. Inam A, Rogers CT, Ramesh R, Remschnig K, Farrow L, Hart D, Venkatesan T, Wilkens B (1990) a-axis oriented epitaxial $YBa_2Cu_3O_{7-x}$-$PrBa_2Cu_3O_{7-y}$ heterostructures. Appl Phys Lett 57:2484–2486
82. Wood RF, Giles GE (1981) Macroscopic theory of pulsed-laser annealing. I. Thermal transport and melting. Phys Rev B 23:2923–2942
83. Singh RK, Narayan J (1989) A novel method for simulating laser-solid interactions in semiconductors and layered structures. Mater Sci Eng B 3:217–230
84. Singh RK, Holland OW, Narayan J (1990) Theoretical model for deposition of superconducting thin films using pulsed laser evaporation technique. J Appl Phys 68:233–247

85. Mihailescu IN, Hermann J (2010) Laser plasma interactions. In: Schaaf P (ed) Laser processing of materials: fundamentals. Applications and developments. In: Springer Series in Materials Science, Springer, Heidelberg
86. van Ingen RP, Fastenau RHJ, Mittemeijer EJ (1994) Laser ablation deposition of Cu-Ni and Ag-Ni films: Nonconservation of alloy composition and film microstructure. J Appl Phys 76:1871–1883
87. Anisimov SI, Bauerle D, Luk'yanchuk BS (1993) Gas dynamics and film profiles in pulsed-laser deposition of materials. Phys Rev B 48(16):12076–12081
88. Singh RK, Narayan J (1990) Pulsed-laser evaporation technique for deposition of thin films: physics and theoretical model. Phys Rev B 41:8843–8859
89. Kaganov MI, Lifshitz IM, Tanatarov LV (1957) Relaxation between electrons and crystalline lattices. Sov Phys JETP 4:173–178
90. Anisimov SI, Kapeliovich BL, Perel'man TL (1974) Electron emission from metal surfaces exposed to ultrashort laser pulses. Sov Phys JETP 39:375–377
91. Wellershoff SS, Hohlfeld J, Glidde J, Matthias E (1999) The role of electron-phonon coupling in femtosecond laser damage of metals. Appl Phys A 69:S99–S107
92. Knight CJ (1979) Theoretical modeling of rapid surface vaporization with back pressure. AIAA J 17(5):519–523
93. Anisimov SI, Imas YA, Romanov GS (1971) Effects of high-power radiation on metals. NTIS Springfield, VA
94. Axente E, Mihailescu IN, Hermann J, Itina TE (2011) Probing electron-phonon coupling in metals via observations of ablation plumes produced by two delayed short laser pulses. Appl Phys Lett 99:081502
95. Serra P, Cleries L, Morenza JL (1996) Analysis of the expansion of hydroxyapatite laser ablation plumes. Appl Surf Sci 96–98:216–221
96. Serra P, Fernandez-Pradas JM, Sardin G, Morenza JL (1997) Interaction effects of an excimer laser beam with hydroxyapatite targets. Appl Surf Sci 109–110:384–388
97. Serra P, Morenza JL (1998) Imaging and spectral analysis of hydroxyapatite laser ablation plumes. Appl Surf Sci 127–129:662–667
98. Serra P, Morenza JL (1998) Fluence dependence of hydroxyapatite laser ablation plumes. Thin Solid Films 335:43–48
99. Miroiu F, Mihailescu IN, Hermann J, Sentis M (2004) Spectroscopic analyses during femtosecond laser ablation of hydroxyapatite. In: SPIE Proceedings of the 7th international conference on optics ROMOPTO, vol 5581, pp. 479–485, September 8–11, 2003, Constanta, Romania
100. Hermann J, Dutouquet C (2002) Local thermal equilibrium plasma modeling for analyses of gas-phase reactions during reactive-laser ablation. J Appl Phys 91(12):10188–10193
101. Jedynski M, Hoffman J, Mroz W, Szymanski Z (2008) Plasma plume induced during ArF laser ablation of hydroxyapatite. Appl Surf Sci 255:2230–2236
102. Myerson A (2002) Handbook of industrial crystallization, 2nd edn. Butterworth-Heinemann, USA
103. Horwitz JS, Sprague JA (1994) Film nucleation and film growth in pulsed laser deposition of ceramics. In: Chrisey DB, Hubler GK (eds) Pulsed laser deposition of thin film. Wiley, New York
104. Amoruso S, Aruta C, Bruzzese R, Wang X (2011) Substrate heating influence on the deposition rate of oxides during pulsed laser deposition in ambient gas. Appl Phys Lett 98:101501
105. Ratner BD, Hench L (1999) Perspectives on biomaterials. Curr Opin Solid State Mater Sci 4(4):379–380
106. Mihailescu IN, Lamolle S, Socol G, Miroiu F, Roenold HJ, Bigi A, Mayer I, Cuisinier F, Lyngstadaas SP (2008) In vivo tensile tests of biomimetic titanium implants pulsed laser coated with nanostructured calcium phosphate thin films. Optoelectron Adv Mater Rapid Commun 2(6):337–341
107. Socol G, Torricelli P, Bracci B, Iliescu M, Miroiu F, Bigi A, Werckmann J, Mihailescu IN (2004) Biocompatible nanocrystalline octacalcium phosphate thin films obtained by pulsed laser deposition. Biomaterials 25(13):2539–2545

108. Tanaskovic D, Jokic B, Socol G, Popescu A, Mihailescu I, Petrovic R (2007) Synthesis of functionally graded bioactive glass-apatite multistructures on Ti substrates by pulsed laser deposition. Appl Surf Sci 254(4):1279–1282

109. Sima F, Ristoscu C, Caiteanu D, Stefan N, Mihailescu CN, Mihailescu IN, Prodan G, Ciupina V, Palcevskis E, Krastins J, Sima LE, Petrescu SM (2011) Biocompatibility and bioactivity enhancement of Ce stabilized $ZrO_2$ doped HA coatings by controlled porosity change of $Al_2O_3$ substrates. J Biomed Mater Res B 96(2):218–224

110. Sima F, Ristoscu C, Stefan N, Dorcioman G, Mihailescu IN, Sima LE, Petrescu SM, Palcevskis E, Krastins J, Zalite I (2009) Shallow hydroxyapatite coatings pulsed laser deposited on $Al_2O_3$ substrates with controlled porosity: correlation of morphological characteristics with in vitro testing results. Appl Surf Sci 255:5312–5317

111. Gauthier O, Bouler JM, Aguado E, Pilet P, Daculsi G (1998) Macroporous biphasic calcium phosphate ceramics: influence of macropore diameter and macroporosity percentage on bone ingrowth. Biomaterials 19:133–139

112. Jelínek M, Kocourek T, Jurek K, Remsa J, Mikšovský J, Weiserová M, Strnad J, Luxbacher T (2010) Antibacterial properties of Ag-doped hydroxyapatite layers prepared by PLD method. Appl Phys A 101:615–620

113. Jelínek M, Weiserová M, Kocourek T, Zezulová M, Strnad J (2011) Biomedical properties of laser prepared silver doped hydroxyapatite. Laser Phys 21(7):1265–1269

114. Socol G, Socol M, Sima LE, Petrescu S, Enculescu M, Sima F, Miroiu M, Popescu-Pelin G, Stefan N, Cristescu R, Mihailescu CN, Stanculescu A, Sutan C, Mihailescu IN (2012) Combinatorial pulsed laser deposition of Ag-containing calcium phosphate coatings. Dig J Nanomater Biostruct 7(2):563–576

115. Capuccini C, Torricelli P, Sima F, Boanini E, Ristoscu C, Bracci B, Socol G, Fini M, Mihailescu IN, Bigi A (2008) Strontium-substituted hydroxyapatite coatings synthesized by pulsed laser deposition: in vitro osteoblast and osteoclast response. Acta Biomaterialia 4:1885–1893

# Chapter 8
# Laser Assisted Bio-printing (LAB) of Cells and Bio-materials Based on Laser Induced Forward Transfer (LIFT)

Bertrand Guillotin, Sylvain Catros and Fabien Guillemot

**Abstract** Laser assisted bio-printing (LAB) is an emerging and complementary technology in the field of tissue engineering envisaging biomimetics applications. LAB allows to print cells and liquid materials with a cell-level resolution, which is comparable to the complex histology of living tissues. By giving tissue engineers control on cell density and organization, LAB potentially holds promise to fabricate living tissues with biomimetic physiological functionality. In this chapter, the physical parameters related to laser induced forward transfer (LIFT) technique, which is implemented in the LAB, are presented. These parameters, such as laser pulse energy and bio-ink viscosity are critical to control the cell printing process. They must be tuned according to each other in order to print viable cell patterns with respect to cell-level histological organization. Processing time is a concern when addressing tissue engineering involving living material like cells. Therefore, concerns regarding the design and technical implementation of LAB based rapid prototyping workstation are discussed. Experimental requirements are described in order to fabricate tissues using LAB. Some typical multi-component printing, 3D printing approaches and bio-printing in vivo are presented.

## 8.1 Laser Assisted Bio-printing for Biomimetic Tissue Engineering

There are several ways for the technological generation of three-dimensional biological structures. As an alternative to the rather straight forward scaffold-based approach of cell seeding on porous templates [1], some authors have suggested that

B. Guillotin · S. Catros · F. Guillemot (✉)
Bioingénierie Tissulaire, INSERM U1026, 146 rue Léo Saignat, Bordeaux 33076, France
e-mail: fabien.guillemot@inserm.fr

B. Guillotin · S. Catros · F. Guillemot
Université Bordeaux Segalen, 146 rue Léo Saignat, Bordeaux 33076, France

V. Schmidt and M. R. Belegratis (eds.), *Laser Technology in Biomimetics*, 193
Biological and Medical Physics, Biomedical Engineering,
DOI: 10.1007/978-3-642-41341-4_8, © Springer-Verlag Berlin Heidelberg 2013

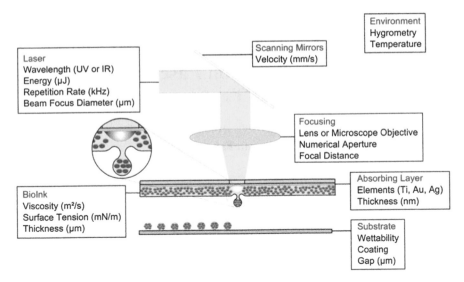

**Fig. 8.1** A typical LIFT experimental set up is generally composed of three elements: a pulsed laser source, a target or ribbon coated with the material to be transferred, and a receiving substrate. The ribbon is a three layer component: a support, transparent to the laser radiation wavelength, coated with a thin absorbing layer (50 nm), coated itself with a transfer layer (50 μm), named bio-ink that contains the elements to be printed like biomaterials, cells or biochemical factors

three-dimensional biological structures can be built from the bottom up by means of the bio-printing. This technology addresses the automated and computer-aided deposition of cells, cell aggregates and biomaterials [2–7]. Ink-jet printers have successfully been applied to pattern biological assemblies [8–10]. Pressure-operated mechanical extruders were used to handle living cells and cell aggregates [3]. At the same time, laser-assisted printing technologies have emerged as alternative methods for assembling and micropatterning biomaterials (i.e. bio-compatible materials) and cells. Laser printing of biological material in general, and living cells in particular, is based on the laser-induced forward-transfer (LIFT) technique (Fig. 8.1), in which a pulsed laser is used to induce the transfer of material from a source film [11–13]. Several set-ups of LIFT have been implemented to print living cells, the main difference being on the choice of the laser source and the hardware (automated or manual) used to positioning the laser beam or the substrate [14, 15]. In this chapter the general term of Laser Assisted Bio-printing (LAB) is used. Under suitable irradiation conditions, and for liquids presenting a wide range of rheologies, the material can be deposited in the form of well-defined circular droplets with a high degree of spatial resolution [16–18].

This chapter introduces the conceptual challenges related to living tissue complexity in the field of tissue engineering and the technical properties that make LAB a suitable tool for biomimetic tissue engineering. It will be discussed that the underlying physics imposes limitations to the tuning range of technical parameters for printing viable cell patterns at cell-level spatial resolution. LAB-based tissue engineering applications for basic research in biology as well as for regenerative medicine are presented.

Living tissue is characterized by its cell types, the biochemical and mechanical properties of the extracellular matrix and the organization of these components. The organs are composed of multiple cell types, which are assembled and interfaced at the micrometer scale. In the liver for example, columns of hepatocytes are interfaced with biliary capillaries on the apical side and with sinusoidal blood vessels on the basal side to form lobules. There are two functional outcomes related to such high density, compartmentalized and integrated cellular organization: (i) homeostasis, in particular metabolic exchange, is optimized; and (ii) functional units are packed together to form organs with a physiological efficiency that is compatible with living tissues. Consequently, miniaturization of tissue engineering processes (i.e. microscale organization of cells) might be necessary to fabricate organotypic structures that compare favorably with the functionality of living organs.

Tissue engineering approaches can be divided into three strategies based on the scale of spatial organization. First, macrosopic strategy can be likened to traditional tissue engineering in which cells are seeded onto a macroporous scaffold. Cells are expected to colonize the inner volume of the scaffold by cell mobility and proliferation, and fluid flow. However, scaffolds do not present the ability to mimic the functional multicellular anisotropy and density of the host tissue. Second, mesostructures or modular blocks, also termed organoids [4, 19, 20], are based on the ability of the cells to self-assemble and their capacity to maintain viability and function, when located within the diffusion limit of nutrient supply. The modular approach enables the production of 3D modules in a variety of shapes (e.g. cylinders) with a lateral diameter between 40 and $1000\,\mu m$ and cell densities of $10^5$–$10^8$ cells/cm$^2$. Furthermore, it allows fabrication of multicellular constructs (e.g. bone-mimicking construct including osteoblasts, osteoclasts and endothelial cells). However, both macroscopic and mesoscopic approaches have not demonstrated the ability to mimic the functional multicellular anisotropy and density of the host tissue. Consequently, it is also conceptually challenging to design an efficient perfusion system that is physiologically interfaced with the engineered tissue and that will branch to the host' vasculature at implantation.

The third strategy is to reproduce the local cell microenvironment. It can be thought as the ultimate target for tissue engineering and cell patterning. It could be defined as the capacity of positioning a single cell into its most suitable environment. Coordinated interactions between soluble factors, different cell types and extracellular matrices (i.e. mechanical and biochemical cues) should be taken into account. Such a cell niche manufacturing approach is unique in its purpose of dealing with tissue complexity and engineering a desired tissue from the bottom up. A microscopic scaffold-free, bottom-up approach to tissue engineering has been proposed [21]. Accordingly, computer-assisted design/computer-assisted manufacturing (CAD/CAM) LAB workstations have been designed and used to print viable cells and to organize them with cell-level resolution, in two dimensional and three dimensional tissue constructs with the aim of mimicking the functional histology of live tissues.

Virtually all cell types can be printed by LAB. Numerous studies have shown successful laser assisted printing of a fairly broad range of prokaryotic cells and eukaryotic cell, which is comprehensively reviewed in [14]. Indeed, the literature does not report altered cell proliferation, cell differentiation or DNA damage due to LAB as compared to conventional cell seeding. Considering human primary cells in particular, the following types have been printed by LAB: human umbilical vein endothelial cells (HUVEC) and human umbilical vein smooth muscle cells (HUVSMC) [22, 23], human mesenchymal stem cells [23–25], adipose-tissue derived stem cells (ADSC) and endothelial colony-forming cells (ECFC) [26], as well as human bone-marrow derived osteo-progenitors (HOP) [27].

## 8.2 Technical Implementation of LAB for Cell and Biomaterial Printing

Major requirements must be considered in order to design a LAB workstation regarding the laser treatment of living cells and the capability for rapid prototyping of biomaterials (Fig. 8.2). An evaluation of the laser source in terms of cell viability and printing resolution leads to the following conclusion:

The wavelength $\lambda$ should not induce alteration of biological materials. Due to the potential denaturation of DNA by UV exposure, and despite short pulse duration and the presence of the light absorbing metal layer, near IR lasers might be preferable to UV lasers.

The laser repetition rate must be considered with respect to high throughput processes. First, to keep energy per spot constant, successive laser spots should not overlap with each other onto the laser-absorbing layer. Moreover, in combination with scanning velocity, the laser repetition rate has to be taken into account to avoid coalescence of vapor bubbles within the thickness of the bio-ink, and reciprocal perturbation of consecutive jets. Moreover, the jet hydrodynamics need to be considered in order to avoid perturbations between two contiguous jets since droplet formation onto substrates occurs in a $100\,\mu s$ timeframe (Fig. 8.2). In that context, most of the LAB set-ups use a nanosecond laser, operating at repetition rates from 1 to 10 kHz.

The beam quality, including divergence, spatial mode, and pulse-to-pulse stability has to be taken into account to ensure the reproducibility, stability and high resolution of the system (Fig. 8.4).

A laser-based workstation dedicated to tissue engineering applications should be designed with the purpose of executing various tasks rather than solely bioprinting. Consequently, the mean laser power should be high enough to perform additional processes like photolithography, photopolymerization, machining, sintering and foaming [28–30].

**(a)**

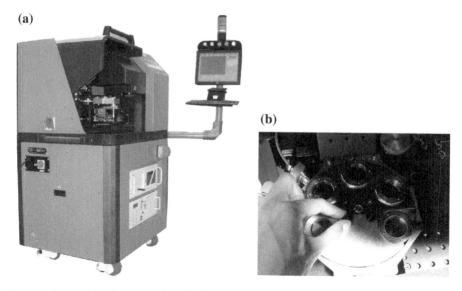

**(b)**

**Fig. 8.2** View of the high-throughput LAB (**a**). High resolution positioning system placed below the carousel holder with a loading capacity of 5 different ribbons (**b**)

## 8.2.1 Droplet Ejection Mechanism

A set-up for a LAB process, which takes advantage of LIFT, comprises three elements: a pulsed laser source, a target coated with the material to be printed and a receiving substrate (Fig. 8.1). The generation of microdroplets by LIFT proceeds through three consecutive steps [12, 31–33]: First, the laser energy is deposited into the skin depth of the absorbing layer, which heats the absorbing layer in its skin depth. Secondly, the heat is transferred to a very thin film of the bio-ink on top of the absorbing layer. Third, at sufficient laser intensity a vapor bubble is generated, which expands and leads to a bio-ink-air interface deformation. Finally, it has been shown that depending on the laser energy, three LIFT regimes can be observed: the subthreshold, the jetting and the plume regime (Fig. 8.4 and [12]). Initially, it was found that the volume of deposited material depends linearly on the laser pulse energy, and that a minimum threshold energy must be exceeded for the occurrence of droplet ejection [31]. As numbered in Fig. 8.1, microdroplet ejection depends on numerous parameters, in order to print viable cells with high printing resolution and high throughput (Fig. 8.3).

A major shortcoming of common LAB is the use of metallic absorbing layers, which are vaporized together with the printed material and may spoil the final tissue engineering product. Recent promising solutions to this issue include the use of a polyimide membrane as the laser-absorbing layer, which is capable of dissipating

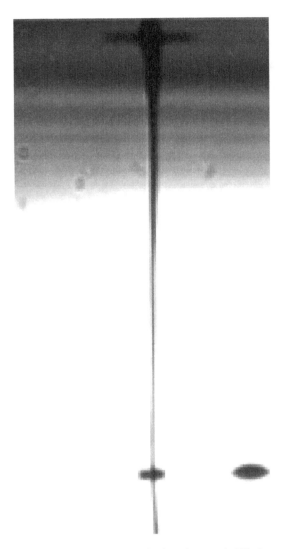

**Fig. 8.3** Mechanism for laser-induced droplet ejection. A vapor bubble is generated (*see II*) by vaporization of the absorbing layer and/or the first molecular layers of the liquid film. At given bio-ink viscosity and film thickness, jetting is observed for intermediary values of laser fluences: T1 < T < T2 (*see IIIb*). For a lower fluence T < T2, the bubble collapses far from the free surface without generating a jet (*see IIIa*). For a higher fluence T < T1, the bubble bursts to the surface, generating sub-micrometer droplets (*see IIIc*). Increasing film thickness or bio-ink viscosity leads to increased threshold T values

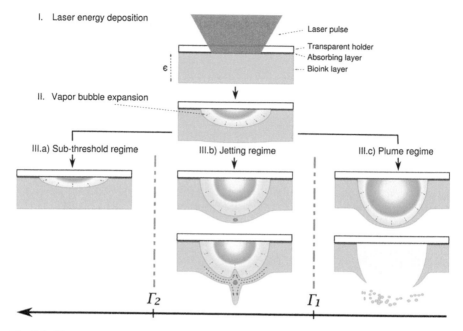

**Fig. 8.4**  Time-resolved image obtained 30 μs after the laser spot deposit showing how a droplet is generated onto a substrate which is positioned face to the bio-ink film

shock energy through elastoplastic deformation [34], or the use of femtosecond laser sources, which enable the generation of the cavitation bubble directly into the bio-ink, and hence avoid the use of a laser-absorbing layer [35].

## 8.2.2 Printing Resolution

The printing resolution and the associated droplet size are mainly determined by the viscosity of the bio-ink and the deposited laser energy. Printing resolution is defined as the number of printed droplets on a given length (e.g. dpi-dots per inch), implying that droplet size decreases as droplet number per length increases. The highest resolution is obtained by printing droplets as small as possible and adjacent to each other on the same length. LAB can print liquids with droplet size in the order of 10 μm. Laser energy deposit can be modulated by tuning the energy of the laser source and/or by cutting the laser beam with a diaphragm aperture stop. The higher the viscosity and/or the lower the energy, the smaller is the droplet diameter (Table 8.1). It is possible to achieve a similar droplet size, with a 0.1 % (w/v) alginate ink (leading to a 27 cps viscosity) printed with a 6 μJ laser energy of (droplet size: $(49 \pm 3.5)$ μm, n = 15), and with a 1 % alginate (w/v) ink (leading to an 110 cps

**Table 8.1** Diameter (in μm) of droplet ejected onto the substrate, depending on alginate concentration in the ink, and laser energy

| Alginate (w/v) (%) | Energy (μJ) | | | |
|---|---|---|---|---|
| | 4.5 | 6.0 | 7.5 | 9.0 |
| 0.1 | 49 ± 4 (n = 15) | 69 ± 4 (n = 15) | c | c |
| 0.5 | 38 ± 3 (n = 15) | 55 ± 5 (n = 15) | 64 ± 5 (n = 15) | 62 ± 6 (n = 15) |
| 1.0 | n.t. | 48 ± 4 (n = 15) | 46 ± 3 (n = 15) | 51 ± 4 (n = 15) |

The different inks were composed of purified water supplemented with 30 % (v/v) glycerol, with varying concentration of alginate. *c* coalescence of contiguous droplets onto the substrate, *n.t.* no transfer of the ink onto the substrate

viscosity) printed at a pulse energy of 12 μJ (droplet size: (51 ± 4.2) μm, n = 15). As a consequence, a wide viscosity range of extracellular matrices can be printed at a comparable resolution.

## *8.2.3 LAB Parameters for Cell Printing*

### 8.2.3.1 Bio-ink Composition

LAB requires cells to be suspended in a liquid bio-ink prior to being printed onto the substrate. In order to print a 3D material containing cells, the bio-ink should gel post printing onto the substrate. With respect to the layer-by-layer 3D building strategy, the gelling process is necessary to stabilize the printed 2D pattern and to support subsequent ink layer for 3D constructs using the layer-by-layer approach (see Sect. 8.2.3.4). Furthermore, the gelling should not harm the cells. Specifically for LAB applications, the bio-ink should harbor properties similar to the physiological extracellular matrix, which is critical for cell homeostasis in vivo [36, 37]. Accordingly, the cells have successfully been printed using various solutions: culture medium alone [22], in combination with sodium alginate [38], thrombin [18], combination of hyaluronic acid and fibrinogen [26] or a combination of blood plasma and sodium alginate [24, 39]. Culture medium supplemented with sodium alginate, or hydrogels like Collagen type I and Matrigel™ can be used as well. Using slightly different laser assisted printing techniques, mouse embryonic stem cells have successfully been printed using gel form of Matrigel™ or 20 % gelatin [40, 41].

Evaporation of the bio-ink is critical because it is typically spread into the target as a 50 μm thin layer. Hence, Othon et al. have proposed to use methyl-cellulose in the bio-ink to help prevent evaporation [42].

### 8.2.3.2 Requirements for High Cell Density Printing with Cell-Level Resolution

In order to achieve microscale cell printing precision, cells should be printed with a minimal volume of surrounding extra-cellular matrix (ECM) or bio-ink. However, because LAB is a LIFT based and nozzle free device, the number of cells in each ejected droplet and hence the printed cell number is statistic [16]. LAB nozzle free set up precludes the cell printing process from clogging issues. Thus, it is possible to use a bio-ink loaded with cell densities in the order of $1 \times 10^8$ cells/ml, which is comparable to the cell confluence observed in living tissue like a parenchyma. To print one single cell in one droplet, bio-ink with a low cell concentration, e.g. $5 \times 10^7$ cells/ml has been used. However, if the cell density is too low on the ribbon, the ejected droplet of ink may not contain any cell [18]. To overcome this problem, at least two strategies can be proposed. (1) Increased laser energy deposit leads to the ejection of bigger droplets. As a result, cells are more likely to be dragged off by draining/capillary effect. (2) Cell density can be increased up to the point cells are touching each other at the surface of the ribbon, i.e. $1 \times 10^8$ cells/ml. In such a condition, the probability of printing droplets that contain cells is as high as possible. If successive droplets are close enough, they may coalesce, thus drawing a continuous line of cells. While larger droplets lead to a higher number of printed cells, the resolution of the cell printing process decreases. If a high resolution (as for single cell printing) is required, the smallest possible droplet must be ejected, which is generally achieved at laser energy just above the cell printing threshold. Accordingly, the LAB virtually prints cells one by one from a high cell concentration bio-ink ($1 \times 10^8$ cells/ml), thus enabling the fabrication of a tissue engineered product with comparable organization and cell density with living tissues in which multiple cell types are in physical contact with each other.

### 8.2.3.3 Viscosity and Laser Energy Influence Cell Viability

Printed cells may not survive the printing process due to exaggerated laser energy deposition or mechanical shock upon impact on the receiving substrate. It has previously been shown that a minimum shock absorbing receiving hydrogel substrate (like Matrigel™) is required for the mechanical shock absorbance of the printed cells [43, 44]. Alternatively in case of insufficient substrate shock absorbing capability, the increase of the bio-ink viscosity by means of sodium alginate improves cell viability [38]. Beside the bio-ink viscosity and the shock absorbing properties of the receiving substrate, the laser energy must be observed and adapted to the maximum radiation dose that the cells may survive. So far, no LAB-induced alteration of cell biology (in terms of phenotype and DNA nicks) has been detected at suitable parameters [25, 41, 42, 45]. According to actual knowledge, the LAB-process is validated for engineering cell containing tissues, however, further studies must be implemented to rule out any genotoxicity in cell based clinical applications of the LAB-process.

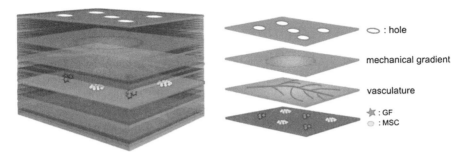

**Fig. 8.5** Schematic principle of the layer-by-layer assembly of complex tissue constructs. These complex tissues feature micropatterns of cells [e.g. mesenchymal stem cells (*MSC*)], biochemical cues [e.g. growth factors (*GF*)], physical cues (e.g. stiffness gradients), and defined shapes (e.g. *holes*). Holes can be processed by laser micro-machining to favor fluid diffusion through the structure, and could also be seeded with endothelial cells

### 8.2.3.4  3-Dimensional Printing: The Layer-by-Layer Approach

Three-dimensional printing of tissue is a challenging task. Tissue engineering applications including hybrid structures for tissue repair demand the management of perfusion and histological complexity. Such applications require a sophisticated architecture with defined properties throughout its entire volume (Fig. 8.5 and [4, 46]). At first sight LAB seems not well suited for building large volume ($cm^3$ size) tissue structures, since the characteristic droplet volume is in the order of 1 pL [47]. Some materials can be used in a layer-by-layer approach to provide volume and or biochemical properties that the bio-ink may not supply, to stabilize the pattern of the printed cells, and support the construct 3D structure in its whole. The layer-by-layer approach is commonly used by the bio-printing approach. The following paragraphs focus on different materials that were used in combination with LAB for the generation of 3D structures, which have potential applications in biological and/or medical fields.

#### Hydrogels as Substrate Layers

For the most part of the literature, hydrogels have been used as substrate to receive cells and to build 3D constructs. Hydrogels have been used for layer-by-layer inkjet printing of cells using fibrin 12, 13 or type I collagen [48]. Sprays of collagen were also used for layer-by-layer cell printing using micro syringe [49, 50]. More recently, LAB was used for the fabrication of a 3D structure with Adipose-tissue Derived Stem Cells in a mixture of blood plasma and sodium alginate [39]. The final structure was then solidified by spraying the $CaCl_2$ cross linker. Using a bio-ink composed of fibrinogen and hyaluronic acid, two arrays of spatially organized ECFC and ADSC were stacked successfully [26]. To each layer of the array, thrombin solution was added, either by wetting or spraying, which resulted in a stable construct made of fibrin.

**Fig. 8.6** Simulated conventional seeding of stacked PCL bio-papers with MG63 cells printed by LAB (**a**), compared to layer-by-layer organization using LAB of MG63 cells and PCL electrospun bio-papers (**b**). The same amount of cells and the same amount of PCL scaffolds have been used in both conditions. *PCL* polycaprolacton

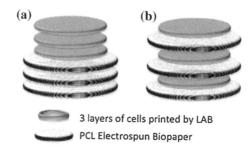

3 layers of cells printed by LAB

PCL Electrospun Biopaper

## Other Bio-Papers

It has been proposed to support the layer-by-layer assembly of biological elements like cells and extracellular matrix, with bio-papers, which are thin films (few hundred microns) of solid biomaterials inserted between successive layers of bio-printed cells (Fig. 8.6). Bio-papers provide the entire structure with mechanical support and may offer to the cells a specific microenvironment (e.g. specific rigidity, porosity) that may not be provided by the liquid bio-ink. The idea of using a bio-paper in a layer-by-layer approach instead of a bulk biomaterial (mm size) is to pattern cells within the material with respect to the diffusion limit of oxygen (200 µm in living tissues). Each layer of the assembly should have a blood supply so that cells within the entire volume of the structure may not suffer from hypoxia. The behavior of three-dimensional hybrid materials built layer-by-layer has been evaluated using electrospun scaffolds of polycaprolacton (PCL) and cells printed by LAB [51]. The electrospinning parameters were adjusted to obtain 100 µm thick films of PCL bio-papers. This study shows that the position of the cells in a three-dimensional tissue engineered product, using a sandwich combining PCL bio-papers and osteosarcoma cell line MG63 printed onto them by LAB, had a significant effect on cell proliferation in vitro and in vivo. PLGA/hydrogel (type I collagen or Matrigel™) bio-papers have been used to print HUVEC [52].

This material is stackable and might be useful for 3D printing. Such bio-paper may be patterned itself, to provide additional control on possible cell migration and differentiation. Alternatively, two different cell types have been co-cultured to stabilize each other in the initial printed pattern [22, 23].

## 8.3 Applications

The micro scale resolution of cell printing by LAB has the potential to deal with the histological complexity of living tissues as shown in the four following examples. Since no clear application of the high resolution cell printing of LAB (as for any other bioprinting technology) has been demonstrated so far, this section is rather speculative. A tradeoff between high resolution cell printing and living tissue complexity and density is postulated. Accordingly potential applications for basic research in cell biology are proposed. Since biological tissues are composed of multiple components in close interactions with each other (cells of different types, proteins and other components of the extracellular matrix), not only three dimensional structures but also multiple cell types and other biological components like extracellular matrix should be considered in a tissue engineered product. Together with the high resolution printing capability of the LAB, it is possible to print different types of cells in close contact with each other, with a high cell concentration, according to a desired spatial organization. The printing resolution achievable by the LAB for multiple cell types and materials is consistent with the study of cell-to-cell, or cell-to-material interactions as well.

### 8.3.1 LAB Engineered Stem Cell Niche

Printing chemical factors onto bio-papers in order to control cell fate (in terms of migration, proliferation and/or differentiation) or to force cells at the desired site of the structure is achievable by LAB. Another approach is the use of LAB for printing patterns of material itself prior to printing cells onto the patterned material [18]. Other studies deal with generating artificial cell niches by co-depositing a suitable combination of stem cells with extracellular matrix components [53]. Indeed, embryonic stem cells have successfully been printed by matrix assisted LAB and have formed embryoid bodies with retained pluripotency [41]. The determinants of stem cell niche renewal, and more generally the biochemical factors that control cell fate and tissue morphogenesis (e.g. hypoxia, ECM rigidity, cell-cell interactions, molecular relationship between adjacent tissues) are still under study. Mechanical and topological determinants of morphogenesis and homeostasis could be studied using bottom-up approaches for engineering tissues with controlled geometric patterns of different types of cells. Future studies may focus on organizing multiple elements like cells, ECM-like materials, and growth factors at different scale of histology.

### 8.3.2 Modelling Capillary Formation

The main limitation especially of thick cellular tissue structures is the time required for the assembly and morphogenesis of a perfused vascular network throughout the entire tissue engineered structure. In certain cases, the assembly and maturation time might be longer than the cell survival time. Micropattern-guided vasculogenesis might help quicken vascular lumen formation as well as branching between the host and the tissue construct. To this aim, endothelial cords have auto-assembled consecutively to endothelial cell alignment with cell-scale accuracy by LAB [22]. Such an approach to modeling endothelial tube formation might be fruitful in the field of vasculogenesis and angiogenesis related research.

### 8.3.3 *Laser Assisted Bio-printed of Tissue Engineering Products*

In vivo transplantation of LAB engineered tissue constructs was demonstrated: Gaebel et al. have used LAB to fabricate a cardiac patch for cardiac regeneration in a rat model of acute myocardial infarction [23]. LAB was used to pattern a co-culture of HUVEC and HMSC onto a 300 μm thick, 8 mm diameter disk of Polyester urethane urea. Healing potential of patterned patch compared similarly to the unpatterned patch, demonstrating the suitability of the LAB procedure for tissue regeneration, and suggesting that patterning may favor faster vasculogenesis and grafting.

Another study hypothesized that the layer-by-layer bio-printing would make the seeding more efficient than the conventional seeding of porous materials. The influence of the three-dimensional organization of MG63 cells and PCL electrospun scaffolds was evaluated regarding cell proliferation in vitro and in vivo. For this purpose, a layer-by-layer sandwich model of assembly was compared to a control hybrid material made of the same amount of material with an alternative 3D arrangement (Fig. 8.6 and [51]). These structures were both evaluated in vitro and in vivo in mouse calvarial defect reconstruction model. These results emphasize the benefit of the layer-by-layer approach to encapsulate cells within a sandwich of PCL bio-papers, either in vitro or in vivo, as far as cell viability and cell proliferation are concerned.

Until now, cell patterning at cell-level resolution has not been studied. Increasing the resolution may help guide faster tissue specific organization like faster vasculogenesis, which may support faster morphogenesis or healing process of vascularized tissues.

### 8.3.4 *In Vivo Printing*

Generally, LAB based studies report on in vitro fabrication, but some preliminary results for in vivo printing exist [54]: Two bone defects were drilled symmetrically in mice calvariae under general anesthesia and the animals were placed inside the LAB workstation. A specific mouse holder was designed in order to position the surface of mice dura mater instead of the quartz substrate (Fig. 8.7a). Then, one defect was refilled by printing thirty layers of a hydroxyapatite slurry. The contra-lateral defect was used as a negative control for bone healing. Three groups, each comprising 10 animals were studied. Animals were sacrificed after 1 week, 3 weeks or 4 weeks. The histological results have shown that the printed material was present in the test defects of all groups (Fig. 8.7b). However, bone repair was inconstant, probably due to the displacement of the printed material after surgery. As a conclusion, in vivo bio-printing has been demonstrated. Future experiment in this model should improve both the sealing procedure of the reconstruction and the biological properties of the printed material for a faster healing process.

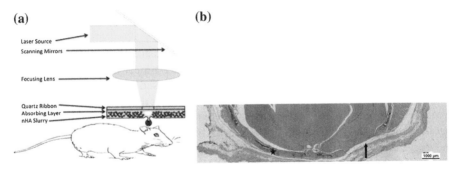

**Fig. 8.7** In vivo printing reconstruction of calvarial bone defect in mouse (**a**). Histological section of mouse calvarial defect (**b**). Complete bone repair on the test side (*star*) was observed in one sample after 3 months. The bone defect control site remains unrepaired (*arrow*)

## 8.4 Conclusions and Perspectives

Laser assisted bio-printing of cells requires taking into account that the droplet ejection mechanism is mainly governed by vapor bubble dynamics. Consequently, the bio-ink should be designed accordingly and spatio-temporal proximity of consecutive laser-induced jets should be considered for optimal printing resolution. Several studies taken together demonstrate the capability of the LAB to print virtually all cell types although many human cell types remain to be validated. These cells can be printed onto numerous biomaterials, either solid or gel, like polymers and nanosized particles of hydroxyapatite. The potential of the LAB to fabricate functional cell containing transplants for tissue repair has been demonstrated, together with the possibility to shunt the transplantation process by operating LAB directly in vivo.

In our opinion, the main issues over the next 5-10 years will concern biological and developmental studies for fundamental research and tissue engineering and repair. Developing tools such as LAB would allow us to create and manipulate the in vitro cell micro environment on demand by controlling intensity and shape of cell patterns and morphogen gradients [55, 56]. Combining LAB with other laser-assisted processes, such as machining and polymerization, should be addressed with specific attention on integrating these different processes in the same workstation to guarantee subcellular resolution. As a direct write method of living cells, the LAB can be combined with other tissue engineering methods [46].

Concerning the layer-by-layer microfabrication of functional tissues that mimic in vivo counterparts, it remains to be determined whether the exact reproduction of the histoarchitecture of living tissue is crucial; in other words, to what extent and resolution cellular self-assembly has to be guided [46]. Future studies involving pattern formation in morphogenesis, specifically the relationship between form and function, should advise this aim. Moreover, the engineering of realistic tissue constructs will help further understanding of tissue physiology and function; this, in turn, will refine tissue engineering strategies and optimize blueprints. To this end, coupling mathematical modeling [57] to the design of micropatterns of cells and morphogens by the abovementioned technologies could be fruitful. Cell-patterning techniques could be promising tools for the fabrication of organotypic tissues, which could elucidate basic cell biology mechanisms, further drug evaluation and toxicity testing in vitro, and become useful platforms within the clinic.

Finally, while cell chip fabrication using LAB can be envisaged, other original applications such as medical robotics should be developed in the coming years, allowing LAB workstations to leave physics laboratories for biological benches.

**Acknowledgments**  We acknowledge financial support from GIS-AMA (Advanced Materials in Aquitaine), ANR (Agence Nationale pour la Recherche), and Région Aquitaine.

# References

1. Hutmacher DW (2000) Scaffolds in tissue engineering bone and cartilage. Biomaterials 21:2529–2543. doi:10.1016/S0142-9612(00)00121-6
2. Mironov V, Boland T, Trusk T et al (2003) Organ printing: computer-aided jet-based 3D tissue engineering. Trends Biotechnol 21:157–161. doi:10.1016/S0167-7799(03)00033-7
3. Jakab K, Norotte C, Marga F et al (2010) Tissue engineering by self-assembly and bio-printing of living cells. Biofabrication 2:022001. doi:10.1088/1758-5082/2/2/022001
4. Mironov V, Visconti RP, Kasyanov V et al (2009) Organ printing: tissue spheroids as building blocks. Biomaterials 30:2164–2174. doi:10.1016/j.biomaterials.2008.12.084
5. Guillemot F, Mironov V, Nakamura M (2010) Bioprinting is coming of age. In: Report from the international conference on bioprinting and biofabrication in Bordeaux (3B'09). Biofabrication 2:010201. doi:10.1088/1758-5082/2/1/010201
6. Klebe RJ (1988) Cytoscribing: a method for micropositioning cells and the construction of two- and three-dimensional synthetic tissues. Exp Cell Res 179:362–373. doi:10.1016/0014-4827(88)90275-3
7. Klebe RJ, Thomas CA, Grant GM et al (1994) Cytoscription: computer controlled micropositioning of cell adhesion proteins and cells. Methods Cell Sci 16:189–192. doi:10.1007/BF01540648
8. Boland T, Xu T, Damon B, Cui X (2006) Application of inkjet printing to tissue engineering. Biotechnol J 1:910–917. doi:10.1002/biot.200600081
9. Nakamura M, Kobayashi A, Takagi F et al (2005) Biocompatible inkjet printing technique for designed seeding of individual living cells. Tissue Eng 11:1658–1666. doi:10.1089/ten.2005.11.1658
10. Saunders RE, Gough JE, Derby B (2008) Delivery of human fibroblast cells by piezoelectric drop-on-demand inkjet printing. Biomaterials 29:193–203. doi:10.1016/j.biomaterials.2007.09.032
11. Brisbane (1971) Pattern Deposit by Laser-Google Patents
12. Young D, Auyeung RCY, Piqué A et al (2002) Plume and jetting regimes in a laser based forward transfer process as observed by time-resolved optical microscopy. Appl Surf Sci 197–198:181–187. doi:10.1016/S0169-4332(02)00322-7
13. Bohandy J, Kim BF, Adrian FJ, (Aug 1986) Metal deposition from a supported metal film using an excimer laser. J Appl Phys 60:1538–1539
14. Schiele NR, Corr DT, Huang Y et al (2010) Laser-based direct-write techniques for cell printing. Biofabrication 2:032001. doi:10.1088/1758-5082/2/3/032001
15. Guillemot F, Souquet A, Catros S, Guillotin B (2010) Laser-assisted cell printing: principle, physical parameters versus cell fate and perspectives in tissue engineering. Nanomed 5:507–515. doi:10.2217/nnm.10.14

16. Barron JA, Krizman DB, Ringeisen BR (2005) Laser printing of single cells: statistical analysis, cell viability, and stress. Ann Biomed Eng 33:121–130
17. Barron JA, Wu P, Ladouceur HD, Ringeisen BR (2004) Biological laser printing: a novel technique for creating heterogeneous 3-dimensional cell patterns. Biomed Microdevices 6:139–147
18. Guillotin B, Souquet A, Catros S et al (2010) Laser assisted bioprinting of engineered tissue with high cell density and microscale organization. Biomaterials 31:7250–7256. doi:10.1016/j.biomaterials.2010.05.055
19. McGuigan AP, Sefton MV (2006) Vascularized organoid engineered by modular assembly enables blood perfusion. Proc Natl Acad Sci 103:11461–11466. doi:10.1073/pnas.0602740103
20. McGuigan AP, Bruzewicz DA, Glavan A et al (2008) Cell encapsulation in sub-mm sized gel modules using replica molding. Plos One 3:e2258. doi:10.1371/journal.pone.0002258
21. Voldman J (2006) Engineered systems for the physical manipulation of single cells. Curr Opin Biotechnol 17:532–537. doi:10.1016/j.copbio.2006.07.001
22. Wu PK, Ringeisen BR (2010) Development of human umbilical vein endothelial cell (HUVEC) and human umbilical vein smooth muscle cell (HUVSMC) branch/stem structures on hydrogel layers via biological laser printing (BioLP). Biofabrication 2:014111. doi:10.1088/1758-5082/2/1/014111
23. Gaebel R, Ma N, Liu J et al (2011) Patterning human stem cells and endothelial cells with laser printing for cardiac regeneration. Biomaterials 32(35):9218–9230
24. Koch L, Kuhn S, Sorg H et al (2009) Laser printing of skin cells and human stem cells. Tissue Eng Part C Methods 091221133515000: doi:10.1089/ten.tec.2009.0397
25. Gruene M, Deiwick A, Koch L et al (2010) Laser printing of stem cells for biofabrication of scaffold-free autologous grafts. Tissue Eng Part C Methods 17:79–87. doi:10.1089/ten.tec2010.0359
26. Gruene M, Pflaum M, Hess C et al (2011) Laser printing of three-dimensional multicellular arrays for studies of cell-cell and cell-environment interactions. Tissue Eng Part C Methods 110629135038006: doi:10.1089/ten.tec.2011.0185
27. Catros S, Fricain J-C, Guillotin B et al (2011) Laser-assisted bioprinting for creating on-demand patterns of human osteoprogenitor cells and nano-hydroxyapatite. Biofabrication 3:025001. doi:10.1088/1758-5082/3/2/025001
28. Duncan AC, Weisbuch F, Rouais F et al (2002) Laser microfabricated model surfaces for controlled cell growth. Biosens Bioelectron 17:413–426
29. Claeyssens F, Hasan EA, Gaidukeviciute A et al (2009) Three-dimensional biodegradable structures fabricated by two-photon polymerization. Langmuir 25:3219–3223
30. Lazare S, Tokarev V, Sionkowska A, Wiśniewski M (2005) Surface foaming of collagen, chitosan and other biopolymer films by KrF excimer laser ablation in the photomechanical regime. Appl Phys Mater Sci Process 81:465–470. doi:10.1007/s00339-005-3260-y
31. Duocastella M, Colina M, Fernandez-Pradas JM et al (2007) Study of the laser-induced forward transfer of liquids for laser bioprinting. Appl Surf Sci 253:7855–7859. doi:10.1016/j.apsusc.2007.02.097
32. Duocastella M, Fernández-Pradas JM, Serra P, Morenza JL (2008) Jet formation in the laser forward transfer of liquids. Appl Phys 93:453–456. doi:10.1007/s00339-008-4781-y
33. Mezel C, Hallo L, Souquet A et al (2009) Self-consistent modeling of jet formation process in the nanosecond laser pulse regime. Phys Plasmas 16:123112. doi:10.1063/1.3276101
34. Brown MS, Kattamis NT, Arnold CB (2010) Time-resolved study of polyimide absorption layers for blister-actuated laser-induced forward transfer. J Appl Phys 107:083103. doi:10.1063/1.3327432
35. Duocastella M, Fernández-Pradas JM, Morenza JL et al (2010) Novel laser printing technique for miniaturized biosensors preparation. Sensors Actuators B Chem 145:596–600. doi:10.1016/j.snb.2009.11.055
36. Engler AJ, Sen S, Sweeney HL, Discher DE (2006) Matrix elasticity directs stem cell lineage specification. Cell 126:677–689. doi:51

37. Engler AJ, Humbert PO, Wehrle-Haller B, Weaver VM (2009) Multiscale modeling of form and function. Science 324:208–212. doi:10.1126/science.1170107

38. Catros S, Guillotin B, Bacáková M et al (2011) Effect of laser energy, substrate film thickness and bioink viscosity on viability of endothelial cells printed by Laser-Assisted Bioprinting. Appl Surf Sci 257:5142–5147. doi:10.1016/j.apsusc.2010.11.049

39. Gruene M, Pflaum M, Deiwick A et al (2011) Adipogenic differentiation of laser-printed 3D tissue grafts consisting of human adipose-derived stem cells. Biofabrication 3:015005. doi:10. 1088/1758-5082/3/1/015005

40. Schiele NR, Koppes RA, Corr DT et al (2009) Laser direct writing of combinatorial libraries of idealized cellular constructs: biomedical applications. Appl Surf Sci 255:5444–5447. doi:10. 1016/j.apsusc.2008.10.054

41. Raof NA, Schiele NR, Xie Y et al (2011) The maintenance of pluripotency following laser direct-write of mouse embryonic stem cells. Biomaterials 32:1802–1808. doi:10.1016/j. biomaterials.2010.11.015

42. Othon CM, Wu X, Anders JJ, Ringeisen BR (2008) Single-cell printing to form three-dimensional lines of olfactory ensheathing cells. Biomed Mater 3:034101

43. Ringeisen BR, Kim H, Barron JA et al (2004) Laser printing of pluripotent Embryonal Carcinoma cells. Tissue Eng 10:483–491

44. Wang W, Huang Y, Grujicic M, Chrisey DB (2008) Study of impact-induced mechanical effects in cell direct writing using smooth particle hydrodynamic method. J Manuf Sci Eng 130:021012. doi:10.1115/1.2896118

45. Hopp B, Smausz T, Kresz N et al (2005) Survival and proliferative ability of various living cell types after laser-induced forward transfer. Tissue Eng 11:1817–1823. doi:32

46. Guillotin B, Guillemot F (2011) Cell patterning technologies for organotypic tissue fabrication. Trends Biotechnol 29:183–190. doi:10.1016/j.tibtech.2010.12.008

47. Guillemot F, Souquet A, Catros S et al (2010) High-throughput laser printing of cells and biomaterials for tissue engineering. Acta Biomater 6:2494–2500. doi:10.1016/j.actbio.2009. 09.029

48. Moon S, Hasan SK, Song YS et al (2010) Layer bylayer three-dimensional tissue epitaxy by cell-laden hydrogel droplets. Tissue Eng Part C Methods 16:157–166.doi:10.1089/ten.tec. 2009.0179

49. Lee W, Debasitis JC, Lee VK et al (2009) Multi-layered culture of human skin fibroblasts and keratinocytes through three-dimensional freeform fabrication. Biomaterials 30:1587–1595. doi:10.1016/j.biomaterials.2008.12.009

50. Lee W, Pinckney J, Lee V et al (2009) Three-dimensional bioprinting of rat embryonic neural cells. NeuroReport 20:798–803. doi:10.1097/WNR.0b013e32832b8be4

51. Catros S, Guillemot F, Nandakumar A et al (2012) Layer-by-layer tissue microfabrication supports cell proliferation in vitro and in vivo. Tissue Eng Part C Methods 18:62–70.doi:10. 1089/ten.TEC.2011.0382

52. Pirlo RK, Wu P, Liu J, Ringeisen B (2012) PLGA/hydrogel biopapers as a stackable substrate for printing HUVEC networks via BioLP. Biotechnol Bioeng 109:262–273. doi:10.1002/bit. 23295

53. Lutolf MP, Blau HM (2009) Artificial stem cell Niches. Adv Mater 21:3255–3268. doi:10. 1002/adma.200802582

54. Keriquel V, Guillemot F, Arnault I et al (2010) In vivo bioprinting for computer—and robotic-assisted medical intervention: preliminary study in mice. Biofabrication 2:014101. doi:10. 1088/1758-5082/2/1/014101

55. Nelson CM, Tien J (2006) Microstructured extracellular matrices in tissue engineering and development. Curr Opin Biotechnol 17:518–523. doi:10.1016/j.copbio.2006.08.011

56. Nelson CM (2009) Geometric control of tissue morphogenesis. Biochim Biophys Acta Bba-Mol Cell Res 1793:903–910. doi:10.1016/j.bbamcr.2008.12.014

57. Baker RE, Gaffney EA, Maini PK (2008) Partial differential equations for self-organization in cellular and developmental biology. Nonlinearity 21:R251–R290. doi:10.1088/0951-7715/21/ 11/R05

# Chapter 9
# Laser-Based Biomimetic Tissue Engineering

**Emmanuel Stratakis, Anthi Ranella and Costas Fotakis**

**Abstract** Tissue Engineering is defined as the technology aiming to apply the principles of engineering and life sciences towards the development of biological substitutes that restore, maintain, or improve tissue function or a whole organ. Its eventual goal is the creation of 3D artificial cell culture scaffolds that mimic the natural extracellular environment features sufficiently, so that cells function in the artificial medium as they would in vivo. Cells in tissue are surrounded by a dynamic cell type-dependent extracellular matrix that provides instructive cues at both the micro- and the nanoscale needed to maintain cell phenotype and behaviour. Cells are thus, inherently responsive to their environment, receptive to micro- and nanoscale features and patterns of chemistry and topography. Lasers are increasingly proving to be promising tools for the controlled and reproducible structuring of biomaterials at micro- and nanoscales. This chapter reviews current approaches for laser based fabrication of biomimetic tissue engineering scaffolds. These include laser processing of natural biomaterials synthesized to achieve certain compositions or properties similar to those of the extracellular matrix as well as novel laser fabrication technologies to achieve structural features on artificial materials mimicking the extracellular matrix morphology on various levels. The chapter concludes

E. Stratakis (✉) · A. Ranella · C. Fotakis
Institute of Electronic Structure and Laser, Foundation for Research and Technology-Hellas,
(IESL-FORTH),P.O. Box 1527, 711 10 Heraklion, Greece
e-mail: stratak@iesl.forth.gr

A. Ranella
e-mail: ranthi@iesl.forth.gr

C. Fotakis
e-mail: fotakis@iesl.forth.gr

E. Stratakis
Department of Materials Science & Technology, University of Crete, 71003 Heraklion, Greece

C. Fotakis
Department of Physics, University of Crete, 71003 Heraklion, Greece

V. Schmidt and M. R. Belegratis (eds.), *Laser Technology in Biomimetics*,
Biological and Medical Physics, Biomedical Engineering,
DOI: 10.1007/978-3-642-41341-4_9, © Springer-Verlag Berlin Heidelberg 2013

with the wealth of arising possibilities, demonstrating the excitement and significance of the laser based biomimetic materials processing for tissue engineering and regeneration.


## 9.1 Introduction

Tissue Engineering is defined as the technology aiming to apply the principles of engineering and life sciences towards the development of biological substitutes that restore, maintain, or improve tissue function or a whole organ [1]. Its primary research aim is, therefore, to provide scaffolds that are biocompatible and capable of integrating living cells that function in the artificial medium as they would in vivo. In this respect, the ideal artificial material for in vitro cell growth is a scaffolding structure that closely mimics the native environment in the body and thus the characteristics of the natural extracellular matrix (ECM).

Biomimetic tissue engineering usually involves four steps: (i) identifying a natural or synthetic material suitable for the fabrication of a scaffold for a specific tissue, (ii) biomimetic structuring of the material to attain an ECM-like architecture, (iii) Further functionalization of the scaffold towards identifying the optimal combination of morphology and chemistry, i.e. surface energy for each specific type of regenerated tissue and (iv) cell seeding into the scaffold for cell culturing in vitro or in vivo.

Regardless the material type used, natural or artificial, of primary importance is the realization of an ECM-like architecture that complies with and provides attachment sites comparable in size and morphology to the ones found in vivo: as presented in Fig. 9.1a, 2D substrates do have limitations, considering that cells in tissue are surrounded by a 3D dynamic cell type-dependent ECM that provides instructive cues at both micro- and nano-scales needed to maintain cell phenotype and behaviour. Cells are thus, inherently responsive to their environment, and receptive to micro/nanosized features and patterns of chemistry and topography [2, 3]. The structural geometry and morphology are thus important parameters for designing artificial cellular scaffolds and a specific scaffold structure has to be tuned for each tissue engineering application. In this context, manufacturing schemes enabling the controlled and reproducible structuring of biomaterials at micro- and nano-scales are very desirable [4]. Lasers are increasingly promising tools for 3D structuring of scaffolding materials with micro-, submicro- and nano-scale resolution. Further advantages of laser structuring include its very high fabrication rate, non-contact interaction, applicability to many types of biomaterial substrates and reproducibility. Furthermore lasers can be easily incorporated into computer-assisted design (CAD) based fabrication systems for complex and customized 3D matrix structure manufacture. Such systems gave rise to a versatile class of scaffold production techniques which are laser-based solid-free-form (SFF) fabrication techniques [5]. The SFF fabrication is a rapid prototyping (RP) technique which facilitates the control over macroscopic properties, such as scaffold shape and microscopic internal architecture.

**(a)**                                                                          **(b)**

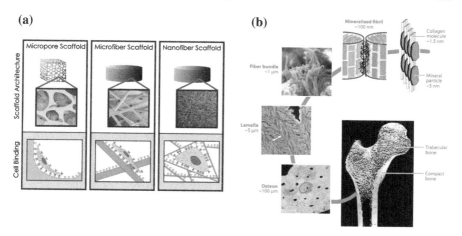

**Fig. 9.1  a** The role of scaffold's length scales on cell behavior. As geometric features become smaller, changes in cell morphology and response can be observed. Cells attached to scaffolds with microscale architectures flatten and spread as if cultured on flat surfaces. On the other hand, cells are more spatially interactive on scaffolds with nanoscale architecture. Nano-featured scaffolds exhibit larger surface areas to adsorb proteins, presenting many more binding sites to cell membrane receptors. Adapted from: Stevens MM, George JH (2005) Science 310:1135–1138. **b** Hierarchical organization of bone over different length scales. Adapted from: Dunlop JWC, Fratzl P (2010) Annu Rev Mater Res 40:1–24

This chapter reviews the state of the art of laser-based fabrication of tissue engineering scaffolds exhibiting biomimetic micro- and nano-topography. Different emerging methodologies are presented comprising laser processing of natural biomaterials synthesized to achieve certain compositions or properties similar to those of the ECM, as well as novel laser fabrication technologies to achieve structural features on artificial materials mimicking the ECM architecture on various length scales. Notwithstanding, further functionalization of the laser structured scaffolds is required towards identifying the optimal combination of biomimetic morphology and chemistry, i.e. surface energy for each specific type of regenerated tissue. Controlling surface energy is a key factor in biomimetic tissue engineering and ECM components should be integrated within scaffolds through covalent conjugation to the scaffolding biomaterials [6–9].

## 9.2  Laser Processing of Biomaterials for Biomimetic Tissue Engineering Scaffolds

### 9.2.1  Overview of the Biomaterials Used for Laser-Based Tissue Engineering Applications

Biomaterials synthesis and development have played a key role in the advancement of tissue engineering applications. Different biomaterials utilized for tissue engineering scaffolds can be artificial or natural and are typically designed with a variety of properties in mind (including chemical, physical, and mechanical). In the following the different biomaterials used for the laser-based biomimetic structuring of tissue engineering scaffolds are presented.

Artificial scaffolds development initially relied upon employing traditional engineering materials, including metallic, semiconducting, ceramic and polymeric ones. For example, titanium, silicon, hydroxyapatite, polyethylene and silicone-based polymers have been used for a wide variety of tissue engineering applications. In general, these materials are designed to be biologically inert to minimize negative consequences of biomaterial-tissue interactions.

Polymeric artificial biomaterials were readily synthesized for tissue scaffolds applications. Initial developments focused on satisfying various criteria, including nontoxic monomers, application-specific polymer chemistry and optimal biocompatibility. Polyesters including poly($\alpha$-hydroxy esters) such as poly(D,L-lactide) (PLA), poly(glycolide) (PGA), and their copolymers (PLGA) have been used extensively as tissue engineering biomaterials because of hydrolytically and enzymatically cleavable ester linkages and monomers that can be readily metabolized [10]. Synthetic biodegradable elastomers are also a class of biomaterials allowing significant control over various synthesis parameters including monomer feed ratios and curing methods. Further advantages are the ability of rapid, facile and scalable synthesis of biomaterials with a potentially wide range of properties [11, 12]. Initial work focused on synthesizing elastomeric scaffolds using naturally occurring monomers that were thermally cross-linked to form polyester networks [13]. However, the aggressive conditions required for polymerization of thermally cross-linked elastomers have hindered their use, especially for tissue engineering strategies that encapsulate growth factors or cells directly into the network. Photo-crosslinkable biomaterials offer significant advantages compared to their thermally cross-linked counterparts, such as (i) rapid and mild cross-linking conditions, (ii) the potential to pattern structures on thin films through photolithography, and (iii) the ability to precisely control cross-linking through chemistry, that provides an additional tool to finely tune physical, mechanical, and chemical properties of polymer scaffolds. Most synthetic routes to obtaining photo-crosslinkable biomaterials begin with a polymeric precursor that is chemically modified through the addition of photoactive groups such as acrylates or methacrylates [14]. For example Poly(ethylene glycol) (PEG), and Poly($\alpha$-hydroxy esters) have been extensively used in soft-tissue engineering applications. Moreover, ring-opening polymerizations of PLA and poly($\varepsilon$-caprolactone) (PCL) monomers have been used to produce photo-crosslinkable networks. Other types of synthetic photo-crosslinkable polymers have been explored for tissue engineering, including acrylated forms of poly(vinyl alcohol) and poly(glycerol-co-sebacate).

Hybrid materials obtained from the integration of biocompatible polymer with bioactive inorganic material is an emerging class of biomaterials used for the fabrication of scaffolds via laser processing. Among the advantages is that such materials might provide tunable mechanical properties, strength and toughness via inorganic (e.g. ceramic) part, plasticity and elasticity via organic (e.g. polymer) part. The sol-gel method has been employed for the synthesis of hybrid organic-inorganic composite materials [15, 16] and particularly sol-gel derived bioactive glass foams have been synthesized in order to be used in bone regeneration [17, 18]. In vitro cell studies in the presence of these foams have shown an increase in osteoblast proliferation and collagen production [19] as well as the stimulation of the formation and mineralisation of bone nodules [20] reinforcing their potential.

Finally, recent biomimetic strategies have focused on the design of biomaterials that mimic and integrate favourably with host tissue in an effort of minimizing the potentially deleterious effects of this interaction. Accordingly, natural biopolymers have been adopted either in pure form or after being modified with various reactive side groups to create synthetic analogues that can be cross-linked by photoinitiation. Natural biopolymers that have been widely used for laser-based tissue engineering applications include proteins, i.e. collagen, gelatine and polysaccharides, i.e. alginate, chitosan, hyaluronic acid, chondroitin sulphate etc. Although naturally occurring biomaterials are most closely simulating the native cellular environment, large batch-to-batch variations upon isolation from biological tissues is the main limitation for their wide application. Poor mechanical performances is also a drawback for transplantation scaffolds made from natural polymers, such as collagen and chitin, which cannot be easily melted upon heating but require the presence of special solvents [21, 22].

## 9.2.2 Laser-Based Methodologies for Biomimetic Tissue Engineering

The primary aim of tissue engineering is the realization of cell culture scaffolds that mimic the 3D architecture of the ECM at different length scales and complies with and provides cellular attachment sites comparable in size and morphology to the ones found in vivo. A multi-scale approach to scaffold structure is especially important in mimicking living systems, because nature often derives properties from multi-scale or hierarchical structures [23]. For example, bone has multiple organizational scales that yield superior mechanical properties, from interacting nanoscale collagen and hydroxyapatite crystals to micro-scale lamella and osteons (Fig. 9.1b). Multiple levels of structural control allow for engineering of unique properties including scaffold size and shape, pore size and geometry, porosity, pore interconnectivity, and surface topology. In this respect, the fabrication of, precisely controlled, porous structures in complex 3D anatomical shapes exhibiting hierarchical surface roughness at meso $(10^{-1}-10^{-3}\,\text{m})$, micro $(10^{-3}-10^{-6}\,\text{m})$ and nano $(10^{-6}-10^{-9}\,\text{m})$ scales is, ideally, required. "Hierarchical" reflects the property that morphological features at scales from the nanometre to millimetre level will determine how well the scaffolding surface mimics the ECM structure, facilitates cell adhesion, growth and proliferation and guides cells to form functional tissue.

Lasers are increasingly proving to be a suitable tool for laser biomaterials processing at multiple length scales. Although, in the far optical field, the minimal achievable structure size is limited by diffraction to the order of wavelength (microscale), the optimal interplay between the laser and material parameters may allow the fabrication of features with dimensions beyond this diffraction limit (nanoscale). In the last decade, this has been accomplished via the application of femtosecond (fs) lasers in biomaterials' processing, which has been increasingly proving to be a powerful approach to overcome the diffraction limit and improve the lateral structure resolution in the fabricated tissue scaffolds [24]. One of the most useful properties of fs laser induced modification is the limited size of the affected volume. Material structuring with laser pulses is induced by optical breakdown, which generates plasma at the focal point of the laser. Because the plasma recombines before thermal diffusion, shock wave propagation and cavitation set in, ablation of the substrate is confined, at least initially, to a small volume. Although the intensity required to initiate breakdown is fairly high the short duration of the pulse allows the threshold intensity to be achieved with a modest fluence. The combination of localized excitation and low threshold fluence can greatly reduce the extent of collateral damage to surrounding areas, so that the size of the affected material can be beyond the diffraction-limited focusing volume. Another key advantage of the use of fs lasers is that due to high peak powers attained nonlinear absorption effects can be driven. This opens new channels for direct nanowriting of transparent scaffolding materials, while three-dimensional structuring becomes feasible. As a consequence, non-linear absorption laser processing techniques provide excellent control over 3D micro- and submicron materials' structuring. Finally, fs lasers allow the unique possibility for hierarchical dual scale, micro- and nanostructuring, which is important for tissue engineering applications [25].

Modern laser-based techniques for biomimetic tissue scaffold fabrication take advantage of the unique features of fs laser-biomaterials interaction and are summarized in Table 9.1. Such techniques rely on different modification processes induced upon absorption of laser photons and subsequent electronic excitation of the irradiated biomaterial. For incident fluences above the material's damage threshold, energy transfer leads to nearly instantaneous mass removal and ablation occurs either by thermal vaporization or by photochemical degradation. At lower fluences, the bonds remain intact and laser energy is converted into heat, while localized heating can be used to sinter or melt the material. On the other hand, photopolymerization takes place when photoinitiator molecules absorb photons and form radicalized species that can initiate polymerization reactions. By moving the laser focus three-dimensionally through the photosensitive biomaterial, complex 3D structures, in some cases with resolution beyond the diffraction limit, can be fabricated. Depending on the principal modification process, laser-based scaffold fabrication methods include selective laser sintering (SLS), stereolithography (SLA), non-linear, two-photon photopolymerization (2PP) and laser ablation texturing (LAT).

It should be emphasized that laser structuring techniques can be readily incorporated to computer aided design and manufacture systems (CAD/CAM systems) for complex and customized 3D scaffold structure design and subsequent

**Table 9.1** Laser-based fabrication methods for biomimetic tissue engineering: Current state of the art

| Fabrication method | Selective laser sintering (SLS) | Stereolithography (SLA) | Direct writing via two-photon polymerization (2PP) | Direct writing via laser ablation micro/nano texturing (LAT) | Laser bio-printing (LBP) |
|---|---|---|---|---|---|
| Advantages | Microporosity induced in the scaffold; A broad range of materials can be used; High mechanical strength; No support structure needed; Fast processing; Potential for double-scale roughness | Rapid response rate; high-form precision; Simple; No special equipment needed; Conductive and mechanical stable; Enables the incorporation of bioactive molecules | Simple;fast; true 3D process; high resolution; Control of external and internal morphology; Precise geometries and pattern; Computer-controlled; Can incorporate biological /bioactive materials | Simple and fast (high fabrication rate); High resolution; Enhanced range of materials can be used; Geometry Control; Double-scale roughness; High aspect ratio; Can process natural polymers | Controlled Cell/Biomaterials positioning; High cell densities possible; |

(continued)

**Table 9.1** continued

| Fabrication method | Selective laser sintering (SLS) | Stereolithography (SLA) | Direct writing via two-photon polymerization (2PP) | Direct writing via laser ablation micro/nano texturing (LAT) | Laser bio-printing (LBP) |
|---|---|---|---|---|---|
| Disadvantages | Material must be in powder form; High temperatures during process (eliminates the possibility of incorporating biological /bioactive materials); Powdery surface finish; Trapped powder inside scaffold; Lack of double-scale roughness.; 3D | Limited choice of photopolymerizable and biocompatible liquid polymer materials; 2.5 D; Lack of roughness control | Low fabrication rate; Multisteps involved; Material has to be photosensitive; Lack of double-scale roughness | Top-down; Limited control over nanoroughness 2.5 D; 3D only at certain architectures; Can be upgraded in 3D using multiple steps | Scale up is difficult due to stacking of 2D layers; Not suitable for constructs in the mm range |

(continued)

**Table 9.1** continued

| Fabrication method | Selective laser sintering (SLS) | Stereolithography (SLA) | Direct writing via two-photon polymerization (2PP) | Direct writing via laser ablation micro/nano texturing (LAT) | Laser bio-printing (LBP) |
|---|---|---|---|---|---|
| Features | Porous Networks; Microporous - Complex architectures | Nano-roughness; Micropores; Nanopores; Craters; Needle like nano-peaks; Pillar or well arrays; Grooves | Micro and submicron pores; Complex 3D architectures | Cones or crater arrays; Micro/Nano-roughness; Grooves; Micropores; Nanopores; Complex architectures | 2D Cell/Biomaterials patterns; Cell/Biomaterials seeding into scaffolds |
| Resolution | $50\,\mu m$ | $1\,\mu m$ | $100\,nm$ | $100\,nm$ | $10\,\mu m$ |
| Materials | Ceramics(HA); Synthetic Polymers (PEEK, PCL); Metals; Different combinations of the above | Polymers; Ceramic Composites; Biomolecule composites; Metals; Semiconductors | Resins; Synthetic Polymers; Biodegradable Polymers; Organic-inorganic hybrids; Ceramic composites | Metals; Ceramics; Semiconductors; polymers; natural polymers; scaffolds; implants | Biomolecules; Cells; Biopolymers; |

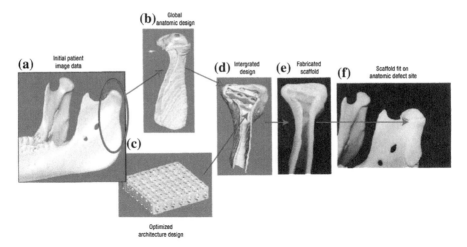

**Fig. 9.2** Image-based procedure for the fabrication of a scaffold fit on anatomic defect site. **a** A coherence tomography (as shown here) or magnetic resonance imaging scan serves as starting point for designing scaffold exterior. **b** The scaffold exterior shape is created with additional features for surgical fixation. **c** Architecture image-design is created using computational topology design. **d** Global anatomic and architecture design are integrated using boolean image techniques. **e** SFF is used to fabricate design from degradable biomaterial. **f** Final fabricated scaffold fits well on the intended anatomic reconstruction site. Adapted from: Hollister SJ (2005) Nat Mater 4:518

reproducible and cost-effective fabrication. SFF methods use a triangular facet surface representation of a structure and build three-dimensional objects layer-by-layer, via structuring or consolidating mater in precisely selected regions. Figure 9.2 summarizes the essential steps followed for SFF-assisted scaffold fabrication. Considering that the SFF methodologies have been reviewed in detail by many articles [26–29], this chapter will only present the most important scaffold architectures prepared via laser-based SFF techniques, without focusing on how computer aided design of those scaffolds had been performed. It is important that such techniques enable the rapid investigation of a wide range of scaffold geometries with a high degree of dimensional control and with fewer limitations on the scaffold exterior shape or the porous architecture [5]. This can give rise to a versatile class of laser-based rapid prototyping SFF scaffold fabrication systems that could potentially be commercialised for mass production and thus attract considerable attention in the following years.

## 9.2.3 Laser Processing of Artificial Biomaterials

The selective laser sintering (SLS) technique employs a high-power laser to sinter thin layers of powdered materials to form solid 3D objects. It has been used for the fabrication of bioceramic scaffolds that can aid the regeneration of hard tissues. Accordingly, pure biopolymer powders such as polyetheretherketon (PEEK) [30], ultrahigh molecular weight polyethylene [31], hydroxyapatite (HA) and physically blended mixtures of PEEK and HA powders have been used [32]. Furthermore fabrication of scaffolds from biodegradable polymers, such as PCL and PLLA has been also been demonstrated. For example, Fig. 9.2 presents PCL a scaffold for replacement of skeletal tissues that mimic mandibular condyle anatomic designs [33]. The scaffolds were seeded with bone morphogenetic protein-7 (BMP-7) transduced fibroblasts. In vivo results show that these scaffolds enhance tissue in-growth, on top of possessing mechanical properties within the lower range of trabecular bone. Compressive modulus (52–67 MPa) and yield strength (2.0–3.2 MPa) were in the lower range of properties reported for human trabecular bone. Other in vivo radiographic studies performed by Kanczler et al. demonstrated that SLS-fabricated polylactic acid scaffolds supported regrowth and bridging of bone gaps in mice [34].

Recent research efforts have been focused on performing SLS with composite nontoxic biocompatible materials that are used in artificial tissues. Several composites, including bioactive ceramics, thermoplastic biodegradable polymers, polymer-coated metals and metals have been processed by means of selective laser sintering. By combining biodegradable polymers and bioactive ceramics, such as HA and b-tricalcium phosphate (b -TCP), composite scaffolds such as HA/PCL, HA/PLGA and b-TCP/PLGA were processed by SLS [35–37]. Duan et al. studied the use of bio-nano-composite microspheres, consisting of carbonated hydroxyapatite (CHAp) nanospheres within a PLLA matrix to produce scaffolds. PLLA microspheres and PLLA/CHAp nanocomposite microspheres of a size of 5–30 μm, suitable for the SLS process, were prepared by emulsion techniques. Finally, biomimetic, SLS-fabricated bionano-composite scaffolds comprising bioactive Ca-P nanoparticles and biodegradable polymers, have been demonstrated to offer a promising approach for bone tissue regeneration [38].

SLA produces computer-designed 3D solid objects in a layer by layer procedure through a selective photo-polymerization of a liquid resin containing precursor and photoinitiator molecules. Polymerization relies on the release of free radicals upon interaction of photoinitiator molecules with UV light. SLA has been initially used to provide HA ceramic scaffolds for orbital floor implants. The building material was a suspension of fine HA powder into a UV-photocurable resin, acting as a binder to hold the HA particles together [39]. Using a UV laser, the resin is burnt out while the HA powder assembly was sintered for consolidation. A similar approach was used by Griffith and Halloran [40] that produced ceramic scaffolds using suspensions of alumina, silicon nitride and silica particles with a photo-curable resin. SLA had also been used for the direct fabrication of 3D biopolymeric scaffolds [39]. Matsuda et al. used a custom made SLA setup to produce computer-aided 3D microarchitectures made of an epsilon-caprolactone biodegradable photopolymer [41]. In another study, Cooke et al. used a biodegradable resin mixture of diethyl fumarate, poly(propylene fumarate) and bisacylphosphine oxide as photoinitiator to produce scaffolds for bone tissue engineering [42]. It has also been successfully utilized to build bone tissue scaffolds from photo-crosslinkable poly(propylene fumarate) with highly interconnected porous structure and porosity of 65 %. The scaffolds were coated by applying acceler-

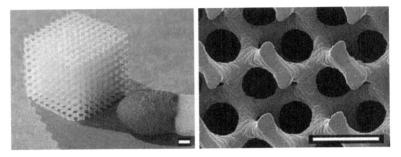

**Fig. 9.3**  Poly(D,L-lactide) (PDLLA) scaffolds with a gyroid architecture fabricated by SLA. Scale bars represent 500 μm. Adapted from: Melchels FP et al (2009) Biomaterials 30:3801

ated biomimetic apatite and arginini-glycine-aspartic acid peptide coating to promote cell behaviour. The coated scaffolds were seeded with MC3T3-E1 pre-osteoblasts and their biologic properties were evaluated using an MTS assay and histologic staining. Later, Dharwala et al. reported the rapid prototyping of tissue engineering constructs using photopolymerizable hydrogels [43]. A more biomimetic complex polymeric scaffold fabricated by SLA is presented in Fig. 9.3. It exhibits a porous and gyroid architecture structured in a resin based on poly(D,L-lactide) macromonomers and non-reactive diluent [44]. It was observed that pre-osteoblasts readily adhered and proliferated well on these scaffolds.

Contrary to SLS, SLA does not involve high processing temperatures, therefore, biological materials such as proteins, antimicrobial agents and viable cells may be directly incorporated within scaffolds following the SLA process. This capability is beneficial for biomimetic tissue engineering applications, in which incorporation of cells or other biomimetic ECM entities into the starting materials may enable the development of more functional scaffolds. For example, Lee et al. fabricated 3D scaffolds for tissue engineering containing poly(lactic-co-glycolic acid) microspheres, which incorporated the growth factor bone morphogenetic protein 2 [45]. Furthermore, Arcaute et al. incorporated dermal fibroblasts within poly(ethylene glycol) that was processed using SLA, and demonstrated cell viability of at least 87 % [46]. However, it should be noted that UV light, monomers, oligomers or photoinitiator molecules may cause damage to deoxyribonucleic acid molecules, cell components or biological materials; these concerns must be addressed before incorporation of cells during the SLA process is translated into clinical use.

2PP has been demonstrated as a technology for the fabrication of 3D structures with submicrometer resolution [47, 48]. When the beam of a fs infrared laser is tightly focused into the volume of a photosensitive biomaterial, the polymerization process can be initiated by two-photon absorption within the focal volume. By moving the laser focus three-dimensionally through the biomaterial, 3D scaffolds with nanometric features can be fabricated. The first materials employed in scaffolds fabrication via 2PP were photocurable commercially available sol-gel hybrid materials such as the SU-8, ORMOCER®, Ormocomp® and the polymer Accura™ SI10 [49–51]. These studies demonstrate the great potential of 2PP technology for the fabrication of 3D scaffolds with submicron resolution with high reproducibility and at a good speed, based on a digitized model. Different cell types such as granulosa cells, endothelial cells and primary hepatocytes cultured within these methods for tissue engineering. Recently, there have been a series of publications on the application 2PP for the development of scaffolds based on specifically designed organic/inorganic hybrid sol-gel materials [48, 52]. Custom made organic/inorganic hybrid materials has been synthesized from some groups, enabling the fabrication of three-dimensional scaffolds with tunable mechanical properties. According to this biocompatible and/or biodegradable hybrid materials have been applied. Particularly, for the fabrication of bioactive, non-biodegradable 3D scaffolds with tuneable shape, size, porosity and mechanical properties, methacryloxypropyl trimethoxysilane (MAPTMS) and methacrylic acid (MAA) were used as the organic photopolymerizable monomers while zirconium n-propoxide or titanium (IV) isopropoxide and the alkoxysilane groups of MAPTMS served as the inorganic network forming moieties. The photopolymerization was occurred due to the addition of the photoinitiator (Irgacure 365) [53, 54]. Altering the metal-MAA molar ratio, Psycharakis and co-workers studied the fibroblasts adhesion and growth in scaffolds with different topologies and mechanical properties [55]. Additionally, hybrid materials such as poly-caprolactone based photopolymers, that are not only biocompatible and biodegradable, but also degrade on a similar time scale as tissue formation [56–58] have been used for the investigation of relationship between scaffold topology and of neuronal cells on 3D scaffolds fabricated using 2PP femtosecond direct laser writing (Fig. 9.4a, ii) [59]. These studies indicate the structuring of reliable microporous scaffolds of complex 3D geometrical shapes that can be used for guided neuronal growth and cell delivery vehicles.

LAT is based on small amount of mass removal from bulk material via laser ablation. Contrary to the use of long laser pulses, where the resolution depends on the micron-sized beam, sub-microscale features may be achieved when texturing with fs lasers. Furthermore, laser micromachining of structures with interior geometries could be performed with fs laser ablation of optically transparent materials [60]. Most important, fs LAT provides the unique possibility for single-step simultaneous structuring at both micro and nanoscales [61]. Cellular scaffolds exhibiting dual scale, micro and nano-dimensions, important for biomimetic tissue engineering applications, can therefore be realized [62]. Finally, LAT can be employed in a two-step process for controlled modification of scaffolds produced by other techniques. Depending on the particular scaffold and process parameters, it may involve texturing at additional length scales, gradient texturing, grain refinement and other functional modification including phase transformations, alloying and mixing of multiple materials and formation of composite layers on the surface without actually affecting the pristine scaffolding material itself. Gradient scaffold morphologies exhibit a continuous spatial change in a given property and allow a continuum of these property values to be tested on a single biological substrate. Therefore, gradient textured scaffolds enable high-throughput screening of cell-biomaterial interactions and enhance traditional tissue engineering techniques. Gradient morphologies could also replicate in vivo physical and chemical gradients in vitro for tissue-engineered

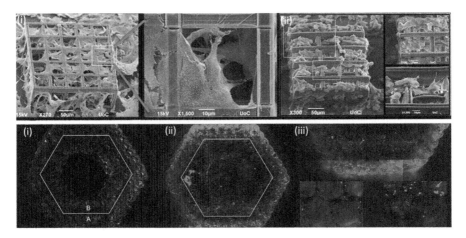

Fig. 9.4 (*Top*) **i** SEM photographs of 3D cell cultures on woodpile microstructures fabricated via two photon polymerization technique. Fibroblast cell cultures on zirconium-based scaffolds after 3 days of culture. **ii** PC12 neuronal cell cultures on biodegradable photocurable PLA scaffolds. Different magnifications are shown. (*Bottom*) Scaffold seeded with cells by means of laser-induced forward transfer (LIFT): **i** dark field image. The white hexagon indicates the border between the two scaffold areas seeded with vascular smooth muscle-like cells (vSMCs) (*A*) and endothelial cells (ECs) (*B*), respectively; **ii** fluorescence image indicating the location of different cell types after the LIFT procedure; **iii** detailed image of the border area. The insets demonstrate that a sharp transition from vSMCs to EC-seeded regions is present along the entire thickness of the scaffold. Adapted from: Ovsianikov A et al (2010) Biofabrication 2:014104

constructs. This is an important capability provided by LAT in view of the fact that currently, most studies on cell-biomaterial interaction are performed on scaffolds with spatially homogeneous properties.

LAT of a large range of biocompatible materials has been used to create scaffolds with controlled pore size and porosity. However, most studies were focused on the application of the fabricated scaffolds for investigating and controlling cell adhesion, orientation and proliferation [63–66]. LAT of electrospun polymers had been also considered in order to improve cell proliferation within the interior regions of such scaffolding materials [67]. Using a fs LAT methodology, Ranella et al. have demonstrated the realization of 3D biomimetic micro/ nano rough textured scaffolds on Si [62]. Tailoring of the morphological features of such scaffolds can be advantageously achieved by tuning the laser structuring parameters resulting in scaffolds of different micro and nanoscale roughness, 3D porosity and surface energy. In particular, the scaffolds prepared at high laser fluences can quantitatively mimic both the structure and the water repellent characteristics of the natural Lotus leaf [61]. This work demonstrated that it is possible to preferentially tune fibroblast cell adhesion and growth, through choosing proper combinations of topography and chemistry of the 3D micro/nano scaffolds. Most important, it is shown that such scaffolds enable primary neurons to grow in 3D, without the use of synthetic ECM coatings or other chemotropic growth factors [68]. Indeed, only few cells survived on a flat Si substrate

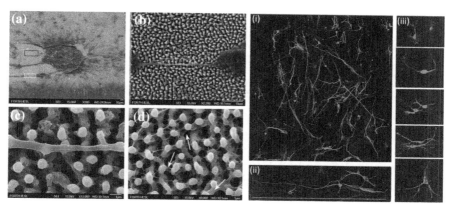

**Fig. 9.5** (*left*) (**a**) Neuronal cluster on the Si spikes area; (**b**) Detail corresponding to white lined inset of (**a**), showing a long neurite that has attached and grown over the spikes; (**c**) Detail corresponding to white lined inset of (**b**), showing protrusions of neurolemma growing over and engulfing the top of the spikes; (**d**) Detail corresponding to *black lined* inset of (**a**), showing the 3D web of cytoplasmic processes growing along the direction vertical to the culture plane. The arrows indicate how multiple processes may initiate from one neurite. (*Right*) (**i**) Primary neuronal culture on Si spikes area, immunostained for neuron-specific immunohistochemistry using beta III tubulin primary antibody (*red*) at 5 days in vitro. The phenomenal neuritic sprouting and extension on the substrate surface is clearly illustrated; (**ii**) Some of the neurons showed extraordinary extension. (**iii**) Examples of different neuronal types and the formation of varicosities were placed on the *right*. Adapted From: Stratakis E et al (2011) Biomicrofluidics 5:013411

used as a control that verified the role of microstructures in absence of synthetic ECM. On the contrary, neuron cells developed into a dynamic cellular aggregate, with long neurites sprouting throughout the microstructured scaffold area (Fig. 9.5). Moreover, a 3D neuron network of nanoscale thin neuritic extensions was formed along the third dimension (perpendicular to the culture plane) utilising the space towards the base of the structures.

LAT has been recently employed for the microfabrication of a 3D biomimetic scaffold for myocardial repair applications [9]. Anisotropy in structural and mechanical properties is a prominent feature of many native tissues and gives rise to unique structure-function relationships that are crucial to maintaining healthy organs. For example, ventricular myocardium is a hierarchical, quasi-lamellar tissue in which functional syncytia of cardiomyocytes (cardiac muscle fibres) are interwoven within collagen. Such structure ultimately results in directionally dependent mechanical and electrical properties, collectively termed cardiac anisotropy. Using excimer laser ablation of a synthetic bioresorbable elastomer (poly(glycerol sebacate)), 3D biomimetic scaffolds with anisotropic structural and mechanical properties were fabricated [9]. The resulting scaffolds were accordion-like, honeycomb-shaped cellular structures with closely matched mechanical properties compared to native adult rat ventricular myocardium, including structural features roughly 200 μm in size and stiffness on the order of 2.1 MPa (Fig. 9.6). Neonatal rat heart cells were seeded on these scaffolds and cultured in vitro for up to one week. It is observed that the laser

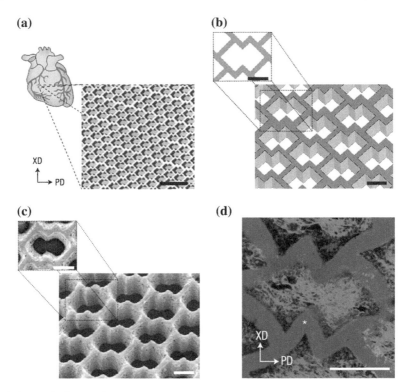

**Fig. 9.6** (**a**), (**b**) Schematic diagrams illustrating the accordion-like honeycomb scaffolds exhibiting anisotropic mechanical properties similar to native myocardium. Preferred (PD) and orthogonal cross-preferred (XD) material directions, respectively corresponding to circumferential and longitudinal axes of the heart are indicated. Scale bars: 1 mm (**a**) and 200 μm (**b**); (**c**) Scanning electron micrographs of the artificial accordion-like honeycomb scaffold made by LAT of poly(glycerol sebacate). Scale bar 200 μm; (**d**) Neonatal rat heart cells were cultured for 1 week on accordion-like honeycomb scaffolds, fluorescently labelled for F-actin (*green*), counterstained for nuclear DNA (*blue*) and imaged by confocal microscopy. Scaffold is indicated by the white asterisk. Adapted from Engelmayr GC Jr et al (2008) Nat Mater 7:1003

textured 3D scaffolds promote heart cell alignment and created a tissue morphology that is similar to native cardiac tissue. Furthermore, tissue constructs formed from these scaffolds exhibited directionally dependent electrophysiological properties as well. The same methodology has been lately applied for the development of scaffolds for heart valve tissue engineering [69].

Classical tissue engineering refers to seeding isolated cells on fabricated scaffolds. A disadvantage of this approach is that cells cannot be placed in defined locations of the scaffold. Moreover, when considering organs the various cell types are not randomly distributed within the tissue but specifically arranged to comply with certain function. In this respect, controlled positioning of cells on the scaffold is therefore desirable. Bioprinting is an emerging technology referring to manufacture complex

3D scaffolds, tissues and organs by printing [70]. It allows cells to be positioned in a controlled way onto scaffolds and/or together with biomaterials. The ultimate aim of such RP approach is building three dimensional biological structures or functional organs, layer-by-layer, from the bottom up [71, 72]. Laser based biomaterials printing methods have been found to be promising in patterning biomolecules and cells onto a substrate. All different techniques work with the Laser Induced Forward Transfer (LIFT) principle that allows printing biomaterials, including cells, with micrometer resolution [73]. Recently, a unique combination of two laser based methods, 2PP for scaffold fabrication and LIFT for cell seeding, has been demonstrated [74]. It is shown that LIFT can be used to deposit multiple cell types precisely within the volume of the 2PP-fabricated scaffold (Fig. 9.4b). Combinations of laser based scaffold fabrication with laser-based bioprinting techniques present a promising approach for the realization of complex biomimetic tissue scaffolds.

In addition to direct scaffold fabrication, laser structuring techniques have been used to fabricate scaffold master moulds for casting biomaterials that are not compatible with such techniques [5, 75]. In this case end scaffolds are fabricated by means of replica moulding methods. This is a simple approach to overcome limitations with regard to scaffolding materials selection. For example, Chu et al. [76] developed a lost-mould technique to produce implants with designed channels and connection pattern. SLA was used to create epoxy moulds designed from negative image of implants. A HA-acrylate suspension was initially cast into the mould. Then the mould and the acrylic binder were removed by pyrolysis while the HA scaffold submitted to a sintering process. The finest channel size achieved was about 366 μm and the range of implant porosity between 26 and 52 %. Moreover, the application of laser-textured solid scaffolds as a template to controllably transfer gradient micromorphologies on different types of polymeric materials, including bioerodible ones, has been recently demonstrated [77]. The original master-substrates were fabricated using ultrafast laser structuring of solid surfaces featured by high-resolution control over the 3D topography at micro-length scales. Microreplication of laser patterned Si surfaces gave rise to 3D polymeric scaffolds, exhibiting controlled gradient roughness, porosity and rigidity. Culture experiments demonstrated that such scaffolds, with the capacity to exquisitely control the pore size, the surface hydrophilicity and rigidity might enable a multi-parametric assessment of the various factors that affect cell behaviour with far-reaching implications for tissue engineering and biomedicine.

## 9.2.4 Laser Processing of Natural Biomaterials

There are several definitions of natural products and the common trend is that a natural product is a chemical compound or substance produced by a living organism. Natural biomaterials are regard as chemical substances that are produced by living organisms with the intent of their application as bio-medical materials. Natural biomaterials provide excellent environments for tissue regeneration, since the different components that make up the extracellular matrix provide a starting point for developing biomimetic scaffolds. The in vivo multiple roles of ECM proteins (collagen, elastin, laminin, fibronectin etc.) and polysaccharides (e.g hyaluronic acid, chitosan, agarose etc.), their biocompatibility and the relative abundance and commercial availability, make these materials attractive for tissue engineering applications [78].

Laser material processing has become increasingly important for the fabrication of scaffolds from natural biomaterials [79–81]. The aim is to achieve an ECM-like morphology complemented by a biomimetic stoichiometry. In the following, the application of laser engineering techniques for biomimetic modification of natural materials will be presented.

The 2PP technique was applied to develop biodegradable 3D scaffolds using natural biomaterials. To this end, there has been some recent studies referred to the synthesis and structuring of biodegradable polymers [56, 59] (Fig. 9.4a, iii), hydrogels and proteins [82]. Specifically, Seidlits and her colleagues reported the use of multiphoton excitation to photocrosslink protein microstructures within three-dimensional, optically transparent hydrogel materials, such as those based on hyaluronic acid [83]. It has also been demonstrated the use of a picosecond pulse green laser for the formation of microfabricated pure protein (bovine serum albumin) microstructures using multiphoton lithography [82]. Besides that the fabrication of sub-micrometer scale 3D structures of bovine serum albumin, avidin and biotinylated bovine serum albumin has been achieved using a femtosecond Ti:Sapphire as well as a picosecond Nd:YAG laser [84]. The authors of this study denote that quite similar grain characteristics and comparable feature sizes were achieved with both laser sources, which demonstrates the utility of the low-cost Nd:YAG microlaser for direct laser writing of protein microstructures [84].

The 2PP technique was also applied to develop precisely defined biodegradable 3D tissue engineering scaffolds fabricated via photopolymerization of gelatine modified with methacrylamide moieties. These studies indicate that the modified gelatine preserves its enzymatic degradation capability after photopolymerization and besides that the developed scaffolds support porcine mesenchymal and primary adipose-derived stem cell (ASC) adhesion, proliferation and differentiation into the anticipated lineage [85, 86].

Recent works have reported on the nanostructuring induced by LAT of collagen [87, 88], gelatine [89] and chitosan [87, 90] with single UV pulses (25 ns, 248 nm) and with 90 fs pulses at 800, 400, and 266 nm [89, 91] (Fig. 9.7). The created foamy layer with nanofibrous properties mimics the nanostructure of the fibrillar network seen in many living tissues and could be developed to construct artificial nanocellular biomimetic materials and for tissue engineering and reconstruction [92]. The micro-foaming transition of biopolymer films was discovered [87] during the study of laser ablation of collagen. In these experiments the absorption of one KrF excimer laser pulse produces a layer of foamy material on the surface of the collagen film. In the foaming process, whose mechanism has been studied and modelled [93] since the discovery, a more selective combination of laser-induced forces is involved to explain the sudden expansion of the laser-excited polymer. It is shown that, for the laser-induced foaming to occur, the target polymer must be moderately absorbing in order to generate a strong pressure wave and the laser absorption must provide

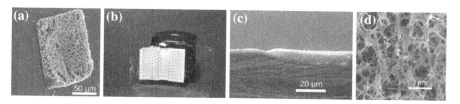

**Fig. 9.7** Examples of LAT induced foams on the surface of (**a**) collagen film; (**b**) gelatin (macroscopic view); (**c**) gelatin (cross-section); (**d**) chitosan (fine nanofibers network). Adapted from: (2009), Journal of Laser Micro/Nanoengineering 4:152

enough gas pressure produced by the ablation products. Irradiation of self-standing films of biopolymers results in the formation of a modified layer with submicrometric structures whose size can be controlled by wavelength selection in the case of fs irradiation [89, 91].

Gaspard et al. have reported on time resolved investigations regarding the effects of transient fs laser-induced foaming by measurements in situ and in real time of the change in the transmittance of a continuous-wave probe HeNe laser through films of collagen and gelatine irradiated with single 90 fs pulses at 800, 400, and 266 nm. Results are compared with those obtained upon irradiation with 25 ns, 248 nm pulses. In this work the temporal evolution of the light transmission through laser irradiated biopolymers collagen and gelatine has been determined in order to separate in time the different processes that lead to the observed nanofoaming structures [92]. KrF laser treatment of collagen and collagen/PVP films caused significant damages on their surfaces leading to the formation of "microfoam" structures as a result of an explosive thermal mechanism assisted by the negative part of laser induced photoacoustic pressure [94]. In the above-mentioned mechanism, the authors considered a purely thermal pathway although the eventuality of reactions of electronic excited states (photochemical pathway) cannot be ruled out. Both collagen and collagen/PVP films behave very similarly. Based on obtained results, has been affirmed that the addition of PVP to collagen causes its resistance to increase the ablation process in a small extent only [94]. KrF laser irradiation was also been used for the generation of microfoam structures on chitosan films [87]. The influence of UV irradiation on the surface of chitosan films has been investigated by Sionkowska et al. who declared that KrF laser treatment caused a significant changes of the surface of chitosan film (forming of the foam) in comparison with UV radiation emitted by incoherent lamp. The deacetylation degree of chitosan calculated from FTIR- ATR data slightly decreases after both, laser irradiation and of UV irradiation from low-intensity lamp [90].

It has been mentioned that gelatine presents some advantages over collagen as its physicochemical properties can be suitably modulated. For instance the thermal and mechanical properties of gelatine, of key importance in the possible applications as a biomaterial, can be enhanced by crosslinking. It is also a much cheaper material than collagen and easier to obtain in concentrated solutions [8]. So, in a number of studies has been reported pulsed laser submicron foam formation in gelatine using single laser pulses of nanosecond (ns) and femtosecond (fs) duration [8, 80, 87, 92, 93]. What seems to be generally accepted is that the morphological characteristics of this layer strongly depend on the biopolymer and on the irradiation wavelength. In collagen, a material of higher mechanical strength, superior internal organization and also of higher water content than gelatine, a nanofibrous network of filaments and interconnected voids is created. In gelatine a nanofoamy layer with uniform bubbles and pores appears as a result of laser irradiation [89].

All these results allow developing the understanding of mechanisms involved in the fs laser processing of biopolymers, of interest in biomedical applications as the obtained laser foam could be used as scaffold to mimic the supramolecular structure and biological functions of the extracellular matrix for three-dimensional tissue regeneration and for development of artificial organs [92].

## 9.3 Conclusions and Outlook

Tissue engineering is an interdisciplinary field involving the combined efforts of biologists, engineers, materials scientists and medical doctors, to achieve biomimetic functional scaffolds. By far, the most challenging aim of tissue engineering is the design of large-scale artificial systems in which biomimetic approaches lead to arbitrary and increasing levels of complexity. This aspect remains in its infancy, so there is a corresponding demand for developing creative solutions. Progress in understanding of the living organisms and their complex functions becomes the inspiration for biomaterials scientists to learn from nature and to advantageously mimic the biological systems on multiple levels. On the other hand, the exciting developments in the fields of nanoscience and micro/nano fabrication technology indicate a revolution in materials design, synthesis and engineering. These emerging technologies will, no doubt, enable the design and fabrication of novel scaffolds incorporating various biomimetic characteristics at the genetic, molecular, nanometer and micrometer scales.

Advanced RP technologies have the potential to incorporate biomimetic approaches in the materials processing and design towards manufacturing of complex artificial scaffolds. In particular, the optical nature of laser direct-write RP techniques provides the unique opportunity to combine machine vision and CAD/CAM automation with non-contact, optically selective, high-throughput precise processing and fabrication. As presented in this chapter, research in this field has shown much progress, considering that laser based RP technologies allow control over the fabrication of 3D porous scaffolds with submicrometer resolution, in a convenient, rapid and cost-effective manner. Despite the remarkable progress, this effort is still far from the ultimate goal of realizing scaffolds that sufficiently mimic the complexity and functionality of natural ECM matrix and there are still many challenges to overcome. Firstly, intense efforts will be required for scaling up the overall processing procedure to include multiple length scales. Although current production techniques are valuable within a limited range of resolutions, future fabrication methods must interlace multiple-scale structures with exceptional controllability. For example, multiple size scales, i.e. nanofeatures imprinted on the surface of a 3D micro-feature, will be required towards mimicking the native extracellular environment. Current laser-based RP methods can manufacture features at scales larger than 100 nm, but there is a need for controllable fabrication at nanometer scale. This is one of the biggest challenges of these technologies towards engineering truly biomimetic hierarchical structured scaffolds and to take full advantage of the potential of nanofeature incorporation.

Another key issue is related to the minimization of thermal damage effects such as melting, burr formation and cracking which are of concern for laser-based processes and limit the functionality of the fabricated scaffolds. As pointed out above, fs lasers present unique capabilities in this respect and should essentially be part of the future laser-based RP schemes for tissue engineering applications. It is encouraging that recent significant improvements in the types of fs sources available have accelerated the acceptance of such lasers as a strong option for large-scale and high-throughput fabrication. On the other hand, research efforts will be necessary for a complete evaluation of the effects of fs laser-biological material interaction. Those studies may include investigation of breakdown products, biological, chemical, corrosion and mechanical properties of the biomaterials processed by fs laser direct write techniques.

In terms of the 3D structural properties aspect, most artificial scaffolds are relatively homogeneous and do not comply with the graded physiochemical properties found in vivo. Therefore, a design challenge for future laser-based tissue engineering lies in development of RP techniques capable of fabricating biomimetic scaffolds with gradient architectures that would encourage heterotypic cell interactions essential to the constructive remodelling process.

Furthermore, there are currently a limited number of biomaterials that are compatible with each laser direct writing technique. The development of novel laser processable and compatible organic and inorganic precursor materials as well as functional biomaterials is thus required for efficient laser based biomimetic tissue scaffold fabrication. Materials development and subsequent fabrication should always be accompanied by detailed in vitro, in vivo and clinical studies to evaluate the functionality of each scaffold.

It should also be mentioned, that the natural ECM contains multiple support proteins and biochemical factors that maintain cell phenotype and response. Incorporating such compounds synergistically within the scaffold architecture is critical for a successful biomimetic scaffold. For this purpose, laser based RP schemes may be complemented with laser assisted biomaterials transfer techniques. Such combined schemes may potentially evolve into future multifunctional RP systems.

Finally, there are other emerging aspects of laser based fabrication techniques which may be exploited for expanding the complexity and novelty of tissue engineered scaffolds. For example, the use of temporally shaped ultrafast pulses may provide an additional route for controlling and optimizing the outcome of processing [95]. Also, the exploitation of filamentation effects [96] which may be produced by using ultrafast lasers open the way for the optimization of sophisticated and demanding processing applications. No doubt, all these techniques require further development before they can become competitive. However, the wealth of arising possibilities in laser based micro and nanofabrication and the number of new approaches to biomaterials synthesis prescribe a future where biomimetic control of artificial biomaterial structure and subsequent functionality can be accomplished with a level of sophistication that we cannot presently envisage.

# References

1. Skalak R, Fox C (1988) Tissue Engineering. In: Proceedings for a workshop held at granlibakken, Lake Tahoe, CA, February 26–29 New York, Alan Liss
2. Stevens MM, George JH (2005) Exploring and engineering the cell surface interface. Science 310(5751):1135–1138
3. Yu LMY, Leipzig ND, Shoichet MS (2008) Promoting neuron adhesion and growth. Mater Today 11(5):36–43
4. Seidlits SK, Lee JY, Schmidt CE (2008) Nanostructured scaffolds for neural applications. Nanomedicine-Uk 3(2):183–199
5. Hollister SJ (2005) Porous scaffold design for tissue Engineering. Nat Mater 4(7):518–524. doi:10.1038/nmat1421PII:nmat1421
6. Cretel E, Pierres A, Benoliel AM, Bongrand P (2008) How cells feel their environment: a focus on early dynamic events. Cell Mol Bioeng 1(1):5–14
7. Khang D, Lu J, Yao C, Haberstroh KM, Webster TJ (2008) The role of nanometer and sub-micron surface features on vascular and bone cell adhesion on titanium. Biomaterials 29(8):970–983
8. Gaspard S, Oujja M, Abrusci C, Catalina F, Lazare S, Desvergne JP, Castillejo M (2008) Laser induced foaming and chemical modifications of gelatine films. J Photoch Photobio A 193(2–3):187–192
9. Engelmayr GC, Cheng MY, Bettinger CJ, Borenstein JT, Langer R, Freed LE (2008) Accordion-like honeycombs for tissue engineering of cardiac anisotropy. Nat Mater 7(12): 1003–1010
10. Chen GP, Ushida T, Tateishi T (2000) Hybrid biomaterials for tissue engineering: a preparative method for PLA or PLGA-collagen hybrid sponges. Adv Mater 12(6):455
11. Bruggeman JP, Bettinger CJ, Nijst CLE, Kohane DS, Langer R (2008) Biodegradable xylitol-based polymers. Adv Mater 20(10):1922
12. Wang YD, Ameer GA, Sheppard BJ, Langer R (2002) A tough biodegradable elastomer. Nat Biotechnol 20(6):602–606
13. Bettinger CJ (2009) Synthesis and microfabrication of biomaterials for soft-tissue engineering. Pure Appl Chem 81(12):2183–2201
14. Ifkovits JL, Burdick JA (2007) Review: photopolymerizable and degradable biomaterials for tissue engineering applications. Tissue Eng 13(10):2369–2385
15. Li HY, Du RL, Chang J (2005) Fabrication, characterization, and in vitro degradation of composite scaffolds based on PHBV and bioactive glass. J Biomater Appl 20(2):137–155
16. Martin AI, Salinas AJ, Vallet-Regi M (2005) Bioactive and degradable organic-inorganic hybrids. J Eur Ceram Soc 25(16):3533–3538
17. Sepulveda P, Jones JR, Hench LL (2002) Bioactive sol-gel foams for tissue repair. J Biomed Mater Res 59(2):340–348
18. Martin RA, Yue S, Hanna JV, Lee PD, Newport RJ, Smith ME, Jones JR (2012) Characterizing the hierarchical structures of bioactive sol-gel silicate glass and hybrid scaffolds for bone regeneration. Philos T R Soc A 370(1963):1422–1443

19. Valerio P, Guimaraes MHR, Pereira MM, Leite MF, Goes AM (2005) Primary osteoblast cell response to sol-gel derived bioactive glass foams. J Mater Sci-Mater M 16(9):851–856
20. Gough JE, Jones JR, Hench LL (2004) Nodule formation and mineralisation of human primary osteoblasts cultured on a porous bioactive glass scaffold. Biomaterials 25(11):2039–2046
21. Mano JF, Vaz CM, Mendes SC, Reis RL, Cunha AM (1999) Dynamic mechanical properties of hydroxyapatite-reinforced and porous starch-based degradable biomaterials. J Mater Sci-Mater M 10(12):857–862
22. Yang SF, Leong KF, Du ZH, Chua CK (2001) The design of scaffolds for use in tissue engineering. Part 1. Traditional factors. Tissue Eng 7(6):679–689
23. Lee J, Cuddihy MJ, Kotov NA (2008) Three-dimensional cell culture matrices: State of the art. Tissue Eng Pt B-Rev 14(1):61–86
24. Kurella A, Dahotre NB (2005) Review paper: surface modification for bioimplants: the role of laser surface engineering. J Biomater Appl 20(1):5–50. doi:10.1177/0885328205052974
25. Stratakis E, Ranella A, Fotakis C (2011) Biomimetic micro/nanostructured functional surfaces for microfluidic and tissue engineering applications. Biomicrofluidics 5(1):013411
26. Hutmacher DW, Sittinger M, Risbud MV (2004) Scaffold-based tissue engineering: rationale for computer-aided design and solid free-form fabrication systems. Trends Biotechnol 22(7):354–362
27. Peltola SM, Melchels FPW, Grijpma DW, Kellomaki M (2008) A review of rapid prototyping techniques for tissue engineering purposes. Ann Med 40(4):268–280
28. Sachlos E, Czernuszka JT (2003) Making tissue Engineering scaffolds work. Review: the application of solid freeform fabrication technology to the production of tissue Engineering scaffolds. Eur Cell Mater 5:29–39; discussion 39–40. doi:vol005a03
29. Yeong WY, Chua CK, Leong KF, Chandrasekaran M (2004) Rapid prototyping in tissue engineering: challenges and potential. Trends Biotechnol 22(12):643–652
30. Schmidt M, Pohle D, Rechtenwald T (2007) Selective laser sintering of PEEK. Cirp Ann-Manuf Techn 56(1):205–208
31. Rimell JT, Marquis PM (2000) Selective laser sintering of ultra high molecular weight polyethylene for clinical applications. J Biomed Mater Res 53(4):414–420
32. Tan KH, Chua CK, Leong KF, Cheah CM, Cheang P, Abu Bakar MS, Cha SW (2003) Scaffold development using selective laser sintering of polyetheretherketone-hydroxyapatite biocomposite blends. Biomaterials 24(18):3115–3123
33. Williams JM, Adewunmi A, Schek RM, Flanagan CL, Krebsbach PH, Feinberg SE, Hollister SJ, Das S (2005) Bone tissue engineering using polycaprolactone scaffolds fabricated via selective laser sintering. Biomaterials 26(23):4817–4827
34. Kanczler JM, Mirmalek-Sani SH, Hanley NA, Ivanov AL, Barry JJA, Upton C, Shakesheff KM, Howdle SM, Antonov EN, Bagratashvili VN, Popov VK, Oreffo ROC (2009) Biocompatibility and osteogenic potential of human fetal femur-derived cells on surface selective laser sintered scaffolds. Acta Biomater 5(6):2063–2071
35. Zhou WY, Lee SH, Wang M, Cheung WL, Ip WY (2008) Selective laser sintering of porous tissue engineering scaffolds from poly(L)/carbonated hydroxyapatite nanocomposite microspheres. J Mater Sci-Mater M 19(7):2535–2540
36. Wiria FE, Leong KF, Chua CK, Liu Y (2007) Poly-epsilon-caprolactone/hydroxyapatite for tissue engineering scaffold fabrication via selective laser sintering. Acta Biomater 3(1):1–12
37. Simpson RL, Wiria FE, Amis AA, Chua CK, Leong KF, Hansen UN, Chandraselkaran M, Lee MW (2008) Development of a 95/5 poly(L-lactide-co-glycolide)/hydroxylapatite and beta-tricalcium phosphate scaffold as bone replacement material via selective laser sintering. J Biomed Mater Res B 84B(1):17–25
38. Duan B, Wang M, Zhou WY, Cheung WL, Li ZY, Lu WW (2010) Three-dimensional nanocomposite scaffolds fabricated via selective laser sintering for bone tissue engineering. Acta Biomater 6(12):4495–4505
39. Levy RA, Chu TMG, Halloran JW, Feinberg SE, Hollister S (1997) CT-generated porous hydroxyapatite orbital floor prosthesis as a prototype bioimplant. Am J Neuroradiol 18(8):1522–1525

40. Griffith ML, Halloran JW (1996) Freeform fabrication of ceramics via stereolithography. J Am Ceram Soc 79(10):2601–2608
41. Matsuda T, Mizutani M (2002) Liquid acrylate-endcapped biodegradable poly(epsilon-caprolactone-co-trimethylene carbonate). II. Computer-aided stereolithographic microarchitectural surface photoconstructs. J Biomed Mater Res 62(3):395–403
42. Cooke MN, Fisher JP, Dean D, Rimnac C, Mikos AG (2003) Use of stereolithography to manufacture critical-sized 3D biodegradable scaffolds for bone ingrowth. J Biomed Mater Res B 64B(2):65–69
43. Dhariwala B, Hunt E, Boland T (2004) Rapid prototyping of tissue-engineering constructs, using photopolymerizable hydrogels and stereolithography. Tissue Eng 10(9–10):1316–1322
44. Melchels FPW, Feijen J, Grijpma DW (2009) A poly(D, L-lactide) resin for the preparation of tissue engineering scaffolds by stereolithography. Biomaterials 30(23–24):3801–3809
45. Lee JW, Nguyen TA, Kang KS, Seol YJ, Cho DW (2008) Development of a growth factor-embedded scaffold with controllable pore size and distribution using micro-stereolithography. Tissue Eng Pt A 14(5):835–835
46. Arcaute K, Mann BK, Wicker RB (2006) Stereolithography of three-dimensional bioactive poly(ethylene glycol) constructs with encapsulated cells. Ann Biomed Eng 34(9):1429–1441
47. Cumpston BH, Ananthavel SP, Barlow S, Dyer DL, Ehrlich JE, Erskine LL, Heikal AA, Kuebler SM, Lee IYS, McCord-Maughon D, Qin JQ, Rockel H, Rumi M, Wu XL, Marder SR, Perry JW (1999) Two-photon polymerization initiators for three-dimensional optical data storage and microfabrication. Nature 398(6722):51–54
48. Farsari M, Chichkov BN (2009) Two-photon fabrication. Nat Photonics 3(8):450–452
49. Almany L, Seliktar D (2005) Biosynthetic hydrogel scaffolds made from fibrinogen and poly-ethylene glycol for 3D cell cultures. Biomaterials 26(15):2467–2477
50. Hsieh TM, Ng CWB, Narayanan K, Wan ACA, Ying JY (2010) Three-dimensional microstructured tissue scaffolds fabricated by two-photon laser scanning photolithography. Biomaterials 31(30):7648–7652
51. Ovsianikov A, Schlie S, Ngezahayo A, Haverich A, Chichkov BN (2007) Two-photon polymerization technique for microfabrication of CAD-designed 3D scaffolds from commercially available photosensitive materials. J Tissue Eng Regen M 1(6):443–449
52. Livage J, Sanchez C (1992) sol-gel chemistry. J Non-Cryst Solids 145(1–3):11–19
53. Ovsianikov A, Viertl J, Chichkov B, Oubaha M, MacCraith B, Sakellari I, Giakoumaki A, Gray D, Vamvakaki M, Farsari M, Fotakis C (2008) Ultra-low shrinkage hybrid photosensitive material for two-photon polymerization microfabrication. Acs Nano 2(11):2257–2262
54. Sakellari I, Gaidukeviciute A, Giakoumaki A, Gray D, Fotakis C, Farsari M, Vamvakaki M, Reinhardt C, Ovsianikov A, Chichkov BN (2010) Two-photon polymerization of titanium-containing sol-gel composites for three-dimensional structure fabrication. Appl Phys a-Mater 100(2):359–364
55. Psycharakis S, Tosca A, Melissinaki V, Giakoumaki A, Ranella A (2011) Tailor-made three-dimensional hybrid scaffolds for cell cultures. Biomed Mater 6(4):045008
56. Claeyssens F, Hasan EA, Gaidukeviciute A, Achilleos DS, Ranella A, Reinhardt C, Ovsianikov A, Xiao S, Fotakis C, Vamvakaki M, Chichkov BN, Farsari M (2009) Three-dimensional biodegradable structures fabricated by two-photon polymerization. Langmuir 25(5):3219–3223
57. Mizutani M, Matsuda T (2002) Liquid photocurable biodegradable copolymers: In vivo degradation of photocured poly(epsilon-caprolactone-co-trimethylene carbonate). J Biomed Mater Res 61(1):53–60
58. Mizutani M, Matsuda T (2002) Photocurable liquid biodegradable copolymers: in vitro hydrolytic degradation behaviors of photocured films of coumarin-endcapped poly(epsilon-caprolactone-co-trimethylene carbonate). Biomacromolecules 3(2):249–255
59. Melissinaki V, Gill AA, Ortega I, Vamvakaki M, Ranella A, Haycock JW, Fotakis C, Farsari M, Claeyssens F (2011) Direct laser writing of 3D scaffolds for neural tissue engineering applications. Biofabrication 3(4):045005
60. Ke K, Hasselbrink EF, Hunt AJ (2005) Rapidly prototyped three-dimensional nanofluidic channel networks in glass substrates. Anal Chem 77(16):5083–5088

61. Zorba V, Stratakis E, Barberoglou M, Spanakis E, Tzanetakis P, Anastasiadis SH, Fotakis C (2008) Biomimetic artificial surfaces quantitatively reproduce the water repellency of a lotus leaf. Adv Mater 20(21):4049

62. Ranella A, Barberoglou M, Bakogianni S, Fotakis C, Stratakis E (2010) Tuning cell adhesion by controlling the roughness and wettability of 3D micro/nano silicon structures. Acta Biomater 6(7):2711–2720

63. Doraiswamy A, Patz T, Narayan RJ, Dinescu M, Modi R, Auyeung RCY, Chrisey DB (2006) Two-dimensional differential adherence of neuroblasts in laser micromachined CAD/CAM agarose channels. Appl Surf Sci 252(13):4748–4753

64. Duncan AC, Rouais F, Lazare S, Bordenave L, Baquey C (2007) Effect of laser modified surface microtopochemistry on endothelial cell growth. Colloid Surf B 54(2):150–159

65. Miller PR, Aggarwal R, Doraiswamy A, Lin YJ, Lee YS, Narayan RJ (2009) Laser micromachining for biomedical applications. Jom-Us 61(9):35–40

66. Patz TM, Doraiswamy A, Narayan RJ, Modi R, Chrisey DB (2005) Two-dimensional differential adherence and alignment of C2C12 myoblasts. Mat Sci Eng B-Solid 123(3):242–247

67. Choi HW, Johnson JK, Nam J, Farson DF, Lannutti J (2007) Structuring electrospun polycaprolactone nanofiber tissue scaffolds by femtosecond laser ablation. J Laser Appl 19(4):225–231

68. Papadopoulou EL, Samara A, Barberoglou M, Manousaki A, Pagakis SN, Anastasiadou E, Fotakis C, Stratakis E (2010) Silicon scaffolds promoting three-dimensional neuronal Web of cytoplasmic processes. Tissue Eng Part C-Me 16(3):497–502

69. Masoumi KL, Johnson JT, Zugates GC, Engelmayr GC (2011) IEEE 37th annual northeast bioengineering conference (NEBEC), 1–2, April 2011

70. Wüst S, Müller R, Hofmann S (2011) Controlled positioning of cells in biomaterials-approaches towards 3D tissue printing. J Funct Biomater 2(3):119–154

71. Guillotin B, Souquet A, Catros S, Duocastella M, Pippenger B, Bellance S, Bareille R, Remy M, Bordenave L, Amedee J, Guillemot F (2010) Laser assisted bioprinting of engineered tissue with high cell density and microscale organization. Biomaterials 31(28):7250–7256

72. Mironov V, Boland T, Trusk T, Forgacs G, Markwald RR (2003) Organ printing: computer-aided jet-based 3D tissue engineering. Trends Biotechnol 21(4):157–161

73. Schiele NR, Corr DT, Huang Y, Raof NA, Xie YB, Chrisey DB (2010) Laser-based direct-write techniques for cell printing. Biofabrication 2(3):032001

74. Ovsianikov A, Gruene M, Pflaum M, Koch L, Maiorana F, Wilhelmi M, Haverich A, Chichkov B (2010) Laser printing of cells into 3D scaffolds. Biofabrication 2(1):014104

75. Gittard SD, Narayan R (2010) Laser direct writing of micro- and nano-scale medical devices. Expert Rev Med Devic 7(3):343–356

76. Chu TMG, Halloran JW, Hollister SJ, Feinberg SE (2001) Hydroxyapatite implants with designed internal architecture. J Mater Sci-Mater M 12(6):471–478

77. Koufaki N, Ranella A, Aifantis KE, Barberoglou M, Psycharakis S, Fotakis C, Stratakis E (2011) Controlling cell adhesion via replication of laser micro/nano-textured surfaces on polymers. Biofabrication 3(4):045004

78. Willerth SM, Sakiyama-Elbert SE (2008) Combining stem cells and biomaterial scaffolds for constructing tissues and cell delivery. doi:NBK27050[bookaccession]

79. Aguilar CA, Lu Y, Mao S, Chen SC (2005) Direct micro-patterning of biodegradable polymers using ultraviolet and femtosecond lasers. Biomaterials 26(36):7642–7649

80. Narayan RJ, Jin CM, Patz T, Doraiswamy A, Modi R, Chrisey DB, Su YY, Lin SJ, Ovsianikov A, Chichkov B (2005) Laser processing of advanced biomaterials. Adv Mater Process 163(4):39–42

81. Vogel A, Noack J, Huttman G, Paltauf G (2005) Mechanisms of femtosecond laser nanosurgery of cells and tissues. Appl Phys B-Lasers O 81(8):1015–1047

82. Ritschdorff ET, Shear JB (2010) Multiphoton lithography using a high-repetition rate microchip laser. Anal Chem 82(20):8733–8737

83. Seidlits SK, Schmidt CE, Shear JB (2009) High-resolution patterning of hydrogels in three dimensions using direct-write photofabrication for cell guidance. Adv Funct Mater 19(22):3543–3551

84. Turunen S, Kapyla E, Terzaki K, Viitanen J, Fotakis C, Kellomaki M, Farsari M (2011) Pico- and femtosecond laser-induced crosslinking of protein microstructures: evaluation of processability and bioactivity. Biofabrication 3(4):045002

85. Ovsianikov A, Deiwick A, Van Vlierberghe S, Dubruel P, Moller L, Drager G, Chichkov B (2011) Laser fabrication of three-dimensional CAD scaffolds from photosensitive gelatin for applications in tissue Engineering. Biomacromolecules 12(4):851–858

86. Ovsianikov A, Deiwick A, Van Vlierberghe S, Pflaum M, Wilhelmi M, Dubruel P, Chichkov B (2011) Laser fabrication of 3D gelatin scaffolds for the generation of bioartificial tissues. Materials 4(1):288–299

87. Lazare S, Tokarev V, Sionkowska A, Wisniewski M (2005) Surface foaming of collagen, chitosan and other biopolymer films by KrF excimer laser ablation in the photomechanical regime. Appl Phys a-Mater 81(3):465–470

88. Wisniewski M, Sionkowska A, Kaczmarek H, Lazare S, Tokarev V (2007) Influence of laser irradiation on the thin collagen films. part I. mechanism of micro-foam structure formation and collagen surface ablation. Polimery-W 52(4):259–267

89. Gaspard S, Oujja M, de Nalda R, Abrusci C, Catalina F, Banares L, Lazare S, Castillejo M (2007) Nanofoaming in the surface of biopolymers by femtosecond pulsed laser irradiation. Appl Surf Sci 254(4):1179–1184

90. Sionkowska A, Kaczmarek H, Wisniewski M, Skopinska J, Lazare S, Tokarev V (2006) The influence of UV irradiation on the surface of chitosan films. Surf Sci 600(18):3775–3779

91. Gaspard S, Oujja A, de Nalda R, Abrusci C, Catalina F, Banares L, Castillejo M (2007) Submicron foaming in gelatine by nanosecond and femtosecond pulsed laser irradiation. Appl Surf Sci 253(15):6420–6424

92. Gaspard S, Oujja M, de Nalda R, Castillejo M, Banares L, Lazare S, Bonneau R (2008) Nanofoaming dynamics in biopolymers by femtosecond laser irradiation. Appl Phys a-Mater 93(1):209–213

93. Lazare S, Bonneau R, Gaspard S, Oujja M, De Nalda R, Castillejo M, Sionkowska A (2009) Modeling the dynamics of one laser pulse surface nanofoaming of biopolymers. Appl Phys a-Mater 94(4):719–729

94. Wisniewski M, Sionkowska A, Kaczmarek H, Skopinska J, Lazare S, Tokarev V (2006) The influence of KrF excimer laser irradiation on the surface of collagen and collagen/PVP films. Int J Photoenergy

95. Klini A, Loukakos PA, Gray D, Manousaki A, Fotakis C (2008) Laser induced forward transfer of metals by temporally shaped femtosecond laser pulses. Opt Expr 16(15):11300–11309. doi:167754[pii]

96. Tzortzakis S, Papazoglou DG, Zergioti I (2006) Long-range filamentary propagation of subpicosecond ultraviolet laser pulses in fused silica. Opt Lett 31(6):796–798

# Chapter 10
# Laser Processing of Natural Biomaterials

Wande Zhang, Peter H. Chung, Aping Zhang and Shaochen Chen

**Abstract** Natural biomaterials have been extensively employed in biomedical applications because of their superior biocompatibility and biodegradability. Among the large family of techniques to process natural biomaterials, laser techniques have been rapidly developed to fabricate biomimetic artificial organs, tissue engineering scaffolds, and other biomedical constructs. Compared to other techniques, laser processing allows more precise control over the geometry and is able to fabricate smaller features with minimal debris generated. The laser processing techniques are generally grouped by three categories: polymerization, ablation, and activation. This chapter introduces several widely used natural biomaterials, including collagen, agarose, hyaluronic acid, and Matrigel$^{TM}$ as well as the recent development in laser processing techniques of those natural biomaterials.

W. Zhang (✉)
Department of Bioengineering, University of California,
San Diego, 9500 Gilman Drive, La Jolla, CA 92093, USA
e-mail: wande117@gmail.com

P. H. Chung
Department of NanoEngineering, University of California,
San Diego, Atkinson Hall, Room 2314, 9500 Gilman Drive #0448,
La Jolla, CA 92093, USA
e-mail: peterchung@ucsd.edu

A. Zhang
Department of Electrical Engineering, Photonics Research Centre,
The Hong Kong Polytechnic University, Kowloon (Hong Kong SAR), China
e-mail: aping.zhang@polyu.edu.hk

S. Chen
Department of NanoEngineering, University of California,
San Diego, SME Building 245B, MC-0448, La Jolla, CA 92093, USA
e-mail: chen168@ucsd.edu

V. Schmidt and M. R. Belegratis (eds.), *Laser Technology in Biomimetics*,
Biological and Medical Physics, Biomedical Engineering,
DOI: 10.1007/978-3-642-41341-4_10, © Springer-Verlag Berlin Heidelberg 2013

## 10.1 Introduction

The rapid development of laser processing techniques has generated much excitement within the field of biomaterials—particularly for those that are naturally derived. Various natural biomaterials, such as collagen, agarose, hyaluronic acid, and Matrigel$^{TM}$, have been treated through polymerization, ablation, or activation. Both in vivo and in vitro experiments have been conducted to demonstrate that biomaterials processed by lasers are effective and efficient for biological applications.

Natural biomaterials comprise biologically- and medically-applied materials derived from living organisms. Natural biomaterials can generally be categorized into three groups: proteins, polysaccharides, and extracellular matrix (ECM)-derived biomaterials. Proteins, including collagen, gelatin, elastin, fibrin, and silk, are composed of amino acid chains bonded by covalent peptide bonds. Polysaccharides, including agarose, hyaluronic acid, and chitin, are composed of sugar chains bonded by glycosidic bonds. ECM-derived biomaterials, such as Matrigel$^{TM}$ and small intestinal submucosa (SIS), contain a large variety of proteins and growth factors and thus are ready for cell culture. Natural biomaterials have been used in medical practice for thousands of years. As early as in ancient Egypt, animal sinew was used as sutures. Compared with synthetic biomaterials, natural biomaterials provide two key advantages: First, since natural biomaterials can be readily recognized by the biological environment, they are more biocompatible and do not invoke an inflammatory response. Second, since natural biomaterials can be recognized by the metabolic system and enzymes, they are more biodegradable and can be beneficial in certain situations where biomaterials are only needed for a temporary amount of time [1–4].

Laser processing of natural biomaterials has been investigated for decades [2, 3, 5–7]. Research in this field can be categorized into three subfields: (1) laser-induced polymerization, where a laser is used to induce cross-linking between natural biomaterial polymer chains; (2) laser ablation, where a laser is used to selectively remove part of the natural biomaterial through thermal or chemical effects; and (3) laser activation, where a laser is used to activate certain moieties on the polymer chains for specific applications. Laser-induced polymerization and laser ablation has been used to make biomimetic artificial organs, tissue engineering scaffolds, and other biomedical constructs. Compared with other material processing methods based on material addition such as inject-printing, laser-induced polymerization allows for more precise control over the geometry and dimensions of the processing area with resolution down to the nanometer scale. Meanwhile, compared to other processing methods that are based on material subtraction, such as etching, the laser is able to completely remove material with minimal generation of debris, thereby minimizing any material property changes that could be caused by the processing tools.

This chapter deals with several natural biomaterials that have widely been investigated, including collagen, agarose, hyaluronic acid, and Matrigel$^{TM}$, which represent all three categories of biomaterials. Each natural biomaterial is described in a separate section with the first part providing background information (such as material properties and current biomedical/biomimetic applications) and the second part featuring a review of the current literature describing the state -of-the-art in laser processing of the biomaterial.

## 10.2  Natural Biomaterials

### 10.2.1  Collagen

The collagen family is composed of over 20 types of proteins and accounts for approximately 25 to 35 % of the human body's total protein weight. Collagen is the most widely used natural biomaterial in biomedical applications. Functionally, collagen serves as an essential component of the connective and supporting tissues that maintain the body's structure and resistance to stretching [8]. Collagen is a fibrous protein whose basic unit is composed of a triple helix–two identical $\alpha 1$ chains and one $\alpha 2$ chain [9]. Collagen offers many properties that make it an excellent candidate for biomedical applications: (1) excellent biocompatibility, as it is biologically derived; (2) availability at a low cost, due to its abundance in tissues; (3) biodegradability (via enzymes such as collagenase) that can be finely tuned by adjusting the degree of cross-linking; (4) ease of functionalization, as a variety of molecules can be attached to the collagen chain to modify its property for different biomedical applications; and (5) formability, as the shape of collagen is easily changed into beads, sheets, or bands for different applications.

Collagen has extensively been used in clinical applications as well as basic biomedical research. In particular, collagen has been used widely in tissue regeneration. For cardiac tissue regeneration, Shi et al. cultured Sca-1 positive stem cells in a collagen scaffold and transplanted the scaffold into C57/BL6 mice to promote cardiac tissue regeneration [10]. To investigate bone tissue regeneration, Lee et al. used collagen/polycaprolactone hybrid fibrous scaffolds in their experiments and found that the collagen-enhanced scaffolds exhibited better mechanical strength and bioactivity than pure polycaprolactone scaffolds [11]. For bladder tissue regeneration, Engelhardt et al. have demonstrated the potential of using collagen/poly(lactic acid-co-epsilon-caprolactone) hybrid scaffolds to promote bladder growth [12]. Kew et al. have shown that collagen fibrous scaffolds are promising candidates for tendon and ligament regeneration [13].

Additionally, collagen has broadly been used as a delivery vehicle for drugs, genes, and cells. Hong et al. attached collagen on polylactide microspheres to promote cell adhesion and proliferation for chondrocyte delivery [14]. Park et al. have shown that collagen/calcium phosphate hybrid structures can promote bone cell adhesion and bone regeneration. Proteins, such as bovine serum albumin, were encapsulated in the hybrid structures and released in a controlled fashion for over a month [15]. Collagen assisted gene delivery experiments were conducted by Adachi et al. They used a collagen/poly(Pro-Hyp-Gly) hybrids as a reliable method to deliver siRNA into cells while preventing fast degradation of the genetic payload [16].

Collagen has also been used in cell culture research. Lam et al. have demonstrated that arachnoid cells isolated from rat brainstems exhibit the same phenotypes in three-dimensional collagen scaffolds as in the in vivo environment, which suggests that collagen structures can be developed as in vitro arachnoid granulation models [17]. Yuan et al. have shown that a 3D collagen microsphere culture system was superior to traditional monolayer cultures [18] in preserving the phenotypic characteristics of rabbit nucleus pulposus cells.

## 10.2.2 Agarose

Agarose is a polysaccharide derived from agar–a cellular component of algae. The agarose molecule is a single or double helix composed of repeating units of agarobiose [19]. Two distinct advantages of agarose make it an excellent candidate for various biological and medical applications. First, the hydroxyl groups in agarose facilitate modifications with a variety of ligands, while other parts of the agarose chain are biologically inert and thus do not interact chemically with nucleic acids and most proteins. Second, the cross-linking of agarose gel extremely enhances its chemical and physical stability, which makes it an ideal candidate for experimental substrate. Due to these main benefits agarose is an ideal source material for different specific biological applications.

In biomaterial research, an important application of agarose is in the field of tissue engineering. Agarose can be blended with various biomaterials to modify its mechanical and biological properties for different applications. Hybrid fibers of agarose and carbon nanotubes have been used as scaffolds for neural tissue engineering [20]. Chitosan-agarose-gelatin hybrid hydrogels and agarose/poly (ethylene glycol) mixtures have been used in cartilage tissue engineering [21, 22]. Hydroxyapatite/beta-tricalcium phosphate/agarose hybrids have been used in bone tissue engineering [23]. In biological research, agarose is widely used as growth media for various microorganisms, such as bacteria, yeast, and fungi [24–27]. Agarose is also one of the primary stabilizing media used in electrophoresis of DNA, polymers, protein, and other macromolecules [28–32].

## 10.2.3 Hyaluronic Acid

Hyaluronic acid (HA), also known as hyaluronan, has been employed as the biomaterial of first choice for numerous biomedical applications [33] from wound healing [34] and tissue engineering [35] to drug delivery [36] and cosmetic products [37]. HA is a naturally occurring biopolymer prevalent throughout the body as an essential constituent of the extracellular matrix (ECM). The only non-sulfated glycosaminoglycan, HA is an unbranched, large molecular weight ($\sim 10^4$–$10^7$ Da), anionic polysaccharide composed of repeating disaccharide units of D-glucuronic acid and N-acetyl-D-glucosamine. Within the ECM, HA interacts with cells via the CD44 antigen and RHAMM (=receptor for hyaluronan-mediated motility) surface receptors and plays a vital role in numerous functions, such as angiogenesis, wound healing, morphogenesis, and tissue reorganization [38–40].

HA features numerous properties particularly attractive for biomedical applications, such as biocompatibility, biodegradability, and non-immunogenicity. Additionally, due to its hydrophilic nature, HA is not protein- or cell-adhesive [41]. This property can be modulated easily via introduction of various moieties at functional sites (e.g. carboxyl and hydroxyl groups) along its polymer backbone. The approaches for chemical modification are extensive and can be used to confer a broad range of functionality to HA [42]. Finally, from a clinical point of view, HA has a well-established history as an FDA-approved biomaterial and continues to be explored across a wide range of developing biomedical applications, particularly within tissue engineering and regenerative medicine [33].

### 10.2.4 Matrigel$^{TM}$

Matrigel$^{TM}$ provides a natural cell growth environment and was first derived in the 1980s from murine Engelbreth-Holm-Swarm (EHS) tumors [43]. Since Matrigel$^{TM}$ contains a large number of basement membrane proteins (e.g. collagen IV and laminin) as well as a variety of growth factors and angiogenic factors, it can been used without further treatment in a variety of applications, such as stem cell differentiation, tissue regeneration, cancer metastasis and angiogenesis. Asakura et al. found that muscle stem cells could differentiate into myocytes, adipocytes and osteoblasts when they were cultured on Matrigel$^{TM}$ [44]. Embryonic stem cell differentiation has also been demonstrated in Matrigel$^{TM}$ [45]. Since Matrigel$^{TM}$ was derived from tumor tissue, many groups used it as platform for cancer cell invasion research. Kramer et al. demonstrated migration and invasion of HT 1,080 cells in Matrigel$^{TM}$ [46]. Later, Fridman et al. found that several cancer cells lines that could not survive in mice were able to generate tumors in Matrigel$^{TM}$ [47].

## 10.3 Laser Processing Methodologies for Biomaterials

### 10.3.1 Laser Processing System Setup

A typical laser system for biomaterial processing consists of a laser source, laser optics, an automated stage, an imaging system, and a computer (Fig. 10.1). A laser source with specific energy level, wavelength, pulse duration, and repetition rate is selected according to the requirements of the application. Generally, a pulsed

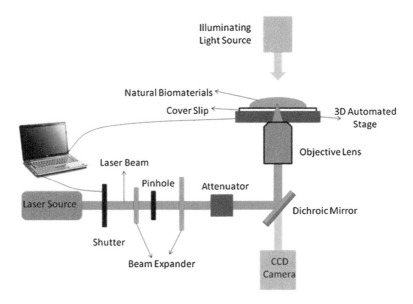

**Fig. 10.1** Schematic of a typical laser processing system for natural biomaterials

solid-state laser with the pulse duration in the range of femtoseconds to nanoseconds is employed in the processing of natural biomaterials. The laser beam is magnified by a beam expander after leaving the laser head. The power of the laser beam can be tuned by an attenuator to accommodate various applications. The beam is then reflected by a dichroic mirror into an objective lens, which is generally mounted in a microscope. The sample holder, usually a piece of coverslip, is fixed on an automated stage above the objective lens. The natural biomaterial is placed on the sample holder and the laser beam is focused into the biomaterials through the objective lens. Being controlled by a computer program, the automated stage is able to locate the laser beam to any position in the biomaterial following an arbitrary scanning path. An imaging system is used to facilitate laser focusing and monitor the entire process in situ.

Laser based three-dimensional patterning is achieved either by (1) using a three-axis piezoelectric translator to move the hydrogels across the laser path or (2) raster-scanning the laser beam across a digital micromirror device (DMD) that acts as a dynamic reflective photomask (based on the method described in [48]). Using this method, they were able to pattern proteins within the hydrogel at sub-μm feature resolutions and thereby guide dorsal root ganglion and hippocampal neural progenitor cells through paths of arbitrary three-dimensional configuration as described in section hyaluronic acid below.

The DMD-based approach is of particular interest due to its versatility (cf. Fig. 10.2). The key advantage of this approach–particularly in comparison to traditional photolithographic techniques–is the ability to generate any pattern rapidly and in a dynamic fashion. By incrementally moving the servo-controlled stage, com-

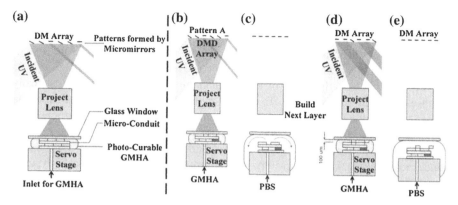

**Fig. 10.2** Fabrication process using the DMD microfabrication system. Schematic of the DMD microfabrication system for the layer-by-layer fabrication of scaffolds (**a**). After fabricating a layer (**b**), the stage moves down 0.5 mm and pulls the structure away from the glass window. After the curing process, PBS (phosphate buffered saline) is pumped into the gap to rinse away any partially polymerized glycidyl methacrylate hyaluronic acid (GMHA) (**c**). To build a new layer above the just-formed structure after the rinsing step (**d**), the stage is moved to bring the just-formed layer 200 μm below the glass slide. Fresh monomer is pumped into the 200 μm gap, and the next layer is then created by UV curing followed by PBS purging again (**e**). Reprinted from Fig. 7 in reference [49] with kind permission from Springer Science+Business Media, LLC

plex structures can be generated layer-by-layer using a set of user-defined images. Additionally, the DMD is amenable to light projection via either a coherent UV or multiphoton laser source [48]. Han et al. demonstrated also the use of a standard (non-coherent) UV lamp source in the DMD-approach [50, 51].

## 10.3.2 Laser-Induced Cross-Linking

In typical applications of photopolymer technology, light induces cross-linking of monomers or oligomers via photopolymerization by the agency of the photoinitiator. The photoinitiator is added to the monomer solution and generates upon illumination free radicals, which stimulate the cross-linking of the monomers. Thus, photopolymerization transforms short-chain liquid-state monomers with low molecular weight to long-chain solid-state polymers with high molecular weight by light. Especially laser light has distinct advantages with respect to photopolymerization due to its spatial and temporal coherence. Since a laser beam is spatially coherent, it can be focused down to a sub-μm spot size to fabricate micro and nanostructures. Meanwhile, due to its temporal coherence, secondary polymerization induced by polychromic effect is eliminated so that the size of focal point is further reduced. Because of its three-dimensional microscale fabrication capability, the laser has been widely used in the cross-linking of natural biomaterials [52–59].

### 10.3.2.1 Processing of Collagen for Biomedical Applications

Although widely used in biomedical applications, native collagen has not extensively been employed as a base material for tissue engineering scaffolds due to its low mechanical strength. Upon in vivo implantation, collagen scaffolds typically collapse under internal pressure. To solve this problem, Chan et al. used an argon laser to enhance the mechanical strength of natural collagen [52]. In their approach, 0.01 % (w/v) Rose Bengal was added to collagen as a cross-linking agent. A 0.2 W pulsed argon laser beam (wavelength 514 nm) with a spot size of 1.5 cm was used to induce cross-linking. Upon characterizing relevant properties, such as elasticity, thermostability, and water-binding capacity of cross-linked collagen, it was found that laser cross-linking was able to strengthen, stiffen, and stabilize collagen membranes without compromising water-binding capability. Furthermore, it was observed that laser scanning was able to generate microscale interconnected pores, which were not present in unprocessed collagen. Iosin et al. have investigated microscale patterning of collagen I based on one and two photon absorption using two different laser systems for the experiments, a Nd:YAG laser at 532 nm and a mode-locked near infrared Ti:Sapphire laser at 740 nm [53]. Eosin Y or Rose Bengal were used as photosensitizers for polymerization. They characterized the relationship between laser power and line width of polymerized collagen. Collagen lines as narrow as 1 $\mu$m were fabricated in this study. The line width depends on the size of the laser beam, the laser power, and the concentrations of proteins and photosensitizer. Additionally, the photochemistry of the processes strongly depends on photosensitizer characteristics, but is independent on the excitation mechanisms (one- or two-photon absorption). Biocompatibility was demonstrated by growing cells on the collagen structures.

   Laser induced cross-linking is able to induce conformational changes in a variety of collagen-rich tissues. This mechanism has been used in tissue regeneration, such as in the treatment of the varicose veins. Traditionally, diseased vessels were removed by surgical operation. However, vein damage frequently occurred during the operation. Frullini et al. have developed an alternative treatment that uses just a blue light emitting diode (LED) to cross-link the collagen in veins [54]. In this approach, riboflavin was used as a cross-linking agent and the blue light from the LED (wavelength range 450–480 nm) was used to activate the polymerization. Venous shrinkage was evident in this study a few minutes after illumination. Additionally, no histological evidence of venous endothelial damage was discovered after the treatment. Similar approaches have also been used outside of the clinical setting in basic biomedical research. Wenzel et al. evaluated the effect of a ruby laser on remodeling of the collagen in the mouse cochlea [55]. Both low (15 J/cm$^2$) and high laser fluence (180 J/cm$^2$) were used in this study. The results revealed that the auditory brainstem response threshold of the cochlea was increased moderately after irradiation at low laser fluence and dramatically at high laser fluence, whereas the increase of auditory capability was linearly related to the increase of laser fluence. Furthermore, collagen density increased after laser irradiation, and a large amount of new collagen fibers was found. This study represents a potential approach for laser-assisted treatment of hearing loss.

Since most biomaterials and tissue are transparent to near infrared (NIR) light, NIR femtosecond lasers can be used to induce two-photon polymerization in collagen/cell mixtures without impairing cell viability. Kuetemeyer et al. used a femtosecond laser to induce two-photon polymerization of collagen in bioartificial cardiac tissue [6] to increase its stiffness, which is necessary for in vivo implantation. In this study, they mixed murine embryonic fibroblasts or neonatal rat cardiomyocytes with collagen I to generate bioartificial tissue. After addition of 0.01 % riboflavin cross-linker, the mixture was raster-scanned by a femtosecond laser beam. Stress-strain measurements one day after laser-scanning showed that the tissue stiffness was increased by 40 %, as compared with control samples, while maintaining cardiomyocytes viability.

### 10.3.2.2  3D Patterning of Agarose for Tissue Engineering

As more research effort has been focused on 3D patterning of biomaterials for tissue engineering applications, 3D patterning of agarose also has been explored. Agarose is well suited as base biomaterial because cells do not adhere to natural agarose and additionally its abundant hydroxyl groups can bond with amino groups and can therefore be easily functionalized. As an example, Wylie et al. patterned agarose, which was modified with coumarin-caged amines, three-dimensionally with a femtosecond laser [7]. In this approach, 6-bromo-7-hydroxy-coumarin, a photocage for amines, was used to functionalize agarose. The femtosecond laser was employed to scan a series of micrometer sized squares at a depth of $40 \mu m$ underneath the surface of the agarose gel. Since the two-photon absorption based femtosecond laser patterning depends non-linearly on the laser intensity, only the laser power in close vicinity to the focal point is high enough to induce chemical modification and to enable three-dimensional structuring, even below the surface of a polymer matrix. A fluorogenic probe was used to verify the production of amines after modification. Fluorescent cross-section images were taken at different depths. It was found that the laser reaction region was from 20 to $60 \mu m$ below the surface when the laser was focused at $40 \mu m$ below the surface. The authors proposed that this approach could provide a means of achieving cell guidance for tissue engineering and regenerative medicine applications, because functionalization can promote cell adhesion in an agarose matrix.

### 10.3.2.3  Laser Patterning of Hyaluronic Acid (HA) for Hydrogel Based Applications

Due to its available functional groups, HA can be easily modified to be photoresponsive. Smeds et al. were among the first groups in the late 1990s to employ this strategy for synthesizing HA hydrogels via laser-initiated photo cross-linking and were able to form hydrogels of various geometries (microcapsules, blocks, and films) [56]. In this early approach, an argon ion laser source (wavelength 514 nm) was used to cross-link a pre-polymer HA solution composed of methacrylated HA (via methacrylic anhydride) and a water-soluble photoinitiator that generated free radicals upon light activation (Eosin Y in the presence of triethanolamine). Since then, this cross-linking approach has broadly been adopted as a method for synthesizing HA hydrogels [57–59] including other light sources such as UV lamps and several commercial photoinitiators. However, laser-based photo cross-linking is still a very specific and unique approach for generating and/or modifying hydrogels where high resolution patterns are required. Musoke-Zawedde and Shoichet used a UV laser to micropattern a hydrogel matrix composed of HA modified with S-2-nitrobenzyl cysteine [60]. UV exposure was used to cleave the 2-nitrobenzyl group to expose thiol groups at patterned locations within the hydrogel. Subsequently, the thiol groups were functionalized with GRGDS (glycine-arginine-glycine-aspartic-acid-serine) oligopeptides (a cell receptor binding sequence) to form patterned gradients to guide neurite outgrowth. Similarly, Seidlits et al. described a general method to three-dimensionally pattern proteins within hydrogels for cell guidance of neuronal cultures [61]. A femtosecond Ti:Sapphire laser at 780 nm wavelength was applied to photoinduced cross-linking of protein microstructures via multiphoton excitation within hyaluronic acid hydrogels.

A DMD approach as described in the laser setup section has been used specifically with HA by Suri et al. to construct three-dimensional scaffolds that can function as nerve conduits (Fig. 10.3) [49]. Authors affiliated with this work (Zhang et al. ; Chen et al.) are now adopting a similar approach that employs a femtosecond laser in place of the UV lamp not only for modification of hydrogels post-polymerization but also for the complete freeform fabrication of an entire HA scaffold.

The use of photocrosslinked HA within a clinical context presents an exciting opportunity, and early demonstrations of this approach have yielded promising results. Smeds et al. extended their initial photocrosslinking investigations to clinical applications that would necessitate in situ polymerization. As described in references [62, 63], a solution of methacrylated HA was applied to corneal lacerations in rabbit models and polymerized in situ using an argon laser beam of low intensity. The lacerations had sealed completely within 6 h, and reformation of the anterior chambers of the treated eyes was observed. Additionally, inflammatory response was minimal based on clinical and histological evaluations, and stromal cell proliferation along with ECM generation was observed at the wound edges.

Due to the native prevalence of HA within cartilage tissue, the in situ polymerization approach was similarly applied in an articular cartilage repair model [57] using the same formulation as described in [62]. In this study, methacrylated HA was used initially to encapsulate chondrocytes in vitro. The chondrocytes remained viable within the polymer and were able to produce significant amounts of cartilaginous matrix. Subsequently, when in situ polymerization of methacrylated HA was performed in osteochondral defect models, cell infiltration into the hydrogel, integration with the native tissue, and cartilaginous matrix generation was observed 2 weeks post-operatively.

More recently, HA has been explored for use in the clinical environment for laser tissue welding. These applications have been investigated by Cohen et al. across various repair models [64–67]. In this approach, HA serves a constituent

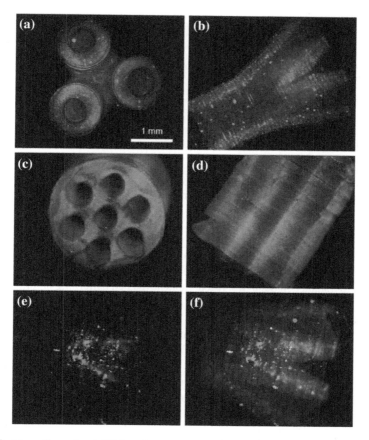

**Fig. 10.3** Three-dimensional HA-based nerve regeneration scaffolds fabricated using a DMD-based dynamic mask system and a UV light source. Branched structures, (**a**) and (**b**), as well as multi-lumen scaffolds, (**c**) and (**d**), can be fabricated. (**e**) and (**f**) demonstrate the viability of Schwann cells seeded inside the scaffold. Reprinted from Fig. 7 in reference [49] with kind permission from Springer Science+Business Media, LLC

of a biomaterial-based solder with the formulation of the solder generally consisting of a 2:1:2 mixture of albumin, indocyanine green dye, and sodium hyaluronic acid. A diode laser module (wavelength at 808 nm) coupled to a quartz silica fiberoptic cable is used for soldering. Exposure time is determined by the color change of the indocyanine dye from green to tan.

This technique has been investigated across animal models for laser facial nerve welding [67], tracheobronchial repair [65], and transluminal repair of esophageal perforations [64]. In the facial nerve and esophageal applications, laser tissue welding yielded comparable or superior outcomes when compared against traditional suture-based closure techniques. In the tracheobronchial model, laser tissue welding demonstrated bond strengths twice that of standard fibrin glue treatments. This approach has recently been explored in human trials for endonasal repair of cerebrospinal fluid leaks with positive outcomes [66].

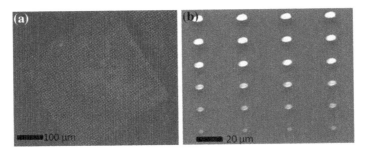

**Fig. 10.4** SEM images of microdots array fabricated with PEGDA/Matrigel™: **a** an array of microdots, the dot size is about 2 μm and the distance between adjacent dots is 20 μm, **b** dose test of microdots fabrication, the laser power was increased from 15 mW (*bottom row*) to 40 mW (*top row*)

### 10.3.2.4 Matrigel™

To better understand cell behavior, researchers have been patterning natural biomaterials with precisely controlled geometries and dimensions. Especially, micro/nano-scale patterns have been attracting more research interest because the biological and chemical stimulations in extracellular matrix often occur at the micro/nano-scale. Despite the wide application of Matrigel™ in biomedical research, literature describing the microscale patterning of Matrigel™ is currently lacking. The authors have explored this niche by micropatterning Matrigel™ using the femtosecond laser-induced two-photon polymerization technique. They used the laser to fabricate microdots arrays to investigate cell migration guided by growth factor gradients. Because Matrigel™ does not have cross-linking groups, poly (ethylene glycol) diacrylate (PEGDA) was incorporated into Matrigel™ to provide a cross-linkable solution. Different ratios of Matrigel™ and PEGDA were tested, and a 1:9 (v/v) ratio of Matrigel™ and PEGDA was found to be the optimum to make fine structures. When the ratio is higher than 1:9 (v/v), aggregation of Matrigel™ was observed in the solution. The aggregation interfered with fabrication due to scattering of the laser beam, which causes uneven cross-linking of the polymer chain. A photoinitiator (Irgacure 819) was added to the solution to facilitate polymerization. By changing the laser power and exposure time, microdots arrays with various sizes were fabricated (Fig. 10.4).

## 10.3.3 Laser Ablation

Natural biomaterials can also be patterned by laser ablation, which is the selective removal of natural biomaterials by vaporization of the material. Due to the Gaussian distribution of the laser energy, only the energy in the center of the focal point is high enough to induce vaporization. Thus, laser ablation is a precise tool in patterning materials. In addition, a laser is able to remove materials without generating debris upon laser induced bond-breaking, which minimize the impact from the ablation tools on the biomaterials. In order to modify the structures underneath of the surface, lasers with near-infrared or infrared wavelength are typically used since most natural biomaterials are almost transparent in the near-infrared spectrum.

### 10.3.3.1  Laser Induced Removal of Collagen

In addition to laser-induced cross-linking of collagen, laser removal has also been extensively explored. Kawata's laboratory was among the first groups to investigate collagen removal using a femtosecond laser [68]. They used the ultrafast laser to create a 3D array of micro-cavities in collagen to a depth of $30\,\mu$m below the surface without damaging the material on top of the microdot. As the laser power was increased, a shorter exposure time, which corresponds to a smaller number of laser pulses, was needed to create the dots. Accordingly, using lower laser powers requires an increased number of pulses or correspondingly a longer exposure time. More importantly, by using larger pulse numbers, they were able to induce the same effect at a laser power well below the ablation threshold of individual pulses, indicating a dose accumulation during the exposure.

Femtosecond lasers are widely used in in vivo applications for the removal of intrastromal collagen due to their excellent capability to penetrate various tissues and their high precision. The treatment of a severe post operational complication, which is called "post-laser in situ keratomileusis ectasia" may be necessary due to removing of central epithelial. In order to investigate the corneal morphologic change after laser treatment, Dong et al. used a femtosecond laser to remove a piece of tissue and successfully created a pocket with a diameter of 7 mm and depth of $80\,\mu$m in rabbit corneas. Subsequently, they injected a 0.1 % riboflavin solution as a collagen cross-linking agent into the corneal pockets and used a UVA beam with a power density of $3\,\text{mW/cm}^2$ to irradiate the pocket for 10 min. They evaluated the topography and pachymetry of the cornea after irradiation and demonstrated that the femtosecond laser was safe and effective in removing collagen-based tissue [69]. It is shown, that the femtosecond laser irradiation minimized epithelial removal during the process.

Other groups have created a variety of three-dimensional microscale structures in collagen by femtosecond laser ablation. Liu et al. used a femtosecond laser to generate microscale holes, grids, and parallel lines in collagen [70] with line widths of about $20\,\mu$m, as depicted in the differential interference contrast (DIC) images shown in Fig. 10.5. The relationships between hole size, laser power and exposure time were characterized. During laser ablation, the thermal effect on the biomaterial may become an issue, because it may impair cell adhesion and proliferation. It was found that only a negligible thermal effect occurred during femtosecond laser ablation, which was demonstrated by culturing mesenchymal stem cells and human HT1080 fibroblasts in the collagen structure resulting in cell viability up to 10 days.

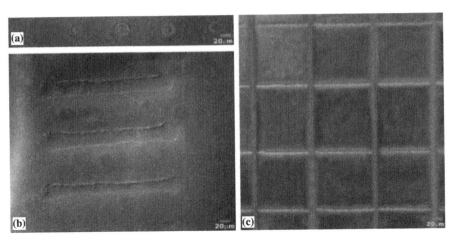

**Fig. 10.5** Various patterns fabricated by a femtosecond laser on collagen surface. Microcraters (**a**); lines (**b**); grids (**c**). Scale bar: 20 μm. Reproduced from Fig. 1 in reference [70] with kind permission from Elsevier

Femtosecond lasers have also been used as cell guidance tracks. Ilina et al. created three-dimensional microtracks in collagen for studying how 3D tracks guide cancer cell invasion and for characterizing the dependence of cancer cell migration on matrix metalloproteinases (MMPs), such as collagenase [71]. A femtosecond laser was chosen in this study due to its three-dimensional microscale patterning ability in collagen due to two-photon polymerization, which is required to develop an in vitro model mimicking the in vivo connective tissue. Furthermore, the laser was able to create structures with minimal debris. Microtracks of various widths and depths were created, and mammary tumor breast cancer cells were then seeded in the collagen 3D microtracks. The results revealed that the cancer cell invasion guided by the microtracks were independent of collagenase degradation. It is suggested that the laser-generated 3D microtrack is a useful model to investigate spatial and chemical requirements of cell migration.

### 10.3.3.2 Laser Ablation of Agarose

Agarose was used as the base material for the fabrication of microchannels in microfluidic systems through ablation by a CAD/CAM based ArF laser fabrication system (wavelength at 193 nm) [72]. The width of the channels ranged from 60 to 400 μm. Since agarose does not support cell adhesion, extracellular matrix gel was filled in the channels and polymerized. Neuroblasts were then seeded in the channels, adhered onto the extracellular matrix gel and proliferated. After 72 h the neuroblasts differentiated into nerve bundles demonstrating the laser as an excellent tool for patterning agarose gel. The use of the laser facilitates an easy and precise control over the geometry, dimensions, and depth of the structures.

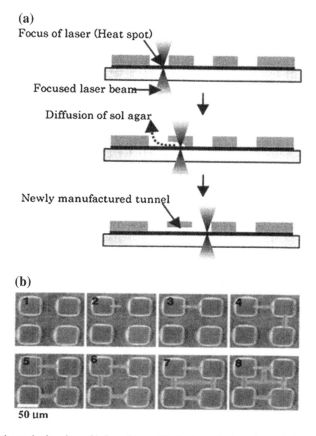

**Fig. 10.6  a** Schematic drawing of infrared laser fabrication of microchannels in agarose. **b** Steps to fabricate microchannels to connect each pair of microchambers. Reproduced from Figs. 4 and 5 in reference [73] with kind permission from Wiley

Moriguchi et al. also used laser-ablated agarose microstructures in combination with imprint lithography for neural cell research [73]. Prior to laser structuring an array of agarose microchambers was fabricated by imprinting a SU-8 mold into an agarose substrate. Subsequently, a 1064 nm Nd:YAG laser was used in order to fabricate microtunnels between two neighboring chambers (Fig. 10.6). The width of the tunnels was approx. 10 μm and could be controlled by the size of the focused laser spot. One neural cell was seeded in each chamber and was cultivated for 48 h. It was observed that the neural cells trapped in the chambers made connections to the neural cells in their adjacent chambers through the microtunnels proving that all fabricated tunnels were open. Similar laser-fabricated agarose structures may serve as a useful platform for the study of cell migration and communication.

### 10.3.3.3 Laser Ablation of Hyaluronic Acid

Barbucci et al. have demonstrated the use of laser ablation to pattern HA modified substrates [74, 75]. HA along with its sulfated derivative (HyalS), was photoimmobilized to aminosilanized glass surfaces via functionalization of the HA and HyalS with a photoreactive group and subsequent UV light treatment. After photoimmobilization, the homogeneous HA and HyalS layers were ablated using an excimer UV laser (wavelength at 248 nm) to create striped patterns with widths ranging from 10 to 100 μm and heights of 800 nm. Cells were cultured on the patterned surfaces to prove the preferential adhesion and proliferation along the ablated patterns for both 3T3 fibroblasts and bovine aortic endothelial cells (BAECs). Results from such studies could provide better understanding of cells interaction with microenvironments.

## 10.3.4 Laser Activation

Activation of natural biomaterials is a unique application of the laser source, though it is currently not being widely used. By attaching or releasing various function groups from the chemical structures of natural biomaterials, the laser beam is able to selectively alter the material surface properties on the microscale. To enhance the laser energy absorption during the laser-induced activation of biomaterials process, Au nanoparticles are commonly encapsulated in natural biomaterials.

### 10.3.4.1 Laser Activation of Agarose

Agarose might be selectively activated for biomolecule immobilization for instance using a He-Cd laser (wavelength at 325 nm) [76]. Vertical chemical channels were created in agarose to guide the growth of axons. Bulk agarose was mixed with S-2-nitrobenzyl-cysteine (S-NBC), and the mixture was irradiated first by UV light to release free nucleophiles. The mixture was then irradiated by the laser to fabricate channels with free sulphydryl (thiol) groups. It was found that at least the exposure duration of 1 s is required for laser activation. The diameter of the channels was about 150 μm, controlled by the size of the laser beam. The cell adhesive peptide fragment glycine-arginine-glycine-aspartic-acid-serine (GRGDS) was conjugated to those channels. The density of conjugated GRGDS was not increased as the laser irradiation time increased further. Ganglia cells were then seeded in the bulk agarose with GRGDS channels. The cell culture results revealed the capability of laser activation to incorporate molecules within agarose for guiding cell migration.

### 10.3.4.2  Laser Activation of Hyaluronic Acid

Recent developments in nanobiotechnology have allowed for more sophisticated functionalization of HA towards a variety of applications. Gold nanoparticles (gold nanorods–GNR) are particularly attractive for laser-based treatments due to the tunability of their absorbance spectra relative to the desired light source. The GNR exhibit an interesting feature, which is their thermally-induced shape change upon laser irradiation. This shape change results in a shift of the spectral absorbance off the resonance with the light source, which in effect self-limits the heat generated by the gold nanorod due to reduced light absorption [77]. Beside this engineering of material properties Matteini et al. demonstrated the incorporation of poly(ethylene)-glycol (PEG) encapsulated GNR within HA for use in in vivo carotid artery closure. This approach provides an improvement over the commonly used indocyanine dye based solders, which are less stable and susceptible to excessive diffusion at the biological site. After topical application of the HA-GNR composite in a rabbit carotid artery model, the site is exposed to a near-infrared diode laser (wavelength at 810 nm). Postoperatively, the arteries demonstrated patency, and integrity of the vascular wall was maintained. Additionally, there was no microgranuloma formation or dystrophic calcification observed, which demonstrated minimal host reaction to the nanoparticles.

## 10.4  Summary

The laser is known as a versatile tool and its application is successfully expanded to the processing of natural biomaterials. Various natural biomaterials, such as collagen, agarose, hyaluronic acid, and Matrigel$^{TM}$, have been treated through laser induced polymerization, ablation, or activation. In vivo and in vitro experiments have been conducted to demonstrate that the laser is both, effective and efficient in processing biomaterials for biological applications. While some drawbacks remain for using natural biomaterials, mainly due to their potential for immunogenic response and their lack of mechanical strength, there has been progress in the development of techniques to address these issues, such as the strengthening of natural biomaterials via crosslinking or the incorporation of other biomaterials [20, 23]. These developments, along with further improvements in laser processing capabilities, will continue to expand the potential applications of natural biomaterials in biomimetic approaches.

**Acknowledgments**  Some results in this chapter were supported by Award Number R01EB012597 from the National Institute of Biomedical Imaging And Bioengineering and grants (CMMI-1130894, CMMI-1120795) from the National Science Foundation.

# References

1. Hubbell JA (1995) Biomaterials in tissue Engineering. Bio-Technology 13:565
2. Lee SH, Moon JJ, West JI (2008) Three-dimensional micropatterning of bioactive hydrogels via two-photon laser scanning photolithography for guided 3D cell migration. Biomaterials 29:2962
3. Lippert T, Chrisey DB, Purice A et al (2007) Laser processing of soft materials. Romanian Rep Phy 59:483
4. Van Vlierberghe S, Dubruel P, Schacht E (2011) Biopolymer-based hydrogels as scaffolds for tissue engineering applications: a review. Biomacromolecules 12:1387
5. Oujja M, Perez S, Fadeeva E et al (2009) Three dimensional microstructuring of biopolymers by femtosecond laser irradiation. Appl Phy Lett 95:3
6. Kuetemeyer K, Kensah G, Heidrich M et al (2011) Two-photon induced collagen cross-linking in bioartificial cardiac tissue. Opt Express 19:15996
7. Wylie RG, Shoichet MS (2008) Two-photon micropatterning of amines within an agarose hydrogel. J Mater Chem 18:2716
8. Di Lullo GA, Sweeney SM et al (2002) Mapping the ligand-binding sites and disease-associated mutations on the most abundant protein in the human, type I collagen. J Biol Chem 277:4223
9. Pollard TD (1990) Molecular cell biology. Nature 346:621
10. Shi CY, Li QG, Zhao YN et al (2011) Stem-cell-capturing collagen scaffold promotes cardiac tissue regeneration. Biomaterials 32:2508
11. Lee H, Yeo M, Ahn S et al (2011) Designed hybrid scaffolds consisting of polycaprolactone microstrands and electrospun collagen-nanofibers for bone tissue regeneration. J Biomed Mater Res Part B-Appl Biomater 97B:263
12. Engelhardt EM, Micol LA, Houis S et al (2011) A collagen-poly(lactic acid-co-epsilon-caprolactone) hybrid scaffold for bladder tissue regeneration. Biomaterials 32:3969
13. Kew SJ, Gwynne JH, Enea D et al (2011) Regeneration and repair of tendon and ligament tissue using collagen fibre biomaterials. Acta Biomaterialia 7:3237
14. Hong Y, Gao CY, Xie Y et al (2005) Collagen-coated polylactide microspheres as chondrocyte microcarriers. Biomaterials 26:6305
15. Park JH, Lee GS, Shin US et al (2011) Self-hardening microspheres of calcium phosphate cement with collagen for drug delivery and tissue engineering in bone repair. J Am Ceram Soc 94:351
16. Adachi T, Kawakami E, Ishimaru N et al (2010) Delivery of small interfering RNA with a synthetic collagen poly(Pro-Hyp-Gly) for gene silencing in vitro and in vivo. Dev Growth Differ 52:693
17. Lam CH, Hansen EA, Hubel A (2011) Arachnoid cells on culture plates and collagen scaffolds: phenotype and transport properties. Tissue Eng Part A 17:1759
18. Yuan MT, Leong KW, Chan BP (2011) Three-dimensional culture of rabbit nucleus pulposus cells in collagen microspheres. Spine J 11:947
19. Rinaudo M (2008) Main properties and current applications of some polysaccharides as biomaterials. Polym Int 57:397
20. Lewitus DY, Landers J, Branch JR et al (2011) Biohybrid carbon nanotube/agarose fibers for neural tissue engineering. Adv Funct Mater 21:2624
21. Bhat S, Tripathi A, Kumar A (2011) Supermacroprous chitosan-agarose-gelatin cryogels: in vitro characterization and in vivo assessment for cartilage tissue engineering. J Royal Soc Interface 8:540
22. Dekosky BJ, Dormer NH, Ingavle GC et al (2010) Hierarchically designed agarose and poly(ethylene glycol) interpenetrating network hydrogels for cartilage tissue engineering. Tissue Eng Part C-Methods 16:1533
23. Sanchez-Salcedo S, Nieto A, Vallet-Regi M (2008) Hydroxyapatite/beta-tricalcium phosphate/agarose macroporous scaffolds for bone tissue engineering. Chem Eng J 137:62
24. Kabanova N, Stulova I, Vilu R (2011) Microcalorimetric study of the growth of bacterial colonies of Lactococcus lactis IL1403 in agar gels. Food Microbiol 29:67

25. Parveen S, Kaur S, David SAW et al (2011) Evaluation of growth based rapid microbiological methods for sterility testing of vaccines and other biological products. Vaccine 29:8012

26. Coban AY, Akgunes A, Durupinar B (2011) Evaluation of blood agar medium for the growth of Mycobacteria. Mikrobiyoloji Bulteni 45:617

27. Hayashi S, Itoh K, Suyama K (2011) Growth of the cyanobacterium synechococcus leopoliensis CCAP1405/1 on agar media in the presence of heterotrophic bacteria. Microbes Environ 26:120

28. Zhang JH, Wang F, Wang TY (2011) A simple and effective superbuffer for DNA agarose electrophoresis. Gene 487:72

29. Bhilocha S, Amin R, Pandya M et al (2011) Agarose and polyacrylamide gel electrophoresis methods for molecular mass analysis of 5-to 500-kDa hyaluronan. Anal Biochem 417:41

30. Giori L, Tricomi FM, Zatelli A et al (2011) High-resolution gel electrophoresis and sodium dodecyl sulphate-agarose gel electrophoresis on urine samples for qualitative analysis of proteinuria in dogs. J Veterinary Diagn Investig 23:682

31. Gerou-Ferriani M, Mcbrearty AR, Burchmore RJ et al (2011) Agarose gel serum protein electrophoresis in cats with and without lymphoma and preliminary results of tandem mass fingerprinting analysis. Veterinary Clin Pathol 40:159

32. Sitaras C, Naghavi M, Herrington MB (2011) Sodium dodecyl sulfate-agarose gel electrophoresis for the detection and isolation of amyloid curli fibers. Anal Biochem 408:328

33. Burdick JA, Prestwich GD (2011) Hyaluronic acid hydrogels for biomedical applications. Adv Mater 23:H41

34. Price RD, Myers S, Leigh IM et al (2005) The role of hyaluronic acid in wound healing—assessment of clinical evidence. Am J Clin Dermatol 6:393

35. Allison DD, Grande-Allen KJ (2006) Review. Hyaluronan: a powerful tissue engineering tool. Tissue Eng 12: 2131

36. Luo Y, Kirker KR, Prestwich GD (2000) Cross-linked hyaluronic acid hydrogel films: new biomaterials for drug delivery. J Controlled Release 69:169

37. Rohrich RJ, Ghavami A, Crosby MA (2007) The role of hyaluronic acid fillers (restylane) in facial cosmetic surgery: review and technical considerations. Plast Reconstr Surg 120:41S

38. Goa KL, Benfield P (1994) Hyaluronic-acid—a review of its pharmacology and use as a surgical aid in ophthalmology, and its therapeutic potential in joint disease and wound-healing. Drugs 47:536

39. Laurent TC, Laurent UBG, Fraser JRE (1996) The structure and function of hyaluronan: an overview. Immunol Cell Biol 74:A1

40. Leach JB, Bivens KA, Patrick CW et al (2003) Photocrosslinked hyaluronic acid hydrogels: natural, biodegradable tissue engineering scaffolds. Biotechnol Bioeng 82:578

41. Toole BP (2004) Hyaluronan: from extracellular glue to pericellular cue. Nat Rev Cancer 4:528

42. Oh EJ, Park K, Kim KS et al (2010) Target specific and long-acting delivery of protein, peptide, and nucleotide therapeutics using hyaluronic acid derivatives. J Controlled Release 141:2

43. Kleinman HK, Martin GR, (2005) Matrigel™: basement membrane matrix with biological activity. Semin Cancer Biol 15:378

44. Asakura A, Komaki M, Rudnicki MA (2001) Muscle satellite cells are multipotential stem cells that exhibit myogenic, osteogenic, and adipogenic differentiation. Differentiation 68:245

45. Xu CH, Inokuma MS, Denham J et al (2001) Feeder-free growth of undifferentiated human embryonic stem cells. Nat Biotechnol 19:971

46. Kramer RH, Bensch KG, Wong J (1986) Invasion of reconstituted basement-membrane matrix by metastatic human-tumor cells. Cancer Res 46:1980

47. Fridman R, Kibbey MC, Royce LS et al (1991) Enhanced tumor growth of both primary and established human and murine tumor cells in athymic mice after coinjection with Matrigel™. J Natl Cancer Inst 83:769

48. Nielson R, Koehr B, Shear JB (2009) Microreplication and design of biological architectures using dynamic-mask multiphoton lithography. Small 5:120

49. Suri S, Han LH, Zhang WD et al (2011) Solid freeform fabrication of designer scaffolds of hyaluronic acid for nerve tissue engineering. Biomed Microdevices 13:983

50. Han LH, Mapili G, Chen S et al (2008) Projection microfabrication of three-dimensional scaffolds for tissue engineering. J Manuf Sci Eng-Trans ASME 130:4
51. Arnold CB, Serra P, Pique A (2007) Laser direct-write techniques for printing of complex materials. Mrs Bulletin 32:23
52. Chan BP, So KF (2005) Photochemical crosslinking improves the physicochemical properties of collagen scaffolds. J Biomed Mater Res Part A 75A:689
53. Iosin M, Stephan O, Astilean S et al (2007) Microstructuration of protein matrices by laser-induced photochemistry. J Optoelectron Adv Mater 9:716
54. Frullini A, Manetti L, Di Cicco E et al (2011) Photoinduced collagen cross-linking: a new approach to venous insufficiency. Dermatol Surg 37:1113
55. Wenzel GI, Anvari B, Mazhar A et al (2007) Laser-induced collagen remodeling and deposition within the basilar membrane of the mouse cochlea. J Biomed Opt 12:7
56. Smeds KA, Pfister-Serres A, Hatchell DL et al (1999) Synthesis of a novel polysaccharide hydrogel. J Macromol Sci-Pure Appl Chem A36:981
57. Nettles DL, Vail TP, Morgan MT et al (2004) Photocrosslinkable hyaluronan as a scaffold for articular cartilage repair. Ann Biomed Eng 32:391
58. Leach JB, Schmidt CE (2005) Characterization of protein release from photocrosslinkable hyaluronic acid-polyethylene glycol hydrogel tissue engineering scaffolds. Biomaterials 26:125
59. Zawko SA, Suri S, Truong Q et al (2009) Photopatterned anisotropic swelling of dual-crosslinked hyaluronic acid hydrogels. Acta Biomaterialia 5:14
60. Musoke-Zawedde P, Shoichet MS (2006) Anisotropic three-dimensional peptide channels guide neurite outgrowth within a biodegradable hydrogel matrix. Biomed Mater 1:162
61. Seidlits SK, Schmidt CE, Shear JB (2009) High-resolution patterning of hydrogels in three dimensions using direct-write photofabrication for cell guidance. Adv Funct Mater 19:3543
62. Smeds KA, Grinstaff MW (2001) Photocrosslinkable polysaccharides for in situ hydrogel formation. J Biomed Mater Res 54:115
63. Miki D, Dastgheib K, Kim T et al (2002) A photopolymerized sealant for corneal lacerations. Cornea 21:393
64. Bleier BS, Gratton MA, Leibowitz JM et al (2008) Laser-welded endoscopic endoluminal repair of iatrogenic esophageal perforation: an animal model. Otolaryngol-Head Neck Surg 139:713
65. Bleier BS, Cohen NM, Bloom JD, et al (2010) Laser tissue welding in lung and tracheobronchial repair an animal model. Chest 138:345
66. Bleier BS, Cohen NA, Chiu AG et al (2010) Endonasal laser tissue welding: first human experience. Am J Rhinol Allergy 24:244
67. Bloom JD, Bleier BS, Goldstein SA et al (2011) Laser facial nerve welding in a rabbit model. Arch Fac Plast Surg 25(3):186–187
68. Smith NI, Fujita K, Nakamura O et al (2001) Three-dimensional subsurface microprocessing of collagen by ultrashort laser pulses. Appl Phy Lett 78:999
69. Dong Z, Zhou X (2011) Collagen cross-linking with riboflavin in a femtosecond laser-created pocket in rabbit corneas: 6-month results. Am J Ophthalmol 152:22
70. Liu YM, Sun S, Singha S et al (2005) 3D femtosecond laser patterning of collagen for directed cell attachment. Biomaterials 26:4597
71. Ilina O, Bakker GJ, Vasaturo A et al (2011) Two-photon laser-generated microtracks in 3D collagen lattices: principles of MMP-dependent and -independent collective cancer cell invasion (vol 8, 015010, 2011). Phys Biol 8:1
72. Doraiswamy A, Patz T, Narayan RJ et al (2006) Two-dimensional differential adherence of neuroblasts in laser micromachined CAD/CAM agarose channels. Appl Surf Sci 252:4748
73. Moriguchi H, Takahashi K, Sugio Y et al (2004) On-chip neural cell cultivation using agarose-microchamber array constructed by a photothermal etching method. Electr Eng Jpn 146:37
74. Barbucci R, Pasqui D, Wirsen A et al (2003) Micro and nano-structured surfaces. J Mater Sci-Mater Med 14:721
75. Barbucci R, Lamponi S, Pasqui D et al (2003) Micropatterned polysaccharide surfaces via laser ablation for cell guidance. Mater Sci Eng C-Biomimetic Supramol Syst 23:329

76. Luo Y, Shoichet MS (2004) A photolabile hydrogel for guided three-dimensional cell growth and migration. Nat Mater 3:249
77. Matteini P, Ratto F, Rossi F et al (2010) In vivo carotid artery closure by laser activation of hyaluronan-embedded gold nanorods. J Biomed Opt 15:6

# Chapter 11
# Future Perspectives

Emmanuel Stratakis, Anthi Ranella and Costas Fotakis

**Abstract** When biomimetics, which involves the translation of mechanisms developed by nature into manmade technology, meets lasers, which come as an emerging technology for high resolution materials processing, a technological push rapidly growing into new and unexplored areas of application in biomimetics is expected. Besides overcoming one of the biggest challenges of manufacturing below the 100 nm scale with laser based solid free form methods, the next years may primarily belong to the field of bioinspired multifunctional materials, since intelligent biological functions and development of responsive biomimetic artificial materials are still in their infancy. Therefore, more efforts will be dedicated to the integration of laser based schemes with other processes to make functional devices or systems and to simultaneously provide unprecedented tools towards understanding the interactions between nanomaterials and complex biological structures. Research on intelligent natural architectures combined with evolutionary optimization is expected to bring up novel bioinspired ideas and approaches for the realization of biomimetic artificial materials that currently remain beyond our grasp. As laser sources continue to bedeveloped, a rich range of technical opportunities in laser based micro- and nanofabrication, nanobiomaterial synthesis and characterization will emerge and suggest a future where control of artificial biomaterial structure and subsequent functionality can be accomplished with a level of sophistication that, presently, we cannot imagine.

E. Stratakis (✉) · A. Ranella · C. Fotakis
Institute of Electronic Structure and Laser, Foundation for Research and Technology-Hellas, (IESL-FORTH), P.O. Box 1527, 71110 Heraklion, Greece
e-mail: stratak@iesl.forth.gr

A. Ranella
e-mail: ranthi@iesl.forth.gr

C. Fotakis
e-mail: fotakis@iesl.forth.gr

E. Stratakis
Department of Materials Science & Technology, University of Crete, 71003 Heraklion, Greece

V. Schmidt and M. R. Belegratis (eds.), *Laser Technology in Biomimetics*, Biological and Medical Physics, Biomedical Engineering, DOI: 10.1007/978-3-642-41341-4_11, © Springer-Verlag Berlin Heidelberg 2013

## 11.1 Future Perspectives of Biomimetics and Laser Technology

Biomimetics involves the transformation of the ideas, concepts and underlying principles developed by nature into manmade technology. Biological systems have, through almost 4 billion years, discovered unique solutions for complex problems which are smart, energy-efficient, agile, adaptable, fault-tolerant, eco-friendly and multifunctional. Such solutions came as a direct consequence of evolutionary pressure which typically forces natural species to become highly optimized and efficient. The adaptation of methods and systems found in nature into synthetic constructs is therefore desirable and nature provides a unique source of working solutions which can serve as models of inspiration for synthetic paradigms.

The superior functions found in natural systems are achieved through a sophisticated control of structural properties at all length scales, starting from the macroscopic world down to the finest detail, right down to the level of atom. It is therefore understood that progress in the field of biomimetics should come as a synergy of the growing knowledge base of biology together with the rapidly developing ability to synthesize, fabricate and manipulate materials' properties at multiple length scales. In this context, several manufacturing technologies have been developed to facilitate the formation of biomimetic patterns on bioinspired materials. Lasers come as an emerging technology for high resolution materials and biomaterials processing, that is rapidly growing into new and unexplored areas of application in biomimetics. A superior advantage of lasers is their ability to be *material independent*, *non-invasive* and *contactless* processing tools that can deliver the desired quantum of energy at a precise location irrespective of the nature and chemical bonding with minimum lateral damage to the immediate environment. Furthermore the possibility of coupling lasers with modern rapid prototyping engineering practices including computer-aided design and manufacturing using robot-controlled and pneumatic/mechanically-driven delivery systems enables creation of 3D components practically of any desired dimension and geometry. The versatility is further extended in the ability of laser to create gradient architectures with controlled porosity, composition, and microstructure, commonly found in natural systems. However, the most distinct advantage of the applications of lasers in biomimetics is that structural or compositional modifications can be realized within multiple length scales.

Although lasers has been proven to be a useful tool in biomimetics, new explorations need to be conducted and new achievements in laser processing capabilities should be realized to prove that this technology can progress future biomimetic artificial systems beyond the current state of the art. In this concept, intense effort will be required for scaling up the overall processing procedure. Current laser production techniques are valuable within a limited range of resolutions, thus future fabrication methods must interlace 3D multiple-scale structures with exceptional controllability. Current solid free form methods can manufacture features at scales larger than 100 nm, but there is a need for controllable fabrication at nanometer scale. This is one of the biggest challenges of these technologies towards engineering truly biomimetic hierarchically structured systems and to take full advantage of the potential of nanofeature incorporation. Controlled structuring at multiple length scales, in a series or, ideally, in a single processing step can be achieved via customized far field fabrication, or through near field optical techniques and/or by sequential application of the two. Besides this, the low throughput is the common challenge pertaining to sub-wavelength processing techniques in particular. A possible solution might be to implement parallel processing by means of multiple optical probes. In such a case, ability to control each probe independently will enhance the throughput of the system. A more versatile control strategy for handling a great number of probes should therefore be developed.

Besides this, though current laser-based fabrication techniques of biomimetic structures are valuable within a limited range of functions, future engineering methods must incorporate multiple functionalities within a single construct. Indeed, the coming years may primarily belong to the field of bioinspired multifunctional materials, since the topics of understanding of intelligent biological functions and development of responsive biomimetic artificial materials are still in their infancy. A major complexity to overcome is the fact that intelligent natural structures do not always function in a unique way but they *adapt* to local functional requirements. Therefore, more efforts will be dedicated to the integration of laser based schemes with other processes and components to make functional devices or systems. For example, issues such as making an integrated device using different materials for various working parts and effectively incorporating desired amounts of delicate functional molecules including smart ones or active biomolecules should be addressed. The development of dynamic feedback-control engineering mechanisms will also be extremely important in this respect and may pave the way for sophisticated material processing.

Along these lines, the capability of manipulating small amounts of molecules including biomolecules, may allow the realization of artificial constructs with exceptional complexity and functionality. Optical tweezers and laser trapping techniques provide innovative, non-contact, means to reach this level of matter control. Laser printing techniques provide additional capabilities in this respect. The optical nature of such techniques affords the opportunity to incorporate machine vision and to employ optically selective material targeting and transfer.

Laser based techniques can be applied not only on synthesis and fabrication of novel biocompatible and/or biodegradable nanobiomaterials with complex interior geometries, but also, on instrumentation of diagnostic tools to analyze their toxicity, to monitor in real time the interactions of biomimetic surfaces with cells/tissues, to detect but still able to manipulate the dispersion of nanoparticles in the body and so on. It could also provide unprecedented tools towards understanding the interactions between nanomaterials and complex biological structures, in strictly quantitative terms.

On the other hand the development of advanced biofabrication methods should work in a synergistic manner with the synthesis of novel, bioinspired, starting materials towards the realization of efficient biomimetic systems. A safe way to ensure the success of such materials is to mimic effective mechanisms and functions that take place in living organisms. The utilization of studies in structural features at macro- as well as at micro- and nano-scales will provide useful information, in this respect.

Nanotechnology has given great impetus forward to the creation of biomimetic surfaces, interfaces, materials and forms that already find application in the biomedical field. Current studies are focused on the modification of chemical behavior, mechanical, electrical and optical properties at the nanoscale, introducing thus the concept of the nanobiomaterial. Owing to the modified properties of nanobiomaterials compared with the respective bulk materials, especially related to the increased surface energy and reactivity, there is a need for specific approaches of its adaptation to existing frames, as well as its biocompatibility and toxicity. In this context, both the interaction of nanobiomaterials with cell/tissue and the effect of size, shape, and surface on their biocompatibility, toxicity and immunogenicity, will be an emerging research field.

Research on intelligent natural architectures, combined with the use of the principles of evolution for optimization, is expected to give rise to bioinspired ideas and approaches for the realization of biomimetic artificial materials that currently remain beyond our grasp. As ultrafast laser sources continue to be developed, a rich range of technical opportunities in laser based micro and nanofabrication, nanobiomaterial synthesis and characterization, will emerge and suggest a future where control of artificial biomaterial structure and subsequent functionality can be accomplished with a sophistication that we cannot presently imagine.

# Index

CPSIA information can be obtained at www.ICGtesting.com
Printed in the USA
LVOW05*0201200115

423468LV00001B/52/P

9 783642 413407